# International Conference on Sustainable Cities (ICSC 2017)



IOP Conference Series: Earth and Environmental Science
Volume 72

Ekaterinburg, Russia
19 May 2017

ISBN: 978-1-5108-4410-0
ISSN: 1755-1307

**Printed from e-media with permission by:**

Curran Associates, Inc.
57 Morehouse Lane
Red Hook, NY 12571

**Some format issues inherent in the e-media version may also appear in this print version.**



Printed by Curran Associates, Inc. (2017)

For permission requests, please contact the Institute of Physics
at the address below.

Institute of Physics
Dirac House, Temple Back
Bristol BS1 6BE UK

Phone: 44 1 17 929 7481
Fax: 44 1 17 920 0979

techtracking@iop.org

**Additional copies of this publication are available from:**

Curran Associates, Inc.
57 Morehouse Lane
Red Hook, NY 12571 USA
Phone: 845-758-0400
Fax: 845-758-2633
Email: curran@proceedings.com
Web: www.proceedings.com

# Table of contents

## Volume 72

**International Conference on Sustainable Cities**
**19 May 2017, Ekaterinburg, Russian Federation**

**Accepted papers received: 21 June 2017**
**Published online: 2 July 2017**

### Preface



### Strategic guidelines for sustainable development of the urban environment to improve the quality of life and restore nature

## Engineering solutions for smart cities



## Energy Efficient solutions for sustainable development

IOP Conf. Series: Earth and Environmental Science **72** (2017) 011001 doi:10.1088/1755-1315/72/1/011001

# Preface

The 2017 2rd International Conference on Sustainable (ICSC 2017) cities was successfully held on May 19, 2017 in Yekaterinburg, Russia. The conference is devoted to new technologies that enable sustainable cities development and promote a stable, comfortable, modern space for life.

There were 3 major sections of the scientific track of the conference:

- Strategic guidelines for sustainable development of the urban environment to improve living standards and restore nature

- Engineering solutions for "smart" cities

- Energy efficient solutions for sustainable development

This volume of IOP Conference Series: Earth and Environmental Science (EES) is a compilation of the accepted papers in ICSC 2017 and represents contributions that were presented in the conference. This year we had over 120 papers submitted for the review and only 29 of them where accepted for the oral presentation.

On behalf of the organization committee I would like to thank all of the conference sponsors, partners and volunteers who made conference possible.

Looking forward to meet you during ICSC 2018.

On behalf of organizing and program committees of the ICSC 2017

Igor Balk, ICSC 2017 co-chair.

# Peer review statement

All papers published in this volume of *IOP Conference Series: Earth and Environmental Science* have been peer reviewed through processes administered by the proceedings Editors. Reviews were conducted by expert referees to the professional and scientific standards expected of a proceedings journal published by IOP Publishing.

IOP Conf. Series: Earth and Environmental Science **72** (2017) 012001 doi:10.1088/1755-1315/72/1/012001

# Economic efficiency of environmental management system operation in industrial companies

**N Dukmasova[1], I Ershova[2], I Plastinina[3] and A Boyarinov[4]**

[1] Department of Environmental Economics, Ural Federal University named after the first President of Russia B. N. Yeltsin, 19 Mira str., Ekaterinburg, Russia
[2] Department of Development of the machine-building production, Ural Federal University named after the first President of Russia B. N. Yeltsin, 19 Mira str., Ekaterinburg, Russia
[3] Department of Environmental Economics, Ural Federal University named after the first President of Russia B. N. Yeltsin, 19 Mira str., Ekaterinburg, Russia
[4] Department of Environmental Economics, Ural Federal University named after the first President of Russia B. N. Yeltsin, 19 Mira str., Ekaterinburg, Russia

E-mail: dykmas-natali@mail.ru

**Abstract.** The article examines the issue of the efficiency of the environmental management system (EMS) implementation in the Russian machine-building companies. The analysis showed that Russia clearly lags behind other developed and developing countries in terms of the number of ISO 14001 certified companies. According to the authors, the main cause of weak system implementation activity is attributed to the lack of interest in ISO 14001 certification on the Russian market. Five-year primary (field) research aimed at the analysis of the environmental priorities of the civilians suggests that the image component of the economic benefits ensures the increase in economic and financial performance of the company due to the increase in customers' loyalty to the products of the EMS adopter. To quantify economic benefits obtained from EMS implementation, a methodological approach with regard to the image component and the decrease in semi-fixed costs due to the increase in the production scale has been developed. This approach has been tested in a machine-building electrical equipment manufacturer in Ekaterinburg. This approach applied to data processing yields the conclusion that EMS gives a good additional competitive advantage to its adopters.

## 1. Introduction

Environmental safety issues have recently been more extensively addressed in the Russian Federation than in the past. Environmental management strategy is part of federal (for the period of 2008–2020) and regional social and economic programs, Russian Federation environmental safety strategy (for the period of 2016–2025), and state environmental stewardship program (for the period of 2012–2020). Russian government aims to reduce emissions by 2.2 times by 2020 compared to 2007, to reduce the number of the cities/towns with high level of air pollution by 2.7 times, and to increase the share of utilized and neutralized waste products and consumption of danger class I–IV wastes in the total amount of wastes by 2.2 times. To attract wide public attention to this issue, in Russia the year 2017 has been declared the Year of Ecology.

The situation is exacerbated by the fact that many large companies in Russia are located within the residential areas due to urbanization, including megalopolises, or such companies are historically town-forming, which means that the cities/towns started their development around such companies (the majority of the Ural towns have been developed according to this principle). Green production will significantly improve the living conditions of the population. EMS implementation is one of the tools that enable achieving this.

As European experience demonstrates, EMS implementation increases economic competitive ability of a company and, thus, its profitability [1–3]. The question is whether this is possible in Russia and its specific conditions?

## 2. Analysis of EMS (ISO 14001) implementation in the Russian Federation

EMS was first implemented in Russia in 1998 (Figure 1). The trail-blazers were some companies operating in mining, processing and metallurgical industries (*GAZPROM* (OAO), *Oil Company Lukoil* (OAO), *NLMK Lipetsk* (OAO), *TMK* (OAO), etc.). Some other chemical and machine-building companies like *Nornickel* (OAO), *Silvinit* (OAO, now part of *Uralkali* (OAO), etc. joined this process in 2005. The very first company in Sverdlovskaya Oblast implemented EMS was *Nizhny Tagil Metallurgical Plant* (OAO), followed by the companies belonging to *UGMK-Holding* (*Ural Mining Metallurgical Company*). The system was introduced in the machine-building companies in Sverdlovskaya Oblast in 2007.

The number of certified companies gradually increased but within the period from 2011 to 2015 there was a decline observed, as due to unstable economic conditions some companies withdrew from the market, while others preferred not to develop EMS (Figure 1).

The main reasons for EMS implementation in the above mentioned companies were external requirements, first of all, the necessity to obtain mandatory environmental management certificates of conformity to be able to operate abroad. If no such a certificate is available, then the goods are not allowed to enter the European market or can be sold only at much cheaper prices. The companies selling their goods in Russia and the neighboring countries do not face such strict requirements for EMS implementation from their partners and buyers.

Therefore, Russia occupies one of the last places in number of the ISO 14001 certified companies.

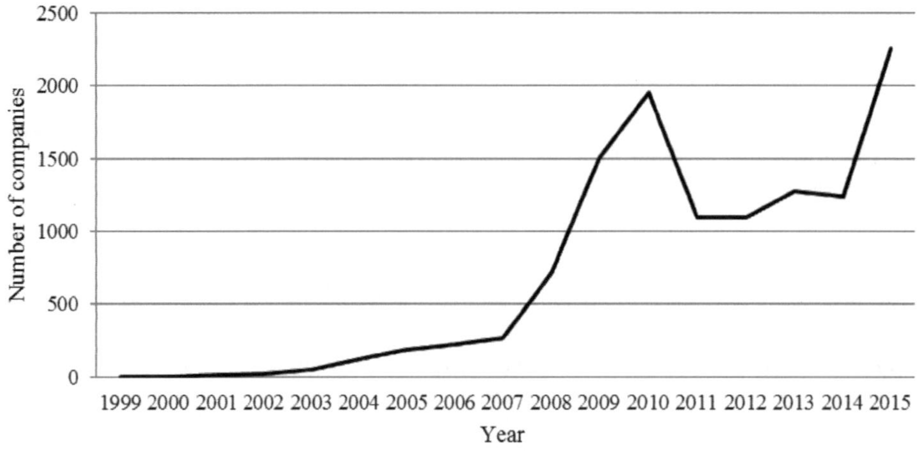


**Figure 1.** Number of Russian companies obtained the certificate

The main cause of weak system implementation activity in Russian companies is the lack of economic interest. State environmental governance in Russia still depends on administrative methods.

According to the sustainable development principle, there are three major concerns that effect evaluation of economic efficiency of the activities (including the sphere of environmental stewardship), i.e. social, environmental and economic concerns. Economic benefits from implementation of environmental activities in Russia are mainly achieved due to reduction of prime costs through lower environmental pollution payments and due to profit improvement through an increase in prices for the goods. Fluctuations in the sales volumes are not normally associated with environmental activities as they are attributed to competition and changes in the market.

The analysis of changes in external economic management of Russian industrial companies shows that nowadays, in conditions of an increased environmental awareness of the consumers and strong influence of non-governmental environmental organizations, competitive recovery through *green* reputation has a significant motivation impact on top management of the companies.

## 3. Methodological approach to definition of EMS implementation

The most widespread approach to evaluate economic efficiency of EMS implementation in the companies is strengthening of their business reputation that leads to the increase in cost of intangible assets defined through goodwill [4–6]. We define economic efficiency in a wider sense of the word, not only through business reputation of the company, but also through its image.

Image, reputation and brand are three notions, very similar with respect to their meaning, therefore it is important to understand the difference and interrelation between them. To form a brand and reputation it is necessary to promote a favorable image, which requires certain efforts from the company. Reputation is based on the created image and is developed due to interaction of target audience and the company. Brand is formed on the basis of a positive image and reputation and is consumer-oriented. If a company wants to be competitive on the market, these notions must become fundamental for it [7, 8].

We agree with the specialists who believe that image is a crucial success factor in different business spheres. It is created due to stable activity of the company, its reliability and honesty in financial relations with the partners.

To improve the image, it is necessary to determine what result we want to get: to increase sales, to enter a new market sector, to attract investors, etc. Nowadays, environmental and social policies of the company also influence its image (for example, implementation of the Quality Management System, EMS, etc.). The choice determines target audience (consumers, bodies of power, environmental departments, etc.) with its specific targets and objectives for quick achievement of the desired image.

The analysis of EMS implementation yielded the entire block of information that is usually not taken into consideration when dealing with conventional economic benefits (Figure 2). As a result, we have formed a proposition that image is involved in economic benefits. It ensures enhancement of economic and financial performance of the company due to increase in customers' loyalty to the products of the EMS adopter. We believe that image component of economic benefits is the indicator that allows assessing economic efficiency of EMS implementation to the fullest extent possible.

To prove our hypothesis, we have carried out primary (field) research aimed at the analysis of environmental benefits and priorities based on random sample of respondents (more than 300 people). The survey showed gradual rise in environmental awareness among consumers.

To develop a practical framework for our suggestion, we have carried out the analysis of current Russian (such documents as *Economic Efficiency of Environmental Costs*, *Economic Efficiency of Environmental Activities*, *Efficiency of Environmental Entrepreneurship*, etc.) and foreign practices for evaluation of environmental and economic efficiency of business activity [5, 6, 9–11]. Available printed sources show that current approaches are based on capital cost recovery principle, reduction of environmental damage, improvement of business reputation, price increase, and improvement of production quality. Nevertheless, none of these approaches takes into account marginal income gained through EMS implementation. Thus, we have proposed a comprehensive framework for evaluation of economic benefits through EMS implementation and operation with due regard to the image component [12, 13].

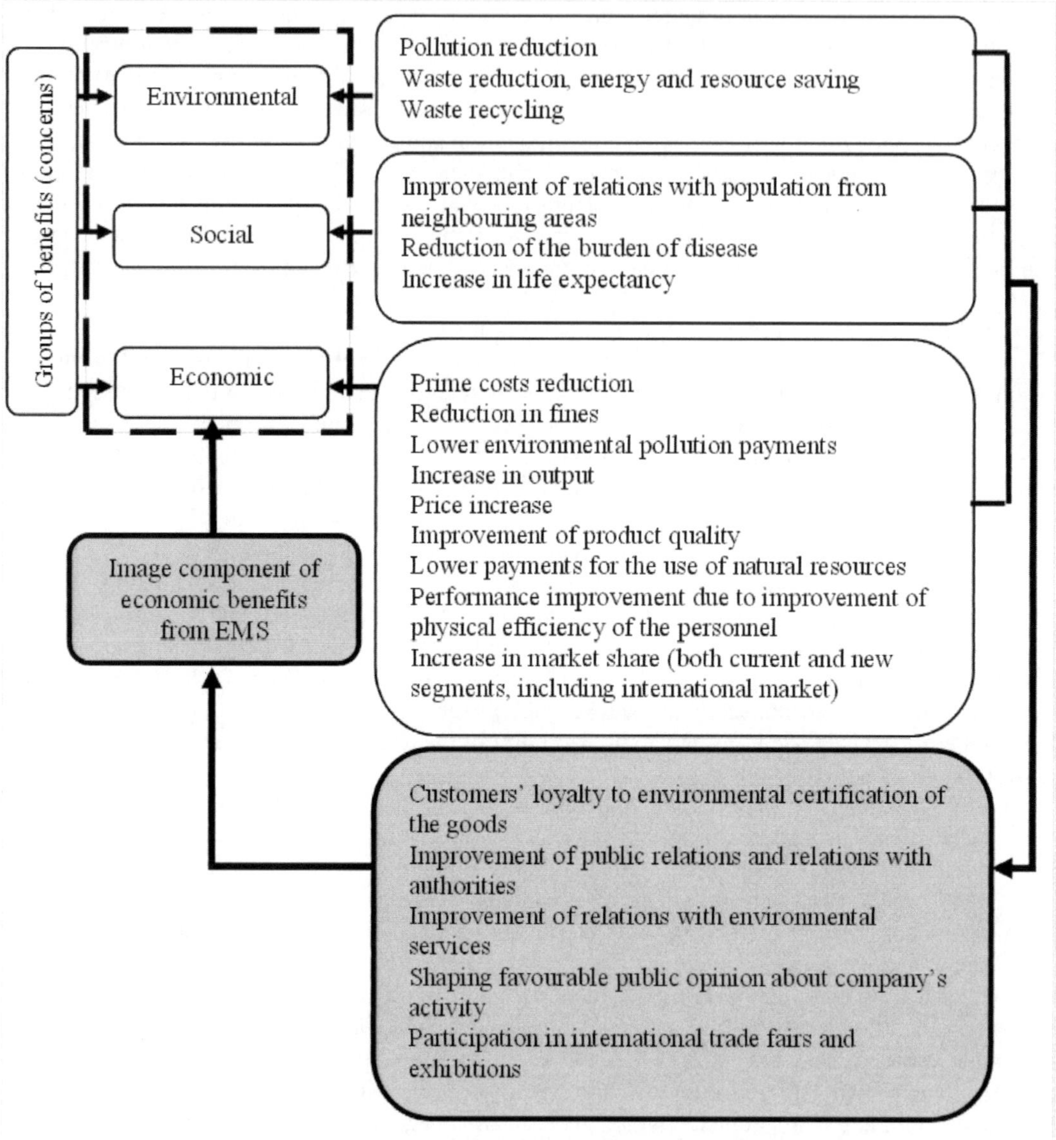


**Figure 2**. EMS implementation and its results

The obtained results show that the following components of the economic benefits can be identified with regard to EMS implementation.

Net profit fluctuation ($\Delta P_n$) due to lower payments for excess pollution ($\Delta P_{excess}$) and reduction of fines for adverse impact on the environment ($\Delta F$) can be calculated as in equation (1):

$$\Delta P_n = \Delta P_{excess} + \Delta F \quad (1)$$

1. Prime cost fluctuation due to lower rated environmental pollution payments (ΔC).

In addition to existing indices, we recommend to calculate the sales gain and prime cost savings due to improvement of the company's image.

2. Marginal income gains from the increase in sales and production volumes and from the savings in semi-fixed production costs during EMS implementation (Pm). In this case we assume that the production costs structure does not undergo any changes when EMS is operated, and additional costs for its implementation are expressed by a separate component. As the increase in sales volumes can be caused by both external (market environment) and internal factors (re-equipment, production modernization, advertising campaign, etc.), marginal income gains must be calculated with application of a correction coefficient a, that determines the influence of the company's image improvement on sales volumes.

Savings obtained from improvement of a company's image can be calculated as in equation (2):

$$P_m = (V - S_{\text{var}}) \cdot a \tag{2}$$

where $V$ = sales volume, th.rubles/year,
$S_{var}$ = semi-variable costs for the total volume of products, th.rubles/year,
$a$ = influence coefficient taking into account image component of the economic benefits.

General formula for calculation of total savings from EMS implementation is as follows (3):

$$E_{EMS} = [(V - S_{\text{var}}) \cdot a - S_{EMS}] + (\Delta P_n + \Delta C) \tag{3}$$

where $E_{EMS}$ = annual savings from EMS operation in a company, th.rubles/year,
$S_{EMS}$ = current costs for EMS operation, th.rubles/year.

Annual efficiency is calculated as follows (4):

$$E_{ann} = \frac{E_{EMS}}{C + E_H K} \tag{4}$$

where $E_{ann}$ = annual efficiency from EMS operation, th.rubles/year,
$C$ = nonrecurrent costs for EMS implementation and operation, th.rubles/year,
$E_н$ = present value index,
$K$ = capital costs for EMS implementation and operation, th.rubles/year.

Crucial element of this approach is allowance for the $a$ coefficient and its definition. The coefficient takes into consideration which part of increased volume of sales is determined by an environmental certificate. The coefficient can be quantified by different methods:

a) Expert evaluation method.
b) Analogue method.
c) Method of regression and correlation analysis.

Our algorithm is based on the regression and correlation analysis. The advantage of such an approach is the possibility to quantify the influence coefficient following the results of the analysis of activities of a large number of companies. However, the obtained results are valid only under stable external conditions, which can be considered a disadvantage of this approach. To eliminate this disadvantage, we suggest that the influence coefficient be calculated for different time periods associated with changes introduced to legislation and changes in market environment.

To apply this approach, it is expedient to improve a reference data base of a company. Calculation of the $a$ coefficient requires data bases covering sales volumes of the industry EMS adopters.

## 4. Research results

To determine the $a$ coefficient, we have carried out the regression and correlation analysis for five Russian machine-building companies within the period from 2005 to 2015. This allowed us to investigate the relationship between the indices under examination. This analysis was based on the reported data which can be freely consulted on the official sites of the machine-building companies.

The following linear relationship was used for the regression and correlation analysis (5):

$$y = ax_1 + bx_2 \tag{5}$$

where $y$ = income index (export index) for the companies,
$x_1$ = ordinal year of certification or assurance of compliance with the existing certificate,
$x_2$ = indices of income (export) fluctuations in the industry,
$a$, $в$ = influence coefficients.

The environmental conformity certificates are issued and re-issued on a regular basis, once every 3–5 years, and since the companies assure compliance with such certificates in different periods of time, the values of $x_1$ in the model range from 1 to 5.

Obtained relationships have been distributed over certain time periods. Within the period from 2005 to 2010, the machine-building companies of Sverdlovskaya Oblast only started to implement EMS, thus the expenses required for such implementation were not covered by the image component of EMS operation. The world economic crisis took place in 2009. It was one of the factors that deteriorated economic activity of the machine-building companies. The period from 2011 to 2013 turned out to be the most stable with regard to both economics and politics. Many companies assured compliance with environmental certificates, which in its turn had its impact on the influence coefficient and its value became positive. The next period, from 2014 to 2016, turned out to be quite difficult for the machine-building industry due to the national trend towards import substitution. The influence coefficient under such unstable conditions raised. It was caused by EMS operation as the consumers preferred the goods in compliance with the requirements of international standards.

The obtained results have been tested against the following criteria: multiple correlation coefficient, standard error, F-test, and Student's t-test, which proves their significance.

Our regression and correlation analysis showed that EMS gives a good additional competitive advantage to the companies operated in stable conditions, and the value of the image component reduces with the increase in number of the companies obtained such a certificate.

Economic efficiency from EMS implementation in the machine-building company in Ekaterinburg calculated on the basis of the approach taking into account the company's image is given in the Table.

**Table**. Calculation of savings and economic efficiency from EMS implementation in a machine-building company in Ekaterinburg

| Year | Approach with regard to the image component | | |
|---|---|---|---|
| | Savings from lower environmental pollution payments, $\Delta P_n + \Delta C$, th.rubles/year | Marginal income gains from the increase in sales volumes, $E_{EMS}$, th.rubles/year | Efficiency with regard to image component, $E_{ann}$, rubles/rubles |
| 2009 | 31.3 | 16,175 | 43.15 |
| 2010 | 3.6 | 18,137.6 | 25.29 |
| 2011 | 9.5 | 17,507.3 | 13.61 |
| 2012 | 31.6 | 17,338.3 | 16.71 |
| 2013 | 28 | 14,338.4 | 9.85 |
| 2014 | 13.2 | 54,227.2 | 47.80 |
| 2015 | 19.1 | 50,869.8 | 44.37 |

The calculations make it evident that our approach yields comprehensive results obtained due to EMS implementation. Application of such an approach will allow top management of a company to evaluate the activities required for EMS operation in a more comprehensive way with due regard to all necessary components.

## 5. Conclusion

1. Nowadays, with the lack of economic interest demonstrated by Russian industrial companies, the main incentive to obtain ISO 14001 certificates is their urge to enter foreign markets and develop relations with foreign partners.

2. We believe that a new component of the economic benefits should be identified and described, i.e. the company's image. This positively influences environmental and social benefits of the business activities of the companies and can lead to improvement of business reputation.

3. Implementation of ISO 14001 EMS allows regulating the impact of production process on the environment (which gradually leads to improvement of its quality), and is also cost-efficient for the company.

4. We have introduced a new approach for calculation of total savings and economic efficiency when EMS is implemented and operated. This approach takes into consideration the image component of economic benefits. This component allows recognizing revenue gains obtained through the increase in prices and production volumes. This has been proved during the analysis of the data obtained for the period from 2006 to 2015 for one of the machine-building companies in Ekaterinburg.

5. Our regression and correlation analysis showed that EMS gives a good additional competitive advantage to the companies operated in stable conditions, and that the value of the image component reduces with the increase in number of the companies obtained such a certificate.

## References

[1] Heras-Saizarbitoria I, Arana G and Boiral O 2016 Outcomes of Environmental Management Systems: the Role of Motivations and Firms' Characteristics *J. Business Strategy and the Environment* **25 (8)** p 545–59

[2] Winter G 2002 *Management – creating your company's own environmental action plan* (Minsk: Tekhnoprint) p 320

[3] Hart Sl 1995 A natural-resource-based view of the firm. *J. Academy of Management Reviev* **20 (4)** p 986–1014

[4] Zhuchenko Yu 2010 Assessment of business reputation of the company *J. Kreativnaya ekonomika* **9-2 (167)** p 32–37

[5] Nishitani K, Kaneko S, Fujii H and Komatsu S 2012 Are firms' voluntary environmental management activities beneficial for the environment and business? An empirical study focusing on Japanese manufacturing firms. Original Research Article *J. of Environmental Management* **105** p 121–30

[6] Alvarez-Garcia J and del RioRama M D 2016 Sustainability and EMAS: Impact of Motivations and Barriers on the Perceived Benefits from the Adoption of Standards *J. Sustainability* **8 (10)** p 1057–64

[7] Mamon N 2010 Methodological approach to image management *J. Management in Russia and foreign countries* **2** p 80–91

[8] Smirnova Yu 2009 Comparative analysis of different approaches to define notions of image and reputation *J. Marketing* **3** p 40–57

[9] Chen Y S 2008 The driver of green innovation and green image – Green core competence *J. of Business Ethics* **81 (3)** p 531–43

[10] Melnyk S A, Sroufe R P and Calantone R 2003 Assessing the impact of environmental management systems on corporate and environmental performance *J. of Operations Management* **21 (3)** p 329–51

[11] Martin-de Castro G, Amores-Salvado J and Navas-Lopez J E 2016 Environmental Management Systems and Firm Performance: Improving Firm Environmental Policy through Stakeholder Engagement *J. Corporate social responsibility and Environmental Management* **23 (4)** p 243–56

[12] Dukmasova N 2006 Problems and perspectives of EMS implementation in Russian companies. *J. Bulletin of Ural State Technical University-UPI. Economics and management* **9 (80)** p 116–22

[13] Dukmasova N and Ershova I 2013 Methodological approach to definition of economic benefits from EMS implementation *J. Bulletin of the Urals Federal University. Economics and management* **6** p 90–97

# Phytofilter – environmental friendly solution for purification of surface plate from urbanized territories

**O Ruchkinova**[1] **and I Shchuckin**[2]

[1] Doctor of Technical Sciences, Professor of Department of Heating, ventilation and water supply and sewerage, Perm National Research Polytechnic University, Russian Federation, Perm

e-mail: xgogax@mail.ru

[2] Senior Lecturer of Department of Heating, ventilation and water supply and sewerage, Perm National Research Polytechnic University, Russian Federation, Perm

e-mail: shchukin-is@yandex.ru

**Abstract**. Its proved, that phytofilters are environmental friendly solution of problem of purification of surface plate from urbanized territories. Phytofilters answer the nowadays purposes to systems of purification of land drainage. The main problem of it is restrictions, connecter with its use in the conditions of cold temperature. Manufactured a technology and mechanism, which provide a whole-year purification of surface plate and its storage. Experimentally stated optimal makeup of filtering load: peat, zeolite and sand in per cent of volume, which provides defined hydraulic characteristics. Stated sorbate and ion-selective volume of complex filtering load of ordered composition in dynamic conditions. Estimated dependences of exit concentrations of oil products and heavy metals on temperature by filtering through complex filtering load of ordered composition. Defined effectiveness of purification at phytofiltering installation. Fixed an influence of embryophytes on process of phytogeneration and capacity of filtering load. Recommended swamp iris, mace reed and reed grass. Manufactured phytofilter calculation methodology. Calculated economic effect from use of phytofiltration technology in comparison with traditional block-modular installations.

## 1. Introduction

Problem of removal and purification of surface plate (SP) is one of the most popular issues of urbanized terrains. Today for purification of land drainage there exist installations of mechanical and physical-and-chemical treatment. Together with moderate effectiveness of purification this installation have high construction and exploitation estimate, what limits the possibility of it extensive use.

Up-to-date systems of purification of surface plate have to answer following purposes: ability to purify wide range of pollutions; environmental friendliness of constructing and exploitation; maximal energy- and resource-saving; aesthetic visual appearance; ability to re-use the land drainage as a source of technical quality water supply. Installations of mechanical and physical-and-chemical purification of land drainage not in full answer following purposes.

Perspective growth option for purification systems land drainage is usage of bioengineered installations as of phytofilters, which are well known in world practice as *raingardens, bioretention filters, stormwater biofilters* (information about a principles of phytofiltration and structures of phytofilters is described in works [1–3]). Statistics of foreign investigations testify that these installations have high technological effectiveness in condition of low-cost construction and exploitation. Nevertheless on the territory of Russian Federation phytofilters nowadays have low distribution. One of the reasons of it are the limits on exploitation of phytofilters in the low temperature conditions (less than +10 ºC), when there is low trend of plants activity and soil microorganisms [4, 5]. This is very important for territories with mild climate, which are characterized the existence of winter season with below zero temperatures and icecap, because, at first, conversion times with near-zero temperature last long and, at second, on that period there appears significant quantity of land drainage.

## 2. Solution for sustaining high effective work of phytofillter and storage purified surface plate in the cold periods

For the purpose of sustaining high effective work of phytofillter in the periods of near-zero temperatures there is offered induction in composition of resin charge components, which are sorbate and ion-selective – peat and zeolite.

The chase of zeoilte as an addition to resin charge components is conditioned of its moderate cost on the one hand, and on the other hand high sorbate and ion-selective abilities in relation to hydrocarbon and heavy metals, which is proven by a large body of researches [6, 7].

Also there is known a positive influence of addiction of zeolites in soil, which influence on plant growth by means of keeping of air-and-water balance of soil [8].

High sorbate size of fen peat related to oil products and heavy metals is mentioned in work [9]. Peat besides its sorbate of oil products and heavy metals capabilities is a source of organic components for plants and soil microorganisms.

Thus in the cold periods purification from dissolved solids is achieved by the processes of sorbate and ion exchange on pointed out materials and in the period of vegetation of plant there take place its phyto- and bioregeneration due to absorption, transformation, stabilization and degradation of cumulated elements under the influence of plants and microorganisms, progressing in resin charge and rhizosphere. Besides the growth of root and top of plants enables the regeneration of void space and therefore capacity of resin charge. Thus the sustainable high quality of SP purification in the whole year is achieved and also long period of work of resin charge before renewal is achieved.

We offer engineering-and-technical solution of storage purified surface plate purposely to its re-use [10]. Realization of solution is provided by use of inert hard-grained bulk material (HGBM), for example, gravel and break-stone, situated under phytofilter (Figure 1). The use of HGBM with water-proof fundament for storage purified SP is cheaper than traditional solutions – reservoirs made of steel, plastic or fiberglass. Besides this construction isn't fracture-prone while frost-heaving phenomenon of watered soils and freezing of water in intergranular space.

## 3. Laboratory investigation of phytofiltration and phytoregeneration processes

For investigation of processes of phytofiltration and phytoregeneration was developed pilot lab phytofiltration mechanism. Filtering columns (Figure 2) are made of soil pipes made PolyVinylChloride, 160 mm diameter 1 m height with load of fixed composition. Columns were filled layer-by-layer like this: 100 mm – sand, 300 mm – complex load ; zeolite; sand in 1 : 2 : 7 proportion to volume; 100 mm – gravel 5–20 mm, 100 mm – gravel 20–40 mm. Substantiation of composition and chemical properties of complex filtering load is written in work [11]. Investigations in dynamic conditions have shown high sorbate and ion-selective capacity of complex filtering load (Table 1).

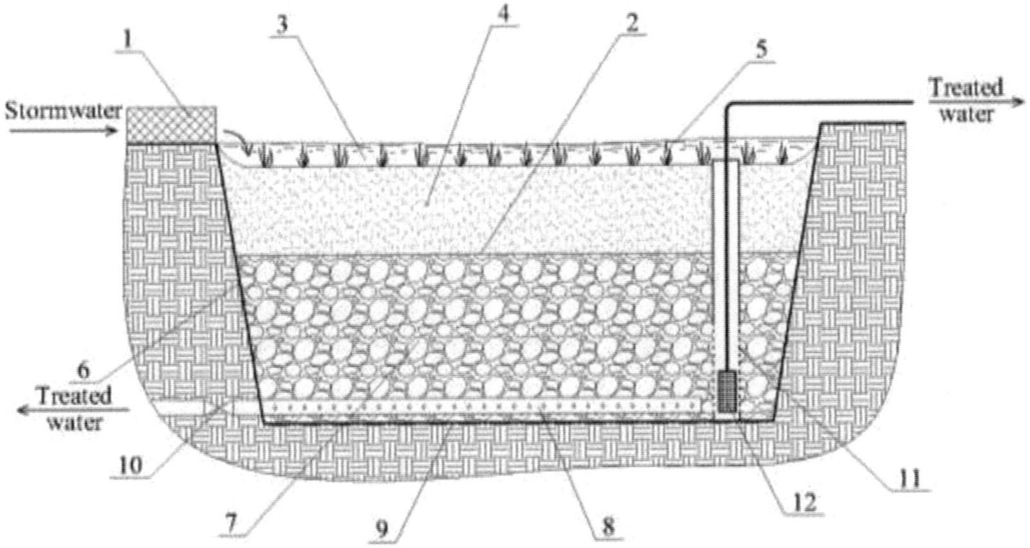


**Figure 1.** Engine for purification and storage of land drainage:
1 – settling chamber or filter (optional); 2 – treatment unit;
3 – bowl, filled with discharged water; 4 – resin charge; 5 – plants; 6 – reservoir with hard-grained bulk material;
7 – hard-grained bulk material; 8 – drain-pipe; 9 – impenetrable fundament;
10 – flow-metering mechanism; 11 – drill-hole; 12 – deep-water pump

**Table 1.** Results of investigation of sorbate and ion-selective capacity of complex filtering load of given composition in dynamic conditions

| Contaminator | Oil-prod. | Cu | Al | Pb | Fe |
|---|---|---|---|---|---|
| Sorbate (ion-selective) volume, mg/g (meq/g) | 40.08 | 0.505 (0.016) | 1.708 (0.061) | 0.738 (0.007) | 0.933 (0.049) |

In the filtering columns were planted common reed grass (*Phragmites commúnis*), mace reed (*Týpha latifólia*) and swamp iris (*Iris pseudacorus)*, 3 columns with plant of every type. Signed types were chosen on the fundament of carried out analysis of 15 plants, which are wide-spread in midlatitudes and meeting next requirements: hydrophily, heavy root system, unpretentiousness, cold resistance. Also there were concerned 3 control columns without plants.

For giving synthetic brine of land drainage individually in every column was used additional reservoir volume of 50 litres and deep-water pump.

As a source of water for watering and preparation of synthetic brine was used centralized aqueduct. For purifying faucet water from pollutions, which could affect the result of experiment (suspended solids, iron compounds, hardness, free chlorine) was foreseen three-level filtration system, containing filters of mechanical, sorbate and ion-selective purification. Filtered faucet water got tanked into supply tank with volume 50 litres, from which it was pumped by deep-water pump through the spreader pipe into filtering columns with period once a twenty-four hours when timer signals. The volume of sprinkling water was 1.5 litres for every filtering column. Excesses of water, which were filtered by the installation were poured in aqueduct. For sampling of purified water were foreseen flexible PolyVinylChloride hoses in the bottom of columns.

For the purpose of plats life support was foreseen the system of illumination with fluorescent lamps OSRAM FLUORA 36W (4 pieces) This lamps a made special for growing up plants and have optimal spectrum of illumination. Switching on and switching off the lamps was automatically by timer. Duration of photoperiod was 14h per day-and-night cycle. Scheme and exterior of phytofiltration mechanism is depicted on Figure 2 and 3.

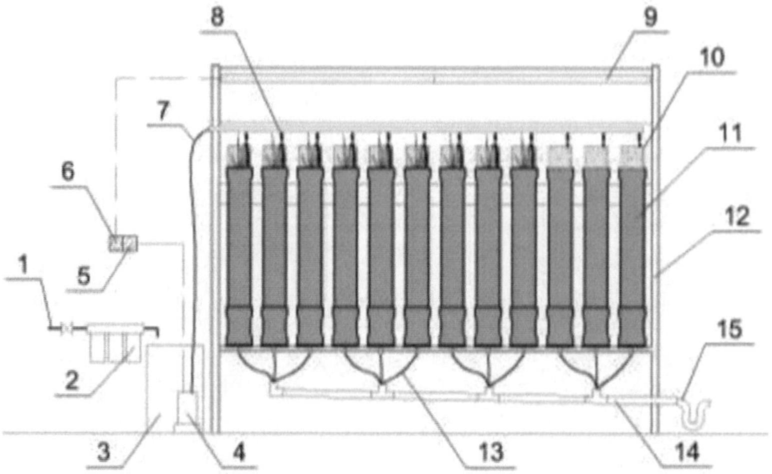


**Figure 2.** Scheme of pilot phytofiltration mechanism:
1 – giving of municipal water; 2 – filtration system; 3 – consumption tank; 4 – deep-water pump; 5 – timer, that controls watering; 6 – timers, that controls illumination; 7 – system of water conveying; 8 – drop bottle; 9 – fluorescent lamp; 10 – clear tube h=250mm.; 11 – phytofiltering column; 12 – timber framing; 13 – sampling hose; 14 – ready-made outflow tube; 15 – water remove in canalization

**Figure 3.** Exterior of pilot phytofiltration mechanism

Phytofiltration effectiveness was investigated by giving in filtering columns surface plate imitator, which contains oil products and heavy metals in concentrations, which are showed in Table 2. In every column in the 7 week period was given 350 litres of SP imitator, which is 70 % of year volume of land drainage in condition of rainfall intense 600 mm per year and square of impenetrable water-gathering territory 40 times bigger than phytofilters square. Samples collection of effluents with determination of degree purification from pollutants took place after giving every 50 litres of SP imitator.

**Table 2.** Composition of surface plate imitator

| Parameter | Reagent for preparation SP imitator | Concentration, mg/litres |
|---|---|---|
| Suspended materials | Sweepings form car road, bolted through bolter № 008 | 500 |
| Oil products, mg/litres | Blend of naphtha Nefras (Нефрас mineral thinner) C2-80/120 and industrial oil И-20A in 1:1 volume proportion | 30 |
| Fe | $FeCl_3$ | 1 |
| Cu | $CuCl_2$ | 0.3 |
| Pb | $Pb(NO_3)_2$ | 0.5 |
| Al | $Al_2(SO4)_3$ | 0.4 |

Results of work of pilot phytofilter mechanism are represented in Table 3.

**Table 3.** Range and central case of concentrations of pollutants in effluents after phytofiltering columns

| Plant type (column numbers) | Range of concentrations of pollutants in effluents, mg/litres<br>Average concentration in effluents, mg/litres | | | | | |
|---|---|---|---|---|---|---|
| | Susp. mat. | Oil-prod. | Cu | Al | Pb | Fe |
| Mace reed (1-3) | 5.2–13.6<br>9.37 | 0.43–1.21<br>0.62 | 0.003–0.006<br>0.005 | 0.01–0.05<br>0.034 | 0.1–0.13<br>0.119 | 0.025–0.108<br>0.075 |
| Reed grass (4-6) | 6.9–10.2<br>8.24 | 0.273–0.618<br>0.452 | 0,001–0,009<br>0.007 | 0.01–0.08<br>0.035 | 0.068–0,17<br>0.097 | 0.024–0.125<br>0.086 |
| Swamp iris (7-9) | 5.8–13.2<br>7.65 | 0.227–1.34<br>0.49 | 0.002–0.006<br>0.005 | 0.01–0.02<br>0.017 | 0.1–0.14<br>0.109 | 0.04–0.062<br>0.051 |
| Control (10-12) | 6.2–16.8<br>14.1 | 0,49–1,36<br>0.86 | 0.002–0.007<br>0.05 | 0.01–0.03<br>0.022 | 0.152–0.161<br>0.164 | 0.032–0.112<br>0.093 |
| MAC 1* | – | 0.05 | 0.001 | 0.04 | 0.006 | 0.1 |
| MAC 2** | – | 0.1–5 | 0.1–4 | 1–10 | 0.2–1 | 2–20 |

*Maximum Allowed Concentration in Russian Federation commercial fishing branch
**Maximum Allowed Concentration in European Union [12]

On the fundament of analysis results of work of pilot phytofiltration mechanism (Table 3) following conclusions were made. Effectiveness of purification of surface plate from suspended materials was amounted 97.2–98.9 %, oil products 95.5–98.9 %, Cu – 97–99.3 %, Al – 80–86.4 %, Pb – 66–86.4 %, Fe – 87.5–97.5 %. The opportunity of drain of surface plate after phytofiltration in the water object is determined by national directives to content of pollutants in water basin. Purified surface plate suits European Union directives to discharge in water basin. For surface plate gathering after phytofiltering in water basins intended to commercial fishing on the territory of Russian Federation there needs deep additional cleaning from oil products and heavy metals wit use of absorbent carbon.

According to offered concept of phytofilter work in the conditions of mild climate, cleaning from salted oil products and heavy metals in the periods of cold temperatures is carried out by the sorbate and ion-selective processes in complex filtering load. For determintaion of purification effectiveness in the cold temperature conditions filtration of SP imitator through the sorbate column at temperature +2 °C, +8 °C and +20 °C took place. Exit concentrations after filtering at pointed out temperatures are shown in Figure 4.

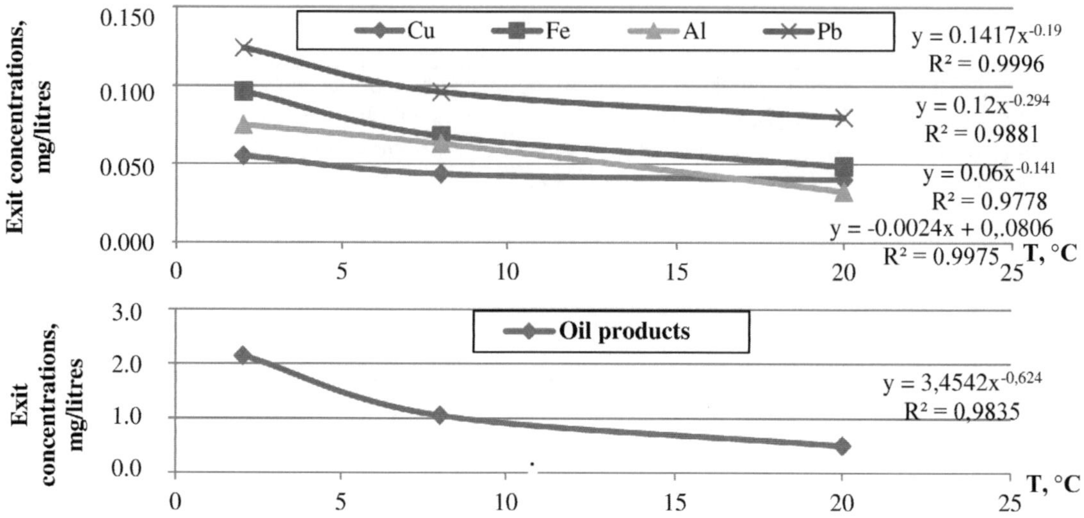


**Figure 4.** Dependence of exit concentrations on temperature

In condition of falling temperature from +20 °C to +2 °C effectiveness of purification from heavy metals have lowered at the level of 4.8–10.6 %, from the oil products – 5.5 %.

For evaluating the speed of phytoregeneration of filtering load there were analyzed dynamics of fluctuations of oil products concentration and motile form of heavy metals in root layer of filtering load in the period of 2 months after breaking the giving surface plate imitator.

Filtering load selection was made from near-root layer of plants, as nearly as possible to the roots.

The results of investigation of phytoregeneration of filtering load under the influence of embryophytes are shown in Figure 5.

There is ascertained that average speed of lowering of the concentration of heavy metals in columns with swamp iris was made – 8.33 mg/kg·month, Al – 4.39 mg/kg·month, Pb – 2.83 mg/kg·month, Fe – 7.77 mg/kg·month, in columns c reed grass: Al – 2.85 mg/kg·month, Pb – 0.58 mg/kg·month, Fe – 7.67 mg/kg·month.. The decrease of concentration of heavy metals in filtering load in columns with mace reed wasn't signed.

Decrease in oil products in filtering load was seen in all columns: with mace reed at 192.41 mg/kg·month, reed grass at 249.3 mg/kg·month, iris at 137.27 mg/kg·month, at control column – 113.60 mg/kg·month. Decrease of oil products concentration in control columns is explained by its degradation by soil microorganisms and vapor of floats. However all columns with plants have shown more than control columns extract of oil products from the load. This can be explained by degradation of oil products by the plants and by stimulation effect by rhizosphere microorganisms, going on at the plant's root layer.

Aiming at determine calculated water volume, which could be accumulated in the volumetric unit hard-grained bulk material, was experimentally estimated water return of river roundstone and granite macadam of different fraction.

Water return of investigated materials more or less didn't depend on fraction and worked out at the average 450 litres/m³ (321 litres/t) for granite macadam and 400 litres/m³ (250 litres/t) for river roundstone.

On the fundament of undertaken studies was developed calculation methodology of phytofilter, which allows depending on square and type of surface of well field, quality of off-taken surface plate, consumers directive to quality and quantity of purified rain waters to define its overall dimensions, periodicity of renewal of filtering load, volume of tank accumulation of purified waters.

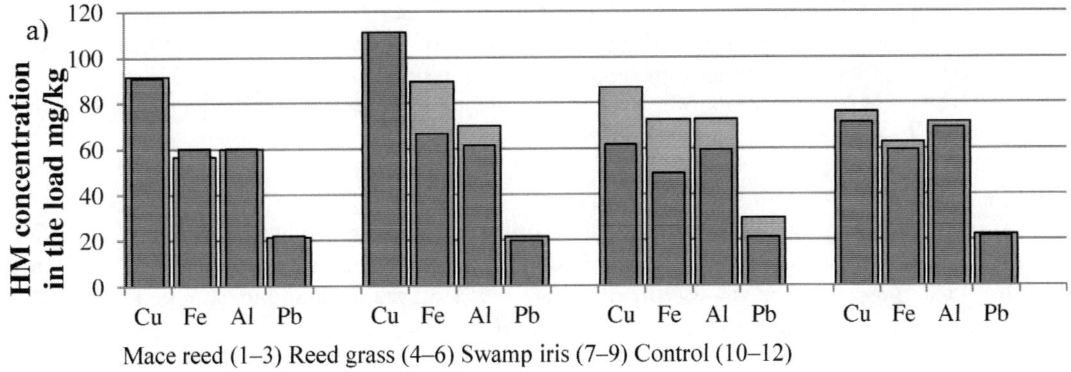


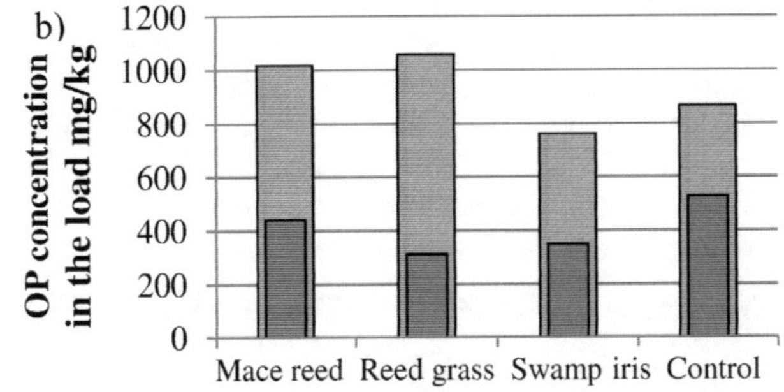


**Figure 5.** Concentration of oil products and heavy metals in filtering load:
a – heavy metals;
b – oil products
■ before phytoregeneration
■ after phytoregeneration (2 month)

## 4. Economic effect from the use of phytofiltering technology and technical solution on storage purified land drainage in hard-grained bulk material

Offered phytofiltering technology was adopted in the process of production of design and estimate documentation in project "Purifying installations for cleaning land drainage for "Yugovskiy integrated plant of milk products"". Economic effect from the use of phytofiltering technology in compare with traditional block-modular installations, estimated by method of life cycle cost estimate equaled 27.09 million of rubles. This effect was reached by lowering cost on handling purifying installations, lowering periods of renewal of sorbate mechanism of additional cleaning, absence of costs for transfer of waste water.

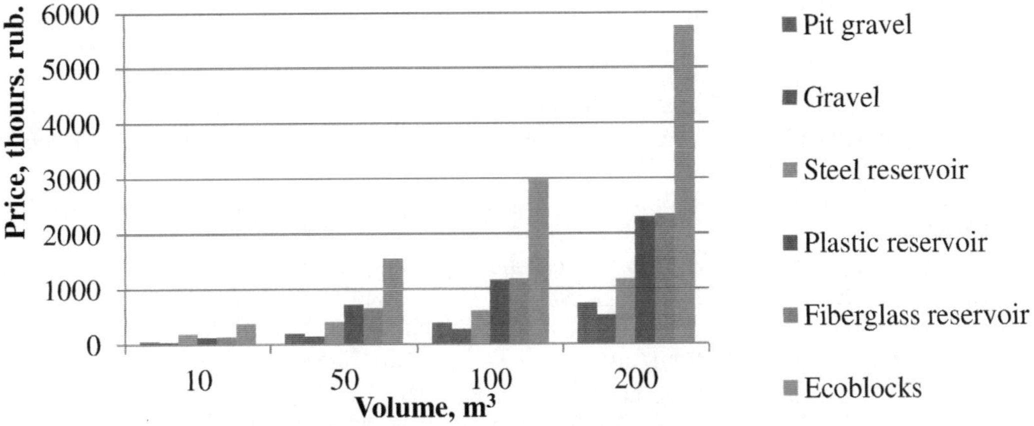


**Figure 6.** Comparison of price of storage purified land drainage

For economic justification of carried technical solution on storage purified land drainage in hard-grained bulk material was carried out calculation materials costs and labor for storage 50 $m^3$ phytofilter purified rainwater (Figure 6). Calculation was carried out in prices for $2^{nd}$ quarter of 2016 year on virtue of catalogues of producers, price of 1 t of gravel of 40–600 mm fraction is 350 rub, pit gravel – 800 rub, water return of pit gravel is 450 litres/$m^3$, gravel – 400 litres/$m^3$.

Calculation has shown that costs of organizing of storage of purified land drainage in comparison with offered solution is lower for 50.8 – 276.8%.

## 5. Conclusion

Manufactured technology of phytofiltration for purification of surface plate, inclusive of conditions of mild climate zone and species composition of plant in temperate latitudes. Developed technical mechanism, which provides the whole-year purification of land drainage and its storage in order to its next re-use. (Patent of Russian Federation No. 2540620).

Experimentally defined optimal composition of filtering load: peat 10 %, zeolite 20–40 %, sand 50–70 % of volume, which provides specified technical characteristics – speed of free-flow filtering 0.3 m/h in condition of head of liquid 0.3 m. Sorbate and ion-selective volume of complex filtering load specified composition in dynamic conditions equaled: oil-products – 40.08 mg/g, Cu – 0.505 mg/g, Al – 1.708 mg/g, Pb – 0.738 mg/g, Fe – 0.933 mg/g.

Estimated dependences of exit concentrations of oil products and heavy metals from temperature, while filtering through complex filtering load specified compositions.

On the fundament of experimental investigations estimated effectiveness of purification at phytofiltering installation: from suspended materirals – 97.2–98.9 %, oil products – 95.5–98.9 %, Cu – 97–99.3 %, Al – 80–86.4 %, Pb – 66–86.4 %, Fe – 87.5–97.5 %.

Estimated influence of embryophytes: swamp iris (Iris pseudacorus), reed grass (Phragmites commúnis) and mace reed (Týpha latifólia) on the process of phytoregeneration and capacity of filtering load. Experimentally defined the speed of phytoregeneration of filtering load from heavy metals and oil products. Swamp iris has made – 8.33 mg/kg·month, Al – 4.39 mg/kg·month, Pb – 2.83 mg/kg·month, Fe – 7.77 mg/kg·month, in columns c reed grass: Al – 2.85 mg/kg·month, Pb – 0.58 mg/kg·month, Fe – 7.67 mg/kg·month, oil products 249.3 mg/kg·month in condition of plant population 50 plants/m2. Determined that effectiveness of regeneration of capacity by plants has down trend like this: swamp iris > reed grass > mace reed.

Manufactured phytofilter calculation methodology, which helps in condition of specified characteristics to calculate needed square of phytofilter, operating time before filtering load renewal, to define volume of hard-grained bulk material for purified water storage.

Economic effect from the use of phytofiltering technology in compare with traditional block-modular installations, estimated by method of life cycle cost estimate equaled 27.09 million of rubles.

## References

[1] Shchukin I 2013 Analysis of existing bioengineering installation of purification ladnd drainage and opportunity of its use in conditions of East Ural *Perm National Research Polytechnic University Messenger* **2** p 122–32

[2] Shchukin I and Melekhin A 2012 The use of filters to purify the land drainage *Water Magazine* **11** p 6–9

[3] Shchukin I 2015 Technology of purification and usage of land drainage from urbanized land in cthe conditions of cold climate *Questions Of Up-to-date Theory and Practice* **2** p 34–40

[4] Antoniadis V and Alloway B 2001 Availability of Cd, Ni and Zn to ryegrass in sewage sludge-treated soils at different temperatures *Water, Air, and Soil Pollution* **132** p 201–14

[5] Blecken G, Zinger Y, Muthanna T, Deletic A, Fletcher T and Viklander M 2007 Biofilter treatment of stormwater: temperature influence on the removal of nutrients *Water Science & Technology* **56** p 83–91

[6] Savin A, Moroxov G, Bondyrev M, Breus V, Nekludov S and Breus I 2011 Bind of fleeting oil

IOP Conf. Series: Earth and Environmental Science **72** (2017) 012002 doi:10.1088/1755-1315/72/1/012002

hydrocarbon by natural zeolite-elay adsorbents *Technology Of Oil and Gas* **3** p 7–15
[7] Chutia P, Kato S, Kojima T and Satokawa S 2009 Arsenic adsorption from aqueous solution on synthetic zeolites *J. Hazard. Matter* **162** p 440–47
[8] Koretskiy A and Ignatova A 2010 Investigation of zeolite influence on chamber of Pegasus for environmental development *Bulletin of the Kuzbass State Technical University* **2** p 92–5
[9] Kasitskaya U and Sarkison N 2007 *Peat resources of Tomsk Oblast and the ways of its use in building* (Tomsk) 292 p
[10] Shchukin I and Melekhin A 2015 Device for Purification of Surface Sewage Waters Patent 2540620 of Russian Federation
[11] Shchukin I and Ruchkinova O 2017 Complex filtering load for phytofilters for land drainage putyfying *Perm Nationla Research University Messenger* **1** p 105–15
[12] RHTU: *MAC of land drainage* Available from: http://enviropark.ru/course/category.php?id=20 [Accessed 13.04.2017]

# Environmental assessment of the measures increasing the sustainability of motor transport

**E Magaril[1], R Magaril[2] and L Abrzhina[1]**

[1] Department of Environmental Economics, Ural Federal University, Mira str., 19, Ekaterinburg, 620002, Russian Federation
[2] Department of Oil and Gas Processing, Tyumen Industrial University, Volodarskogo str., 38, Tyumen, 625000, Russian Federation

E-mail: larisa.abrzhina@mail.ru

**Abstract**. Ensuring the sustainable development of motor transport, whose emission of greenhouse gases and other contaminants is one of the main sources of air pollution, requires informed decision-making on the implementation of effective environmental protection measures. A simplified methodological approach to a comparative assessment of the effectiveness of these environmental protection measures has been developed. The offered approach has been approved at estimating application for motor fuels of the multipurpose additive, which increases both the energy efficiency and environmental safety of motor transport.

## 1. Introduction

Despite the advantages that modern society derives from the use of transport, technological progress achieved and the tightening of emissions standards, automotive transport remains one of the main sources of pollution by toxic substances and greenhouse gases, detrimental to human health. Vehicle emissions create increased concentrations of pollutants, especially in cities where their contribution to air pollution in the breathing zone is significantly higher than that from other sources.

Apparently, emissions from vehicles of of benzo(α)pyrene and its analogues – polycyclic aromatic hydrocarbons, which are the strongest of all carcinogens, is the most serious problem. Emissions of polycyclic aromatic hydrocarbons are maximal in idle running, i.e. near traffic lights, and thus they come directly to the breathing zone of people, while their content in exhaust gases exceeds the maximum allowable concentrations (MAC) level by an order of magnitude [1, 2]. Due to the high boiling point (310 °C at 1.33 kPa for benzo(α) pyrene) and chemical inertness, these hydrocarbons are adsorbed on the road surfaces, soil, etc. Figure 1 shows dynamics of average annual concentrations of benzo(α)pyrene on average throughout Russia in 2011–2015; figure 2 shows dynamics of lung cancer incidence in 2004–2014. In large cities, the concentration of benzo(α)pyrene exceeds the MAC to a significantly greater degree. There is an increase in the average incidence (by about 20 % over 10 years) despite a decrease in the level of benzo(α)pyrene concentrations over the past 5 years, which is explained by both the cumulative effect of harmful influence and synergistic effect of various factors.

To reduce the environmental footprint of the transport sector various instruments and methods are used, including upgrading road infrastructure [5], rational organization of road traffic [6, 7], improvement of the engines and cars' design [8, 9], and the use of fuels with improved environmental characteristics [10–14]. Methods to improve the sustainability of motor transport are characterized by their differences in timing, efficiency and costs for their implementation.

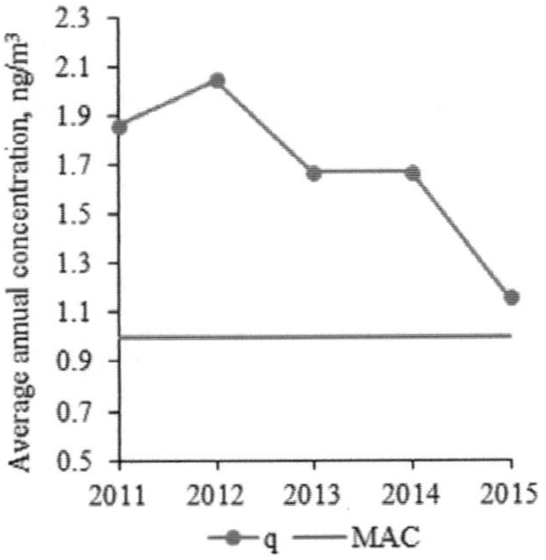


**Figure 1.** Dynamics of average annual concentrations of benzo(α)pyrene (q, ng/m$^3$) in Russian cities, relative to the level of daily average maximum allowable concentration, MAC (according to [3])

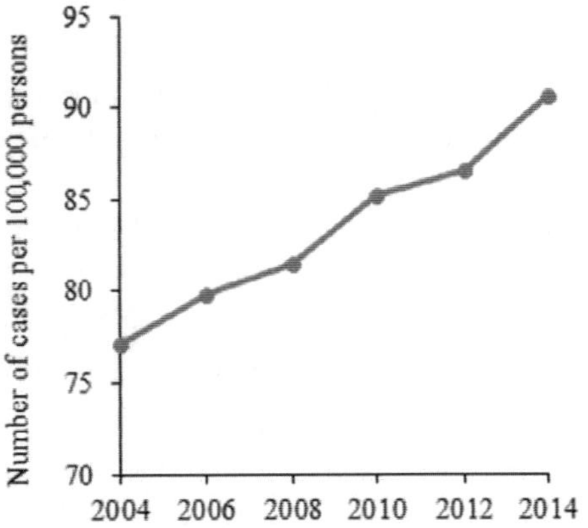


**Figure 2.** Dynamics of lung cancer incidence in Russia (according to [4])

In order to rapidly implement highly efficient technological solutions that increase the environmental sustainability of motor transport it is necessary to compare possible alternative scenarios with an assessment of externalities and costs. Given the difficulty of fully accounting for externalities, it seems important to use simplified procedures for assessing the feasibility of implementing the environmental measures that can identify the dominant effects of implementation.

## 2. Methodical approach to accounting for externalities of the motor transport operation and assessment of the effectiveness of the environmental protection measures

Exhaust gases of vehicles contain a significant enumeration of toxic substances that directly enter human's breathing zone. At the same time, the contribution of various components of the exhaust gases to the total toxicity varies significantly; thus the ratio of gross emissions without taking into account the toxicity of the components does not adequately characterize the level of the negative impact as a whole, nor does it allow the contribution of the individual pollutants to be readily estimated to the total toxicity of the exhaust gases. Conversion into the conditional mass units while taking into account the relative environmental hazard allows one to solve this problem and calculation of the equivalent mass $G$, conventional ton, can be made using the formula (1):

$$G = \sum_{i=1}^{N} G_i \cdot K_i , \qquad (1)$$

where $G_i$ is the actual mass of the i-th pollutant, entering the atmospheric air during the reporting period, calculated in tons; $K_i$ is the coefficient of the relative environmental hazard of the i-th pollutant, calculated as the reciprocal value of the maximum allowable concentration of the i-th pollutant ($MAC_i$), conventional ton per ton.

The relative environmental hazard of emission components should be taken into account when calculating the atmospheric air damage (for example, in Russia this approach was used in [15], equation (2)):

$$D_{ps}^{el} = d_{sp} \cdot G \cdot K_e \cdot J_d \qquad (2)$$

where $D_{ps}^{el}$ is the environmental damage from polluting toxic substances emission to the atmospheric air, monetary units per year; $d_{sp}$ is the index of specific atmospheric air damage caused by emission of the mass unit of a pollutant, monetary units per conventional ton; $K_e$ is the coefficient of the environmental situation and environmental significance of the state of the atmospheric air for a given district; $J_d$ *is* a deflator index used for damage index calculation from the current price to the base (comparable) one.

In the environmental assessment of implementation of the specified environmental measure relative to the basic situation, the effectiveness of the environmental protection measure $e_{ps}$, %, concerning the reduction of emissions of polluting toxic substances and prevention of the corresponding environmental damage during the reporting period can be determined according to the formula (3):

$$e_{ps} = \frac{D_{ps0}^{el} - D_{ps1}^{el}}{D_{ps0}^{el}} \cdot 100\% = \frac{G_0 - G_1}{G_0} \cdot 100\% = \frac{\Delta G}{G_0} \cdot 100\% \tag{3}$$

where $D_{ps0}^{el}$, $D_{ps1}^{el}$ is the environmental damage from the atmospheric air pollution by the emissions of polluting toxic substances, and $G_0$, $G_1$ is the equivalent mass of the emissions, accordingly, in the base case and after implementation of the environmental protection measure; $\Delta G$ – the prevented equivalent mass of the emissions of polluting toxic substances after implementation of the environmental protection measure.

Thus, in order to assess possible scenarios aimed at reducing the damage from emissions of polluting toxic substances a relative change in the equivalent mass of the pollutants should be taken into account. It is thus possible to evaluate the impact of the environmental protection measure on the emissions of the individual pollutant and prevention of the corresponding environmental damage caused by the specified component of the emissions.

For a more complete accounting for externalities of the motor transport operation when evaluating the atmospheric air damage it had been previously suggested to take into account the negative effects caused by not only emissions of polluting toxic substances, but also carbon dioxide emissions as well as the consumption of atmospheric oxygen. Formulas have been proposed for calculating the environmental damage from carbon dioxide emissions $D_{CO_2}^{el}$, and from consumption of oxygen during combustion of fuel, $D_{O_2}^{el}$ [16]. When all the constants are combined, these harmful contributions can be estimated, respectively, according to the equations (4) and (5):

$$D_{CO_2}^{el} = d_{CO_2} \cdot a \cdot J_d \cdot G_f \tag{4}$$

where $d_{CO_2}$ is the specific environmental damage from the emission to the atmospheric air of 1 ton of carbon dioxide, monetary units per ton; $G_f$ is the actual mass of the fuel consumed during the reporting period, calculated in tons; a is the constant.

$$D_{O_2}^{el} = b \cdot J_d \cdot G_f, \tag{5}$$

where $b$ is the constant. In the environmental assessment of the environmental protection measure's impact on the environmental damage from carbon dioxide emissions, and on the environmental damage from oxygen consumption, relative to the basic case, the effectiveness of environmental protection measures $e_{CO_2}$, % to reduce carbon dioxide emissions during the reporting period, and efficiency $e_{O_2}$, %, to reduce consumption of oxygen, at the different absolute values of the corresponding damages prevented, will be equal according to the equations (6) and (7):

$$e_{CO_2} = \frac{D^{el}_{CO_{20}} - D^{el}_{CO_{21}}}{D^{el}_{CO_0}} \cdot 100\% = \frac{G_{f_0} - G_{f_1}}{G_{f_0}} \cdot 100\% = \frac{\Delta G_f}{G_{f_0}} \cdot 100\% \quad (6)$$

$$e_{O_2} = \frac{D^{el}_{O_{20}} - D^{el}_{O_{21}}}{D^{el}_{O_0}} \cdot 100\% = \frac{G_{f_0} - G_{f_1}}{G_{f_0}} \cdot 100\% = \frac{\Delta G_f}{G_{f_0}} \cdot 100\% \quad (7)$$

Thus, in order to assess possible scenarios to reduce damage from carbon dioxide emissions and from consumption of oxygen in the combustion process, the fuel consumption changes should be taken into account.

This approach is also applicable when comparing alternative options for the environmental protection measures.

## 3. Environmental assessment of application of the multifunctional additive to motor fuels as a method to increase environmental sustainability of motor transport

For different countries their differing priorities can mean their directions to improve the environmental sustainability of vehicles can differ. Meanwhile, the use of highly effective additives that improve the properties of hydrocarbon fuels is a quick and inexpensive way to solve this problem and therefore, it is promising for both developed and developing countries.

The multifunctional all-purpose additive for fuels [12, 17] has a complex positive effect on the environmental and performance characteristics of vehicles using gasoline and diesel fuel at the very low cost of the method equal to $ 1per ton of gasoline and $ 3 per ton of diesel fuel. To assess the environmental effectiveness of the additive application using the proposed simplified approach, the significant factors, which should be taken into account, are the effect of the additive on pollutant emissions and fuel consumption, Table 1.

**Table 1.** Effect of the multifunctional additive application in optimal concentrations on the polluting substances' emissions and fuel consumption

| Characteristics | Vehicles with gasoline engines | Vehicles with diesel engines |
|---|---|---|
| Reduction of emission, on average | | |
| CO | –23 % | –16 % |
| $NO_x$ | –23 % | –21 % |
| $C_mH_n$ | –22 % | –36 % |
| benzo(a)pyrene and its analogs | –95 % | – |
| $PM_{2.5}$ | – | –45 % |
| Fuel consumption, on average | –10 % | –6 % |

In the retrospective assessment of the application of the environmental protection measure proposed in Russia, we assume that in the reporting period (2010–2015) the emissions of the standard pollutants (CO, $NO_x$, $C_mH_n$, $PM_{2.5}$) of all vehicles meet the Euro-3 standards. The data on fuel consumption were used to determine the annual amount of emissions [18], the relative hazard coefficients $K_i$, conventional tons per ton, of the specified pollutants were taken to be equal, according to the procedure [15], to 0.4 for CO, 16.5 for $NO_x$, 0.7 for $C_mH_n$, 33.5 for $PM_{2.5}$. Figure 3 demonstrates the effect of the additive on the equivalent mass of the sum of these pollutants. Figure 4 shows the change in the case of the individual pollutants, taking into account the relative toxicity of emissions.

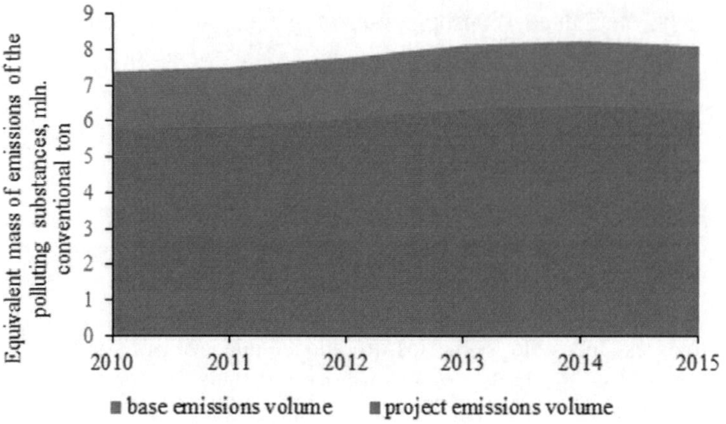


**Figure 3.** Change to the equivalent total emissions of CO, $NO_x$, $C_mH_n$, $PM_{2.5}$ after implementation of the environmental protection measure relative to the base scenario

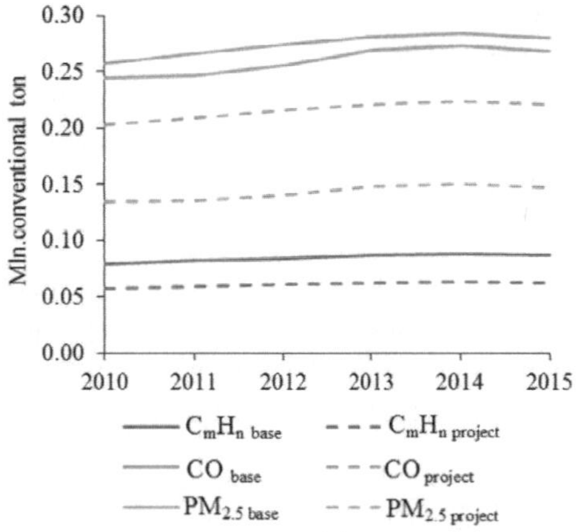


(a)

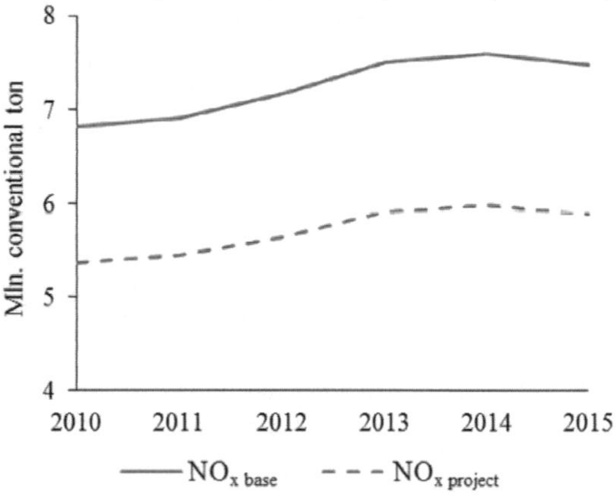


(b)

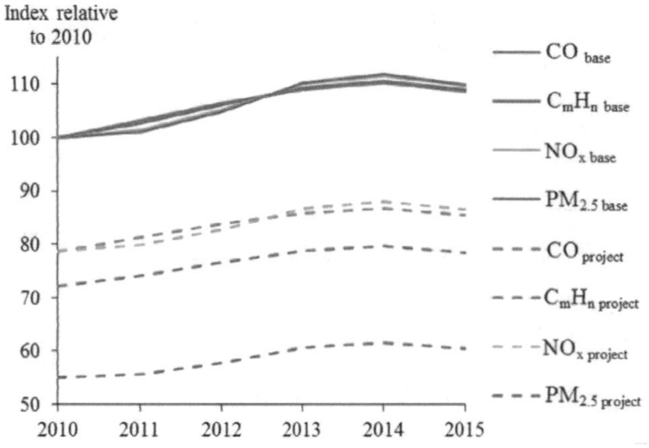


(c)

**Figure 4.** Dynamics of the equivalent mass of the emission components: (a) – CO, $C_mH_n$, $PM_{2.5}$, (b) – $NO_x$, and of the equivalent mass indexes (c) after implementation of the environmental protection measure relative to the base scenario

It should be noted that the relative toxicity of benzo(α)pyrene ($C_{20}H_{12}$),which is the strongest carcinogen, is significantly higher than that of the gaseous toxic components of emissions and particulate matter, and according to the procedure of [15] it is assumed to be 12,500 conventional tons per ton. It is advisable to calculate the effectiveness of the implementation of this environmental measure for the reduction of the amount of CO, $NO_x$, $C_mH_n$, $PM_{2.5}$ and benzo(α)pyrene emissions separately. Figure 5 shows the calculated effectiveness of the multifunctional additive to fuels relative to the reduction of CO, $NO_x$, $C_mH_n$, $PM_{2.5}$, carbon dioxide and oxygen consumption (taking into account fuel consumption data [18]), as well as relative to the reduction of benzo(α)pyrene ($e_{C_{20}H_{12}}$).

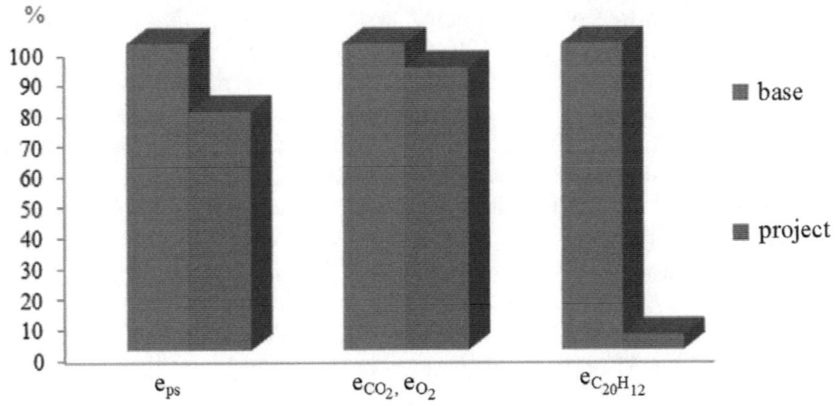


**Figure 5**. Effectiveness of the multifunctional additive application in preventing environmental damage

Thus, the assessment carried out demonstrates that the environmental protection measure under consideration significantly increases the environmental sustainability of vehicles. Given the 95 % reduction in benzo(α)pyrene emissions, and its high relative hazard, which is 3–5 times greater than the relative hazard of emissions of other pollutants under consideration, this effect of the additive application is of the highest environmental significance. A high level of environmental compatibility, substantial fuel-savings and low cost of implementation are significant arguments for the extensive use of the environmental protection measure proposed.

## 4. Conclusion

The need to make informed decisions on the selection of optimal measures that will improve the environmental sustainability of motor transport while taking into account the complexity of fully accounting for external factors makes it important to develop simplified approaches to assess the effectiveness of the implementation of the relevant environmental protection measures. An approach was proposed for assessing the effectiveness of environmental protection measures in the operation of motor vehicles and also considering externalities caused by emissions of polluting toxic substances, carbon dioxide and the consumption of oxygen during fuel combustion. The proposed approach allows the effectiveness of the environmental protection measure to reduce emissions of polluting toxic substances and corresponding environmental damage to be evaluated by calculating the relative change in the equivalent mass of emissions. The effectiveness of environmental protection measures at reducing carbon dioxide emissions and oxygen consumption during fuel combustion and to prevent appropriate environmental damage is proposed to be estimated from the calculation of the relative change in fuel consumption. The use of the proposed approach for assessing the effectiveness of use of the multifunctional fuel additive made it possible to give recommendations for its wide application. The developed simplified approach can be used in assessing the effectiveness of the environmental protection measures both for motor transport, and for stationary sources of atmospheric pollution.

**Acknowledgement**
The work was supported by Act 211 Government of the Russian Federation, contract No. 02.A03.21.0006.

**References**

[1] Magaril E 2013 The influence of carbonization elimination on the environmental safety and efficiency of vehicle operation *Inter-national Journal of Sustainable Development and Planning* **8(2)** p 231–45
[2] Magaril E 2015 Carbon-free gasoline engine operation *International Journal of Sustainable Development and Planning* **10 (1)** p 100–8
[3] Yearbook of the state of atmospheric pollution in cities throughout Russia in 2015 Available from: http://www.meteorf.ru/product/ infomaterials/ezhegodniki/ [Accessed 30th April 2017]
[4] Russian Cancer portal Available from: http://www.oncoforum.ru/o-rake/statistika-raka/rak-legkogo-prognozy-i-statistika.html [Accessed 30th April 2017]
[5] Cantarella GE, Vitetta A 2006 The multi-criteria road network design problem in an urban area *Transportation* **33 (6)** p 567–88
[6] Polimeni A and Vitetta A 2014 Vehicle routing in urban areas: an optimal approach with cost function calibration *Transportmetrica B-Transport Dynamics* **2 (1)** p 1–19
[7] Yin Y F and Lawphongpanich S 2006 Internalizing emission externality on road networks *Transport Res D-Tr E* **11 (4)** p 292–301
[8] Gibbs L, Anderson B, Barnes K et al 2009 *Motor gasolines technical review* (FTR-1) (Chevron Corporation, USA)
[9] Mark J and Morey C 2000 *Rolling smokestacks: cleaning up America's trucks and buses* (UCS, Massachusetts)
[10] Boulter P G and Latham S 2009 *Emissions factors 2009: Report 5-a review of the effects of fuel properties on road vehicle emissions* TRL Report PPR358 (TRL Limited, Wokingham)
[11] Magaril E 2014 The solution to strategic problems in the oil refining industry as a factor for the sustainable development of automobile transport *WIT Transactions on Ecology and the Environment* **190 (2)** p 821–32
[12] Magaril E 2015 Increasing the efficiency and environmental safety of vehicle operation through improvement of fuel quality *International Journal of Sustainable Development and Planning* **10(6)** p 880–93
[13] Anderson L G 2015 Effects of using renewable fuels on vehicle emissions *Renew Sust Energ Rev* **47** p 162–72
[14] Pearson R J and Turner J W G 2014 Using alternative and renewable liquid fuels to improve the environmental performance of internal combustion engines: key challenges and blending technologies. In: Folkson R (ed) Alternative fuels and advanced vehicle technologies for improved environmental performance *Towards zero carbon transportation* (Woodhead Publishing Ltd, UK) p 52–89
[15] Vershkov L, Groshev V and Gavrilov V 1999 *Prevented Environmental Damage Identification Methodology* (Moscow: State Committee for Environmental Protection)
[16] Magaril E, Abrzhina L and Belyaeva M 2014 Environmental damage from the combustion of fuels: Challenges and methods of economic assessment *WIT Transactions on Ecology and the Environment* **190 (2)** p 1105–15
[17] Magaril E and Magaril R 2016 Improving the environmental and performance characteristics of vehicles by introducing the surfactant additive into gasoline *Environmental Science and Pollution Research* **23(17)** p 17049–57
[18] Analytical agency "Autostat" Available from: http://www.autostat.ru [Accessed 30th April 2017]

IOP Conf. Series: Earth and Environmental Science **72** (2017) 012004 doi:10.1088/1755-1315/72/1/012004

# Reagent removal of manganese from ground water

**G Brayalovsky, E Migalaty and O Naschetnikova**

Department of Water Management and Technology, Ural Federal University named after the first President of Russia B. N. Eltsin, 17, Mira str., Ekaterinburg, 620002, Russia

e-mail: o.b.naschetnikova@urfu.ru

**Abstract**. The study is aimed at the technology development of treating drinking water from ground waters with high manganese content and oxidizability. Current technologies, physical/chemical mechanisms and factors affecting in ground treatment efficiency are reviewed. Research has been conducted on manganese compound removal from ground waters with high manganese content (5 ppm) and oxidizability. The studies were carried out on granular sorbent industrial ODM-2F filters (0.7–1.5 mm fraction). It was determined that conventional reagent oxidization technologies followed by filtration do not allow us to obtain the manganese content below 0.1 ppm when treating ground waters with high oxidizability. The innovative oxidation-based manganese removal technology with continuous introduction of reaction catalytic agent is suggested. This technology is effective in alkalization up to pH 8.8–9. Potassium permanganate was used as a catalytic agent, sodium hypochlorite was an oxidizer and cauistic soda served an alkalifying agent.

## 1. Introduction

Almost all ground water treatment plants operate on the conventional "aeration-filtration" scheme (simplified aeration). However, plants' long-time operating experience has shown that the basic aeration treatment technology followed by light granular sorbent filtration fails to provide quality target water in terms of manganese ions [1]. Reducing the manganese content by means of conventional treatment methods takes place under increased pH values, significant excess of iron over manganese and oxidation not more than 3 $mgO_2/l$ [1]. Currently, the issue of high oxidizability in ground treatment employing reagent methods is insufficiently explored and remains valid.

In ground treatment in noncentralised water supply environment can be realized by means of development, research and adoption of up-to-date high-performance water treatment technology employing reliable small water purification units [1, 2]. Hence, the following trends are considered urgent nowadays: development and application of supplementary techniques in order to enhance manganese removal processes; search for optimal doses and reagent combinations; and research of modern catalytic material properties [3].

The Paper presents the studies conducted on in ground treatment using granular sorbent industrial ODM-2F filters; the fraction was 0.7–1.5 mm, the bed depth equaled 1 m and filter velocity was 10m/hr. Reagents were introduced directly to filter system.

## 2. Physical/chemical characteristics of manganese state in ground waters

Soluble manganese content in oxygen-deficient underground and surface waters of the Sverdlovsk region reaches 5 ppm [4]. Russian sanitary norms restrict maximum admissible manganese content in domestic water up to 0.1 ppm, while in some European countries this value is allowed to be lower than 0.05 ppm. The excess of manganese content deteriorates organoleptic properties of water. Manganese being higher than 0.1 ppm causes spots on sanitary wares and a film on pipelines. The latter is peeled off in the form of dark sediment [5].

Manganese in ground waters is presented as ion $Mn^{2+}$, salts of which are soluble. To remove manganese from water one should transfer it to insoluble state by oxidizing $Mn^{2+}$ to $Mn^{3+}$ and $Mn^{4+}$. Oxidized manganese compounds are cleaved with the formation of practically insoluble hydroxides $Mn(OH)_3$ and $Mn(OH)_4$. The latter when precipitated through crushed diatomaceous rock (granular sorbent) exhibits catalytic properties, i.e. accelerates $Mn^{2+}$ oxidation process by dissolved oxygen [6].

As compared to air-oxidized ferrous bicarbonate whose pH value equals seven, manganese oxidizes within the range of high pH values and oxidation-reduction potentials. Optimum oxidation value for manganese bicarbonate oxidation is pH = 10. The lower pH values can only be kept by applying stronger oxidizing agents (chlorine, ozone, etc.) with their high excess in the reaction [7].

In addition, manganese is generally found in ground waters together with ferrous iron, but it is the iron that is removed from water utilizing the basic aeration treatment technology. Due to strong oxidizing agents, manganese oxidizes and sorbs only after iron since the latter is an easily oxidizable element. Therefore, joined purification of iron and manganese on the same stage at filter plants is feasible provided their concentration equals minimum values up to 2–5 ppm and strong oxidizing agents are employed.

Sanitary norms [8] on manganese removal directly in deferrization without adding reagents are outdated and fail to meet present requirements of the processes described. Application of various schemes to purify water from manganese or iron or both depends on contaminant concentration, plant capacity and local conditions.

## 3. Experimental investigation

Ground waters of the Sverdlovsk region are mainly polluted by manganese, although it is non-typical in some cases when the content of manganese exceeds that of iron.

The research of manganese removal in the city Verkhnaya Pyshma conducted by the Department of Water Management and Technology in Ural Federal University proved the concentration of manganese to be 4.5 ppm, ferrum – 0.6 ppm, silicon – 17 ppm, oxidizability – 5.5 $mgO_2/l$ and pH = 6.8. According to the suggested reagent purification scheme, corresponding reagents were added into water, then it was immediately filtered through light granular sorbent at 10 m/hr. Preliminary research showed that mixing such water with oxygen or chlorine combined with alkalization regardless of the reagent dose did not lead to manganese retention.

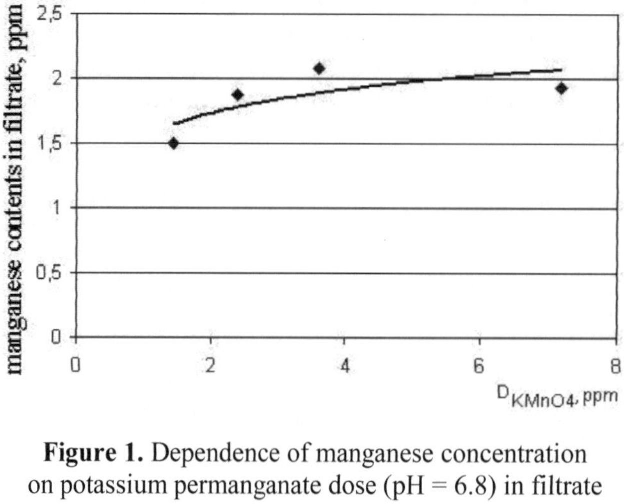


**Figure 1.** Dependence of manganese concentration on potassium permanganate dose (pH = 6.8) in filtrate

IOP Conf. Series: Earth and Environmental Science **72** (2017) 012004 doi:10.1088/1755-1315/72/1/012004

Then, tests were carried out to oxidize manganese with potassium permanganate followed by filtration and the results are presented in Figure 1. Introducing even low doses of potassium permanganate reduced the concentration of manganese in filtrate up to 1.5–2 ppm. However, the purification efficiency was not affected by the increase of the dose. This results from permanganate being used for oxidation of organic matters. Once the concentration of potassium permanganate is higher than 20 ppm, sharp increase of manganese content in filtrate occurs because of permanganate migration to filtrate. Given contact time of high dose of reagent with water is increased in the following filtration, the water of drinking quality is likely to be obtained by means of such process. This proves the occurrence of additional processing operation and the necessity to mount the reactor prior to fast filter.

For manganese to oxidize effectively, the pH value of water to be purified should be 8.0–8.5 [2]. Previous experiment was repeated but water was additionally alkalified with caustic soda with pH up to 8.5. The results are presented in Figure 2.

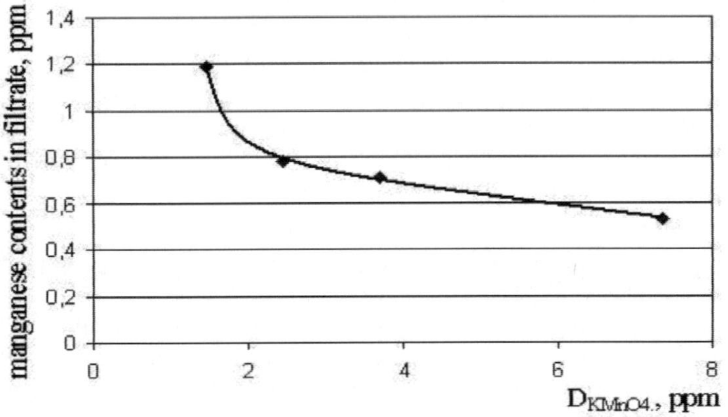


**Figure 2**. Dependence of manganese concentration on potassium permanganate dose (pH = 8.5) in filtrate

As shown in Figure 2 the increase of potassium permanganate dose results in the decrease of manganese content in filtrate. However, the dependence is not linear; the higher the dose, the slower the growth of the purification efficiency.

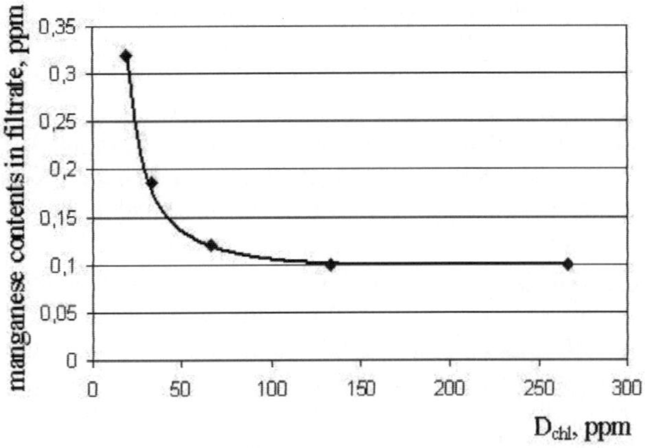


**Figure 3**. Dependence of manganese concentration on available chlorine dose (pH = 8.5) in filtrate

While filtering water mixed with potassium permanganate, the sorbent was modified: its granules were covered with the layer of manganese hydroxide performing several functions simultaneously – being an oxidation reaction catalytic agent, oxidizing manganese and sorbing organic compounds. Water was purified in such modified sorbent; meanwhile the oxidizing agent (sodium hypochlorite) and the alkalifying agent (cauistic soda) were added into water. Figure 3 shows the results of the experiment.

On the graph one can see that manganese is removed to meet drinking sanitary norms when chlorine dose is higher than 100 ppm in modified sorbent. Residual chlorine accounts for 80 ppm necessitating additional processing operation on its removal. In addition, organic matters and silicon compounds causing progressive decrease of purification efficiency gradually contaminate the catalytic layer (the layer of manganese hydroxide on a granule surface).

## 4. Technological guidelines

To avoid shortcomings of the above-listed processes, the manganese removal technology based on the application of granular sorbent with renovating catalytic layer was developed. The technology implies adding into water alkalifying and oxidizing (sodium hypochlorite) agents. Besides potassium permanganate constantly added to water accounts for 2–5 ppm. Potassium permanganate reacts instantly with manganese and organic matters creating insoluble suspension of manganese hydroxide – catalytic agent of manganese oxidation reaction. Sorbent is constantly modified by new doses of catalytic agent. In such sorbent, the contaminated catalytic agent tends to be covered with a new film of manganese hydroxide enabling oxidizing soluble manganese even with low amount of available chlorine. The sorbent should be regenerated every 6 hours by backwash as the growing catalytic layer is squeezed out to filtrate.

Based on this technology the manganese removal plant with the capacity of 300 $m^3$ per day was created in Verkhnaya Pyshma. Potassium permanganate available dose accounted for 3 ppm, chlorine dose was 5 ppm. Manganese concentration after purification was less than 0.005 ppm. The operation of the plant for a year has proved the reliability and efficiency of the suggested technology.

## 5. Conclusions

Having studied the manganese removal from ground waters with high oxidizability and manganese content (about 5 ppm), we found that the conventional oxygen or chlorine treatment of such waters even in alkaline environment fails to remove manganese irrespective of the oxidizing agent doze. Introducing insignificant doses of potassium permanganate reduces by half the manganese content in filtrate when precipitated through crushed diatomaceous rock (granular sorbent) due to the modification of the sorbent's surface. Given that, manganese hydroxide is used as the oxidation reaction catalytic agent and the dissolved organic compound sorbent. However, the introduction of potassium permanganate alone to serve as an oxidizer fails to decrease manganese in filtrate to the level required by Sanitary Regulations and Norms [8]. The pH value of water to be purified should account for 8.0–8.5, the potassium permanganate dose has to be 2–5 ppm and that of sodium hypochlorite is limited to 5 ppm for manganese to oxidize effectively. We developed the oxidation-based manganese removal technology with continuous introduction of a reaction catalytic. The ground treatment catalytic is potassium permanganate, while sodium hypochlorite serves as the principal oxidizing agent. As compared to conventional technologies (oxidation with oxygen or/and potassium permanganate followed by filtration), the technology discussed allows reducing oxidizability being above 5 $mgO_2$/l and manganese content accounting for 4 ppm in source water to meet the requirement of Sanitary Regulations and Norms [8].

This innovated technology tested on actual ground waters of the Ural region proved to be highly efficient for manganese removal since the concentration of manganese after purification did not exceed 0.005 ppm, thus confirming to the requirements for quality standard drinking water in the Russian Federation.

**Acknowledgments**
The most pleasurable thanks to Pyrkova Tamara, Foreign Languages Department, the Institute of Social and Political Sciences, Ural Federal University named after the first President of Russia B. N. Eltsin.

**References**

[1] Kulsky L A and Strokach P P 1986 *Technology Purification of natural waters* (Kyiv: Higher School) p 352
[2] Nikoladze G I 1978 *Deferribution of natural and circulating waters* (Moscow: Stroiizdat) p 160
[3] Brayalovsky G B et al 2016 Application of coagulants in the process of iron removal and desalination of groundwater *Water management in Russia 3* p 92–6
[4] 2014 Problems and Ways to Increase the Efficiency of Water Sector Enterprises *Resolution of the Regional Congress of Vodokanals of the Sverdlovsk Region* Available from: https://er.ru/projects/chistaya-voda/userdata/files/2014/11/28/vodokanal.doc [Accessed 14th May 2017]
[5] 1994 *Manual on drinking water quality control* (Geneva: WHO) 1 p 257
[6] Novikov V K and Mikhailova E M 1990 *Methods of purification of natural waters from manganese, iron and other pollutants Overview* (Moscow: Institute for the Economy of Housing and Communal Services named after Pamfilova K D) p 52
[7] Kulakov V V et al 1998 *Iron and manganese removal* (Khabarovsk: Pub. DVGUPS) p 100
[8] Sanitary Regulations and Norms 2.04.02-84* 1996 *Water Supply Public Utilities* (Moscow: GP CPP) p 128

# Simulation of freezing and thawing of soil in Arctic regions

**N Vaganova**[1] **and M Filimonov**[2]

[1] Laboratory of Applied System Analisys, Graduate School of Economics and Management, Ural Federal University and IMM UrB RAS, S. Kovalevskaya str., 16, Ekaterinburg, 620990, Russia
[2] Ural Federal University and IMM UrB RAS, S. Kovalevskaya str., 16, Ekaterinburg, 620990, Russia

E-mail: vna@imm.uran.ru

**Abstract**. A mathematical model and a numerical method for modeling long-term forecasts of the influence of various climate scenarios on the temperature fields in permafrost soils is developed. This approach allows, in particular, to estimate long-term changes in permafrost, depending on the temperature of air varying, on the precipitations and other factors affecting the degradation of permafrost. Computations are presented for obtaining the temperature in the subsurface layer of the soil, depending on the soil moisture for given values of average monthly temperatures and the intensity of solar radiation.

## 1. Introduction

We will consider a permafrost soil as base ground, which may conserve its temperature below zero during long time period. In spite that permafrost zones take about 25 % of the land of the globe, these areas are very susceptible to external influences [1–4]. In Russia, permafrost takes approximately 10 bln. km$^2$ that is more than 60 % of the territory. As a result of climatic and technogenic influences, changes in permafrost and, in particular, in the frozen ground have a significant impact on economic [5–8]. For example, in the cities of Norilsk District, a bearing capacity of soil foundations reduces when the temperature rises and permafrost loss of their stability during thawing, and dozens of buildings have to be distructed. Due to frost heaving in the Yamburgskoye gas condensate field, about 8000 gas pipeline piles were cut in the course of three years. About 40 % of all engineering structures in the cryolithozone are under deformation. A special attention for the sustainable development of the Arctic region is devoted to long-time prognosis of changes in permafrost soil, which related to various scenarios of climate warming and changes in the upper layer of the soil. Degradation of permafrost leads to considerable difficulties in construction and operation of various engineering structures, some of which are already possible to be in an emergency condition although there were built with the presence and preservation of the bearing properties of frozen ground. Due to the climat dynamics in Western Siberia, the permafrost zone boundary is estimated with the shifting to the North by 200–400 km [6]. In large areas it leads to changes in the grounds and to numerous accidents and destruction of various structures. Therefore, obtaining an accurate prediction of the possibility and rate of permafrost degradation will allow to take an opportunity to minimize the destruction of engineering infrastructure, for example, by thermal stabilization of the soil, or by application of new matherials and methods in constructing in the Arctic.

## 2. Mathematical Model of Heat Distribution

### *2.1. Heat Equation with a Phase Transition*

For simulation of heat propagation in the upper layer of soil, the ideas related with modeling of underground pipelines, especially the thermal fields and the thermal traces at the day surface, are considered, which were used to find a possible damage in underground pipelines shells [9, 10], and for modeling an open geothermal systems [11]. Numerical methods for solving problems of process description permafrost degradation are currently the most effective and universal method of research [12]. The difference methods and the corresponding numerical codes for solving boundary value problems for the heat equation are presented in a large number of papers. Fundamentals of finite difference methods are detailed in many monographs [13, 14]. In the numerical simulation of the propagation of thermal fields in the upper layer of soil, we used the idea of binding of parameters of the model and numerical algorithms to geographic coordinates [5].

We will consider an upper layer of ground surface directly illuminated by the sun. The computational domain is a three-dimensional box, where $x$ and $y$ axes are parallel to the ground surface and the $z$ axis is directed downward. We assume that the size of the region $\Omega$ is defined by positive numbers $L_x$, $L_y$, $L_z$: $-L_x \le x \le L_x$, $-L_y \le y \le L_y$, $-L_z \le z \le 0$. Let $T=T(t,x,y,z)$ be soil temperature at the point $(x,y,z)$ at the time moment $t$. Then the following thermal diffusivity equation may be considered:

$$\rho\left(c_v(T)+k\delta(T-T^*)\right)\frac{\partial T}{\partial t}=\operatorname{div}\left(\lambda(T)\operatorname{grad}T\right), \tag{1}$$

where $\rho=\rho(x,y,z)$ is density [kg/m$^3$], $T^*=T^*(x,y,z)$ is temperature of phase transition,

$c_v(T)=\begin{cases} c_1(x,y,z), & \text{for } T<T^*, \\ c_2(x,y,z), & \text{for } T>T^*, \end{cases}$ is specific heat [J/kg K],

$\lambda(T)=\begin{cases} \lambda_1(x,y,z), & \text{for } T<T^*, \\ \lambda_2(x,y,z), & \text{for } T>T^*, \end{cases}$ is thermal conductivity[W/mK],

$k=k(x,y,z)$ is specific heat of phase transition, $\delta$ is Dirac $\delta$-function.

The coefficients included in equation (1) may vary at different points in the computational domain because of heterogeneity of the soil and different layers.

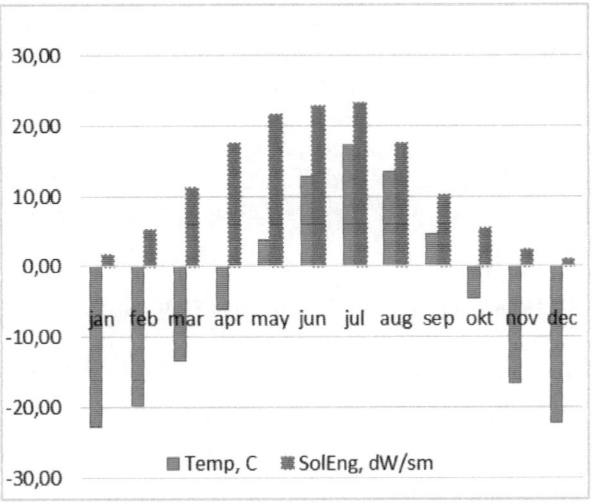


**Figure 1**. Monthly average temperature (blue) and solar radiation (red)

### *2.2. Boundary Conditions at the Soil Surface*

It is necessary to consider basic features of thermal fields forming. The ground surface $z = 0$ is the main zone of formation of the natural thermal fields. On this surface, the equation of balance of flows is used as a boundary condition, with taking into account the main climate factors: temperature of air and solar radiation. In Figure 1 the monthly average temperature and solar radiation are shown.

Thus it is necessary to solve equation (1) in the area Ω with initial condition

$$T(0,x,y,z) = T_0(x,y,z) \tag{2}$$

and boundary conditions

$$\alpha q + b(T_{air} - T|_{z=0}) = \varepsilon\sigma(T^4_{z=0} - T^4_{air}) + \lambda \left.\frac{\partial T}{\partial z}\right|_{z=0}, \tag{3}$$

$$\left.\frac{\partial T}{\partial x}\right|_{x=\pm L_x} = 0, \quad \left.\frac{\partial T}{\partial y}\right|_{y=\pm L_y} = 0, \quad \left.\frac{\partial T}{\partial z}\right|_{z=-L_z} = 0. \tag{4}$$

Condition (2) determines the initial distribution of soil temperature at the time moment from which we plan to start the numerical calculation. Condition (3) is obtained from the balance the heat fluxes at the ground surface $z = 0$. $Tair = Tair(t)$ denotes the temperature in the surface layer of air, which varies from time to time in accordance with the annual cycle of temperature; $\sigma$ = 5,67·10$^{-8}$W/(m$^2$K$^4$) is Stefan-Boltzmann constant; $b = b(t,x,y)$ is heat transfer coefficient; $\varepsilon = \varepsilon(t,x,y)$ is the coefficient of emissivity. The coefficients of heat transfer and emissivity depend on the type and condition of the soil surface. Total solar radiation $q(t)$ is the sum of direct solar radiation and diffuse radiation. A part radiation goes to the soil, let suppose it is equal to $\alpha q(t)$, where $\alpha = \alpha(t,x,y)$ is the part of energy that is formed to heat the soil, which in general depends on atmospheric conditions, angle of incidence of solar radiation, i.e. latitude and time. The boundary conditions for another object to simulations may differ due to the essential heat fluxes probably presenting [5].

The bottom surface ($z = -Lz$) and lateral faces ($x = \pm Lx$, $y = \pm Ly$) are assumed that the heat flux is equal to zero. The numbers *Lx, Ly, and Lz* are choozen large enoght to avoid the boundary conditions influence on the inner domain. In the model we use the snow cover and other factors are taken into account by variating the functions $\alpha(t,x,y)$, $b(t,x,y)$, $\varepsilon(t,x,y)$. The proposed model and developed algorithm and codes allow to choose these functions so that the temperature distribution in the ground in the deep of 10 meters corresponds to the preset temperature, i.e. adapts the model and computations to a specific geographic location.

The basic parameters that observed in connection with the thermal fields in soil are cyclic annual repeating of temperature vertical profiles and the active layer thickness (ALT). In verification and estimation on numerical results these parameters are key for the model reliability.

## 3. Results of Numerical Simulations

For solving the problem (1)–(4) an implicit finite-difference method is used [5, 15]. The computational area is a cube with sides equal to 50 meters. As a basic soil we will use a loam with the following parameters. Thermal conductivity: frozen – 1.93 W/(m K), melted – 1.57 W/(m K), volumetric heat: frozen – 2150 kJ/(m$^3$ K), melted – 3490 kJ/(m$^3$ K), volumetric heat of phase transition – 1.415·105 kJ/(m$^3$ K). The background temperature of permafrost is -3 °C, except for the layer of seasonal thawing (freezing) of soil (ALT).

Let consider the profiles of temperature in upper soil layer. In Figures 2–7 the horizontal and vertical axis correspond to the deep and the temperature, respectively. In the Figures 2, 4 and 6 temperatures on spring and summer are presented, in Figures 3, 5 and 7 temperatures on autumn and winter are shown. The numbers in the figures correspond to the month number: 1 is January, etc.

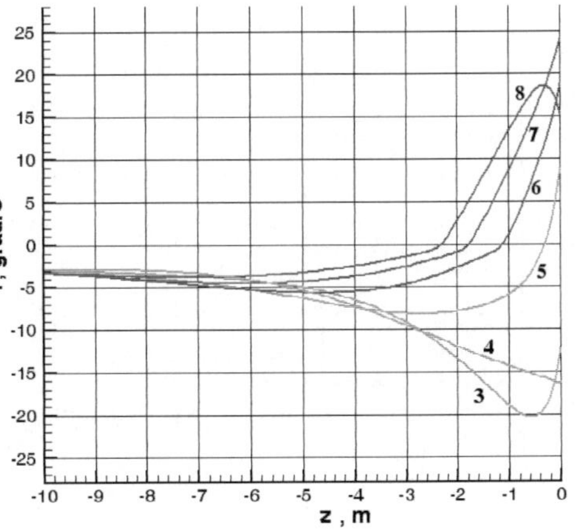


**Figure 2**. Temperatures in the soil on Spring and Summer. Humidity of the soil is 20 %

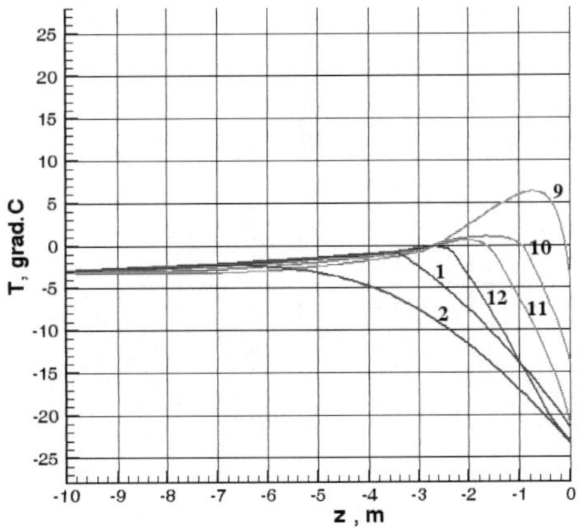


**Figure 3**. Temperatures in the soil on Autumn and Winter. Humidity of the soil is 20 %

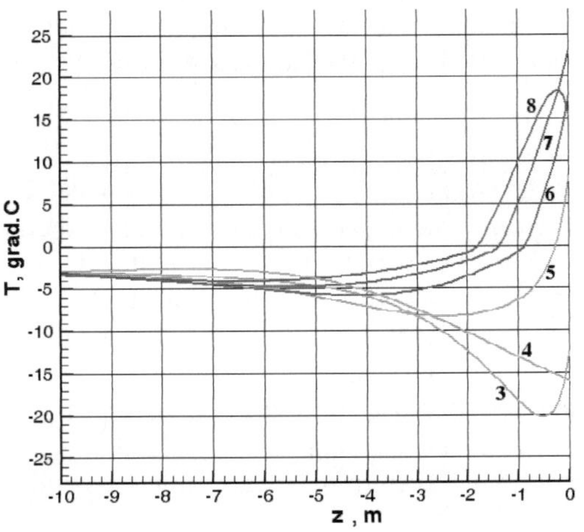


**Figure 4**. Temperatures in the soil on Spring and Summer. Humidity of the soil is 50 %

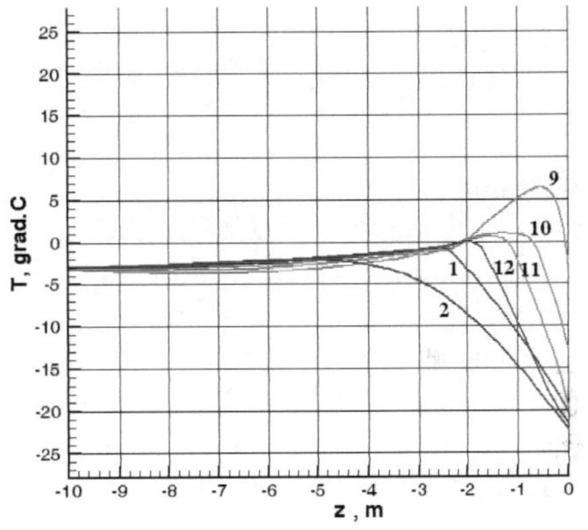


**Figure 5**. Temperatures in the soil on Autumn and Winter. Humidity of the soil is 50 %

The soil properties are dependent with the soil humidity. Comparison of ALT for soil with different humidity shows that wet soil shows that freezing level reaches 2.5 and 2 meters for humidity of 10 % and 20 %, respectively, and this difference it is necessary to take into account for a riprap platforms construction. As a rule, to construct platforms on permafrost to prevent thawing the surface of soil 2 meters riprap (concrete slab, sand layer, and Penoplex) is used that approximately equals to the thickness of the depth of seasonal thawing of the upper layer (ALT). In Figures 6 and 7 the presence of 2 m layer of riprap are presented. The riprap consists of three layers: penoplex (0.06 m), sand (1.7 m) and the concrete slab on the top (0.3 m). Parameters: cement slab with density 2500.0 kg/m$^3$, thermal conductivity 1.69 W/(m K), specific heat 0.84 kJ/(kg K); sand with density 1600.0 kg/m$^3$, thermal conductivity 0.47 W/(m K), specific heat 0.84 kJ/(kg K), penoplex has density 80.0 kg/m$^3$, thermal conductivity 0.032 W/(m K), specific heat 1.53 kJ/(kg K). The riprap using allows to conserve the soil in a frozen state. But it is necessary to note that the problem of the permafrost degradation is related with a proper constructing of engineering facilities [16], in particular, with using of pipe foundations or additional devices, helping the soil to stay frozen [17–18].

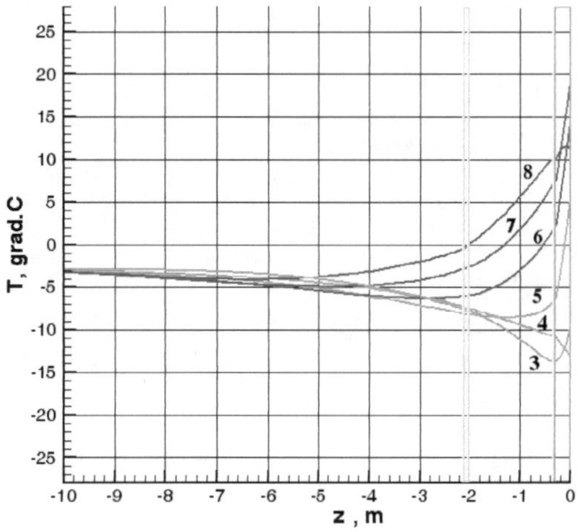

**Figure 6**. Temperatures in the soil with a riprap (colored layers) on Spring and Summer. Humidity of the soil is 20 %

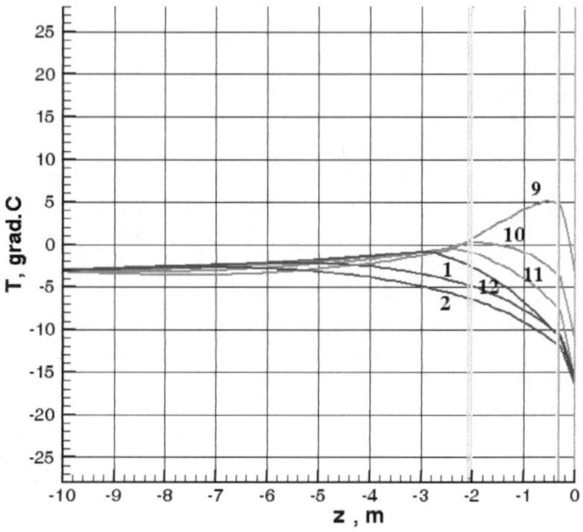

**Figure 7**. Temperatures in the soil with a riprap (colored layers) on Autumn and Winter. Humidity of the soil is 20 %

## 4. Conclusion

A mathematical model of heat distribution in an upper layers of soil is presented. This model takes into consideration a number of most important climatic and thermophysical parameters. The presented approach allows to carry out numerical simulations, which estimate distribution of temperature and cyclic annual movement of freezing/thawing points in the soil. This model allows also to simulate long-term changes in the soil, depending on the air temperature varying, on the precipitations and other factors affecting the degradation of permafrost. This approach allows to predict the long-term dynamics of permafrost degradation in a selected specific region in the Arctic, to minimize the impact of such changes for engineering facilities with various technological methods, to determine an optimal strategy for the development of different buildings and engineering constructions and safe operation modes in Arctic And Subarctic regions.

**Acknowledgment**
This work was supported in part by a grant from RFBR No. 16-01-00401 and the program UrB RAS 15-7-1-13.

**References**

[1] Zhang T, Barry R, Knowles K, Heginbottom J and Brown J 1999 Statistics and characteristics of permafrost and ground ice distribution in the Northern Hemisphere *Polar Geography* **23 (2)** p 132–154
[2] Jorgenson M, Yoshikawa K, Kaveskiy M, Shur Y, Romanovsky V, Marchenko S, Grosse G, Brown J and Jones B 2008 Permafrost characteristics of Alaska *Proc. 9th International Conference on Permafrost* p 121–22
[3] Zhao L, Wu Q, Marchenko S and Sharkhuu N 2010 Thermal state of permafrost and active layer in Central Asia during the international polar year *Permafrost and Periglacial Processes*, **21 (2)** p 198–207

[4] Yu W, Gu M, Chen L, Lai Y, Yi X and Xu L 2014 Influence of urbanization on permafrost: a case study from Mohe County, northernmost China *The Cryosphere Discuss* **8** p 4327–48 doi: 10.5194/tcd-8-4327-2014

[5] Vaganova N, Filimonov M 2015 Computer simulation of nonstationary thermal fields in design and operation of northern oil and gas fields *AIP Conf. Proc.* **1690** (020016) doi: 10.1063/1.4936694

[6] Nelson F, Anisimov O and Shiklomanov N 2002 Climate Change and Hazard Zonation in the Circum-Arctic Permafrost Regions *Natural Hazards* **26** p 203–25

[7] Reto Knutti and Jan Sedlacek, "Robustness and uncertainties in the new CMIP5 climate model projections," *Nature Climate Change*, vol. 3, pp. 369–373, 2013, doi: 10.1038/nclimate1716.

[8] Zhang Y, Chen W and Cihlar J 2003 A process-based model for quantifying the impact of climate change on permafrost thermal regimes *J. Geophys. Res.* **108** (D22) p 4695 doi: 10.1029/2002JD003354

[9] Bashurov V, Vaganova N, Kropotov A, Pchelintsev M, Skorkin N and Filimonov M 2012 Nonlinear model of a pipeline in a gravity field with an ideal fluid moving through it *Journal of Applied Mechanics and Technical Physics* **53** (1) p 43–48 doi:10.1134/S0021894412010063

[10] Vaganova N 2014 Mathematical model of testing of pipeline integrity by thermal fields *AIP Conf. Proc.* **1631** p 37–41 doi: 10.1063/1.4902455

[11] Vaganova N and Filimonov M 2015 Simulation and Numerical Investigation of Temperature Fields in an Open Geothermal System *Lect. Notes Comput. Sci.* **9045** p 385–91 doi: 10.1007/978-3-319-20239-6_44

[12] Westermann S, Langer M, Boike J, Heikenfeld M, Peter M, Etzelmüller B and Krinner G 2016 Simulating the thermal regime and thaw processes of ice-rich permafrost ground with the land-surface model CryoGrid 3 *Geosci. Model Dev.* **9** p 523–46 doi: 10.5194/gmd-9-523-2016

[13] Samarsky A A and Vabishchevich P N 1995 *Computational Heat Transfer Volume 2 The Finite Difference Methodology* (Chichester, Wiley) 432 p

[14] Yanenko N 1971 *The method of fractional steps (The solution of problems of mathematical physics in several variables)* (Berlin etc.: Springer-Verl.)

[15] Vaganova N and Filimonov M 2015 Parallel splitting and decomposition method for computations of heat distribution in permafrost *CEUR Workshop Proceedings* **1513** p 42–9

[16] Vaganova N and Filimonov M 2016 Different shapes of constructions and their effects on permafrost *AIP Conf. Proc.* **1789** 020019 doi: 10.1063/1.4968440

[17] Filimonov M and Vaganova N 2013 Simulation of thermal stabilization of soil around various technical systems operating in permafrost *Applied Mathematical Sciences* **7 (144)**, p 7151–60 http://dx.doi.org/10.12988/ams.2013.311669

[18] Filimonov M and Vaganova N 2012 Simulation of thermal fields in the permafrost with seasonal cooling devices *Proc. ASME.* **45158** vol 4, p 133–41 doi: 10.1115/IPC2012-90287

IOP Conf. Series: Earth and Environmental Science **72** (2017) 012006 doi:10.1088/1755-1315/72/1/012006

# Permafrost thawing from different technical systems in Arctic regions

**M Filimonov**[1] **and N Vaganova**[2]

[1] Ural Federal University and IMM UrB RAS, S. Kovalevskaya str., 16, Ekaterinburg, 620990, Russia
[2] Laboratory of Applied System Analisys, Graduate School of Economics and Management, Ural Federal University and IMM UrB RAS, S. Kovalevskaya str., 16, Ekaterinburg, 620990, Russia.

E-mail: fmy@imm.uran.ru.

**Abstract**. A new three-dimensional model of thermal interaction in a "heat source-and-soil" system is proposed to study the process of permafrost degradation from various engineering facilities operating in the Arctic regions, taking into account a number of physical and climatic factors that affect the heat distribution. On the base of the proposed model, a software complex was developed to predict long-term dynamics of permafrost thawing in the upper layer of soil, and this approach was used in the design of 11 northern Russian oil and gas fields and is in a good agreement with numerical rezults and experimental data. Numerical calculations are presented for illustration the possibility of carrying out long-term forecasts for the determination of permafrost zone defrosting during operation of production wells in northern oil and gas field.

## 1. Introduction

Permafrost soils, which conserve a negative temperature deeper than the active layer thickness (ALT), takes about 25 % of the globe land [1–6]. In Russia, permafrost zones are extremely important in view to both oil and gas field exploitation and a general development of the Arctic. Thawing of iced soils due to warming of the climate, or various industrial impacts, will be followed by subsidence of the earth's surface and dangerous permafrost geological processes, called thermokarst, and as a result, it leads to disruptions and accidents [7–11].

The irrational design and functioning of cluster sites in Arctic oil and gas fields leads to significant additional and financial costs, which under the conditions of low oil prices reduces the profitability of development and operation of new and existing oil and gas fields. In this regard, it is relevant to reduce the cost of designing and optimizing the operation of new and functioning oil and gas fields. Therefore the problem of reducing the intensity of thermal interactions in the "heat source – permafrost" zones has a particular importance for solving problems of energy saving, environmental protection, safety, cost savings and enhance operational reliability of various engineering structures.

Further, we consider a three-dimensional model which allows to describe heat distribution in upper layers in permafrost soils with taking into consideration not only most significant climatic factors (seasonal changes in temperature and intensity of solar radiation due to geographic location of the field) and physical factors (different thermal characteristics of non-uniform ground that change over the time) [12] but also engineering construction features of the production wells [13]. This model may be used for simulations including other types of technical systems such as ripraps, tanks, pipelines, flare systems and others [14].

## 2. Mathematical model and basic heat flows

First, let consider heat exchange on a flat ground surface directly illuminated by the sun. Let the initial time be $t_0=0$, and the ground is a box $\Omega$ and has a temperature $T_0(x,y,z)$ (Figure 1). The computational domain is a three-dimensional box, where $x$ and $y$ axes are parallel to the ground surface and the z axis is directed downward. We assume that the size of the region $\Omega$ is defined by positive numbers $L_x$, $L_y$, $L_z$: $-L_x \le x \le L_x$, $-L_y \le y \le L_y$, $-L_z \le z \le 0$. To simulate the propagation of heat in this volume the following mathematical model is suggested.

*Heat Transfer Equation with a Phaze Transition*

Let $T = T(t,x,y,z)$ be soil temperature at the point $(x,y,z)$ at the time moment t. The main heat flow associated with climatic factors on the surface $z = 0$ is shown in Figure 1.

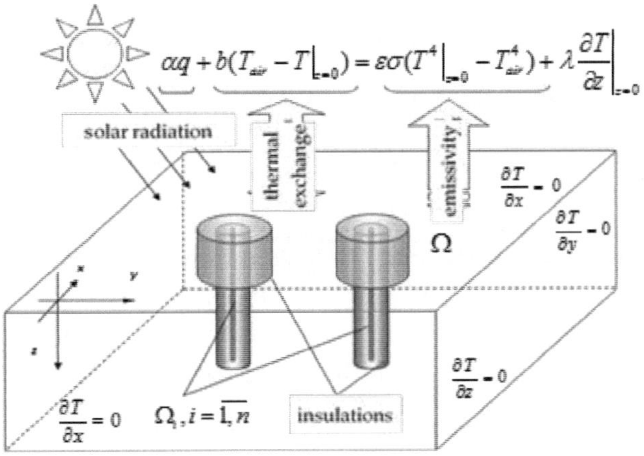


**Figure 1**. An area for simulation

We will consider an equation of the contact (diffusion) heat conductivity with inhomogeneous coefficients is used as a basic mathematical model with including localized heat of phase transition – an approach to solve the problem of Stefan type, without the explicit separation of the phase transition. The heat of phase transformation is introduced with using Dirac δ-function as a concentrated heat of phase transition in the specific heat ratio. The obtained discontinuous function then "shared" with respect to temperature, and does not depend on the number of measurements, phases, and fronts. Thus, the modeling of thawing in the soil is reduced to the solution in $\Omega$ of the equation

$$\rho\left(c_v(T) + k\delta(T - T^*)\right)\frac{\partial T}{\partial t} = \mathrm{div}\left(\lambda(T)\ \mathrm{grad}\ T\right) \tag{1}$$

where:

$\rho = \rho(x,y,z)$ *is* density [kg/m$^3$],

$T^* = T^*(x,y,z)$ is temperature of phase transition,

$c_v(T)= \begin{cases} c_1(x,y,z), \text{ for } T < T^*, \\ c_2(x,y,z), \text{ for } T > T^*, \end{cases}$ is specific heat [J/kg K],

$\lambda(T) = \begin{cases} \lambda_1(x,y,z), \text{ for } T < T^*, \\ \lambda_2(x,y,z), \text{ for } T > T^*, \end{cases}$ is thermal conductivity [W/m K],

$k = k(x,y,z)$ is specific heat of phase transition, $\delta$ is Dirac $\delta$-function.

The coefficients included in equation (1) may vary at different points in the computational domain *Ω* because of heterogeneity of the soil and possible presence of engineering structures.

*Boundary Conditios*

The ground surface z = 0 is the main zone of formation of the natural thermal fields. On this surface the equation of balance of flows is used as a boundary condition, with taking into account the main climate factors: air temperature and solar radiation. On the surfaces Ωi, bounding the objects in Ω, a set of the temperatures Ti(t), i = 1,...,n, is given. The bottom surface (z = -Lz) and lateral faces (x = ±Lx, y = ±Ly) of the parallelepiped Ω (Figure 2) is assumed that the heat flux is equal to zero.

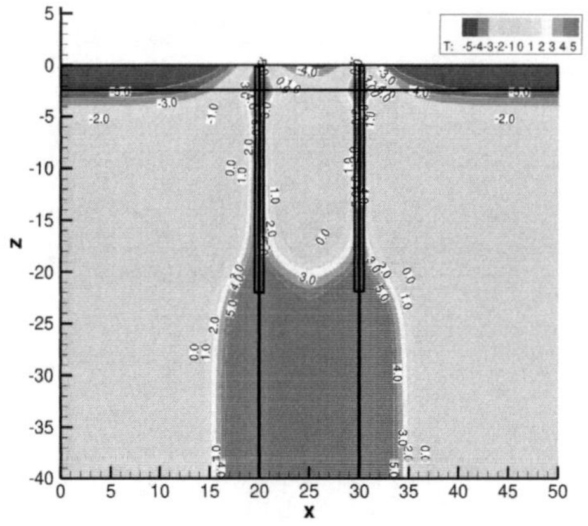


**Figure 2.** Thermal field around the wells for the moment when the left well has stopped

Thus it is necessary to solve equation (1) in the area *Ω* with initial condition

$$T(0,x,y,z) = T_0(x,y,z) \tag{2}$$

and boundary conditions

$$\alpha q + b(T_{air} - T|_{z=0}) = \varepsilon\sigma(T^4_{z=0} - T^4_{air}) + \lambda \frac{\partial T}{\partial z}\bigg|_{z=0}, \tag{3}$$

$$T|_{\Omega_i} = T_i(t), \quad i = 1,...,n, \tag{4}$$

$$\frac{\partial T}{\partial x}\bigg|_{x=\pm L_x} = 0, \quad \frac{\partial T}{\partial y}\bigg|_{y=\pm L_y} = 0, \quad \frac{\partial T}{\partial z}\bigg|_{z=-L_z} = 0. \tag{5}$$

Condition (2) determines the initial distribution of soil temperature at the time moment from which we plan to start the numerical calculation. Condition (3) is obtained from the balance the heat fluxes at the ground surface *z = 0*. Conditions (4) appear in the case if in the ground there are different objects, which temperature is different from the surrounding soil.

## 3. Numerical results

On the base of ideas in [15–17] a finite difference method is used with splitting by the spatial variables in three-dimensional domain to solve the problem (1)–(5). We construct an orthogonal grid, uniform, or condensing near the ground surface or to the inner surfaces. The original equation for each spatial direction is approximated by an method to solve a system of linear differential algebraic equations is used. Heterogeneity of the coefficients in (2) is taken into account by the corresponding heat flux at each point of the difference grid. The grid is constructed in such a way that grid points take place in layers of insulating materials or in the considered engineering constructions (for example, in a concrete slab in riprap layers). In addition, if an engineering construction is a heat source (for example, a well), we introduce dummy nodes of the inner boundary, that allows to accurately determine the position and intensity of heat generated by this source.

A similar approach is used in the study of thermal fields on the day surface from an underground pipeline [18] and a geothermal system [19] without of the phase transition.

Note, that for the considered problems an analytical methods of investigation, such as [20], is difficult and unavailable, but, in a case it is possible to construct an exact solution [21], applicable for numerical methods verification. Complexity of analytical methods application is also related with nonlinear boundary condition (3), which application is determined in [22–24].

On the base of the model (1)–(5) a program complex «Wellfrost» is developed, which was approved for 11 oil and gas fields located in the permafrost zone. In comparing of the numerical and experimental data it was observed that boundary of permafrost thawing from an injection well, operated for 3 years, coincides in approximation up to 5% with experimental data. This fact allows to suggests that the developed algorithm and software package «Wellfrost» may serve to test various other techniques. In the following we present some numerical rezults for some Russian northern oil and gas fields.

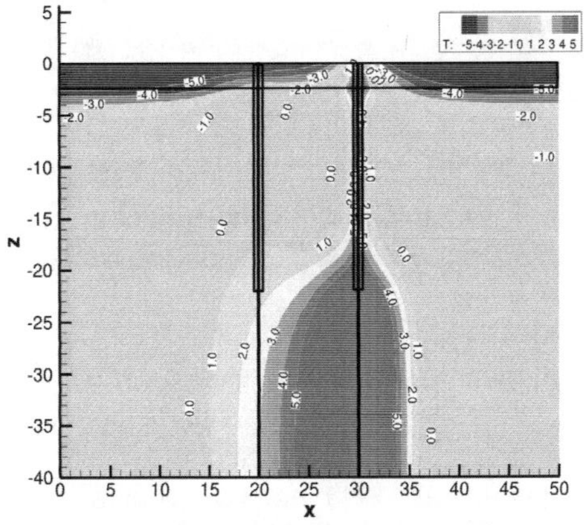


**Figure 3**. Thermal field around the wells after 1 year after the left well has stopped.

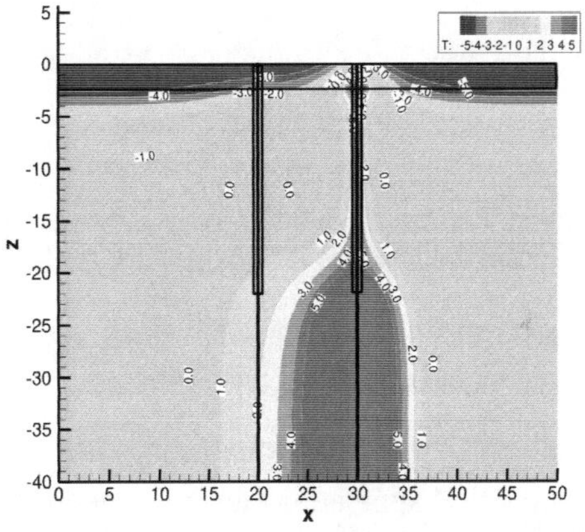


**Figure 4**. Thermal field around the wells after 2 years after the left well has stopped.

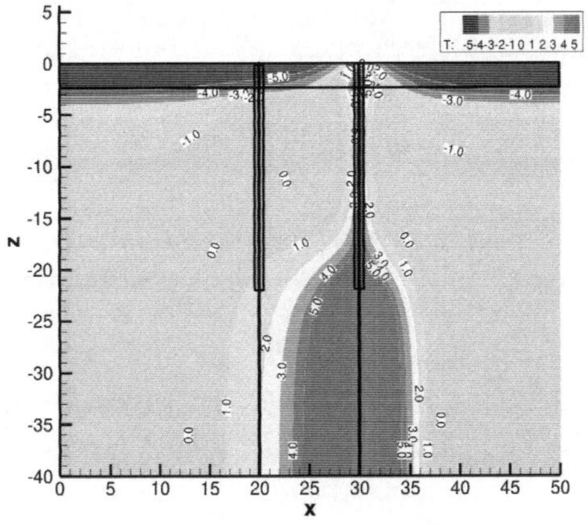


**Figure 5**. Thermal field around the wells after 3 years after the left well has stopped.

Let consider a system of two wells with thermal insulations in the upper layer in the deep to 12 m. The wells are drilled in a loam soil with the temperature -0.7 ºC. The well diameter is 0.18 m, distance between the wells is 10 m. The oil heated up to 45 ºC leads to the soil thawing around the wells. In Figure 2 distribution of temperature in the vertical slice along the wells is shown for 6 years of exploitation. The fronts of thawing from the wells (zero isotherm) in the bottom of the area are jointed. In the upper layers of soils one can see the ALT (approx. 2 m) and the part conserved frozen and stable due to the thermal insulation of the wellheads. The figures 3–5 show the cooling process when the left well has stopped. The zero isotherm turns back to the left wellhead. The front of thawing movement is present in figures 6 and 7. The numeration of lines corresponds to the years of exploitation. The results of computations allow to estimate of thermal influence of the inserted engineering constructions and to preven some damages caused by permafrost thawing and degradation.

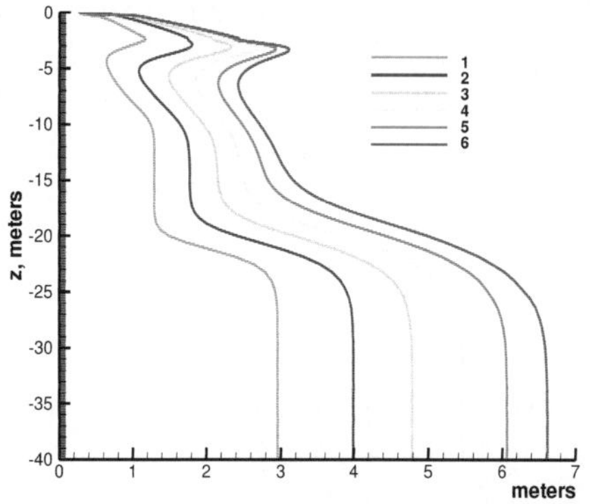


**Figure 6**. Front of thawing movement during 6 years from the well in exploitation.

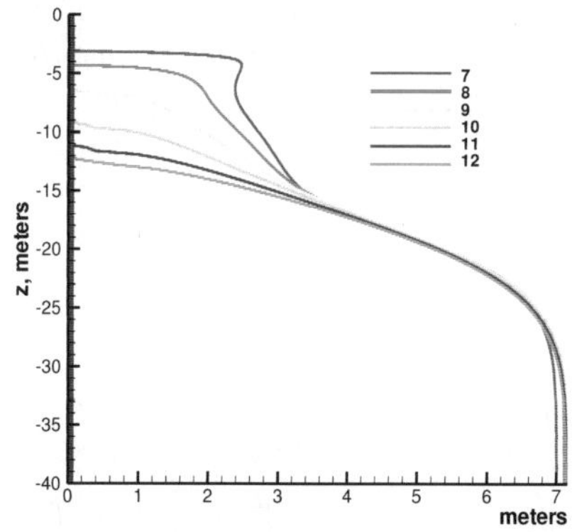


**Figure 7**. Front of thawing movement during 6 years to the stopped and cooling well.

## 4. Conclusion

1. Based on the mathematical model, a complex of computational codes is developed with a cloud interface for numerical three-dimensional modeling of non-stationary thermal fields in the upper layer of permafrost soils from various technical systems located in the northern oil and gas field.

2. Numerical simulations for well sites allow to improve safety and efficiency of northern oil and gas fields due to an optimal location of wells and other technical systems in this area and have a significant economic effect during the design stage. For example, in 2012, the simulations allowed to reduce by 50% the size of a well pad for placing the producing wells for one of the Russian northern oil and gas fields.

3. A basic feature of the developed package of computational codes is a possibility of adaptation of the model parameters to the selected geographical location of the area under investigation. For this purpose, an original iterative algorithm is developed that allows to take into account both the varying thickness of the snow, and the number of sunny days per year, and the feature of the topsoil layer, etc.

**Acknowledgment**

This work was supported in part by a grant from RFBR No. 16-01-00401 and the program UrB RAS 15-16-1-10.

## References

[1] Zhang T, Barry R, Knowles K, Heginbottom J and Brown J 1999 Statistics and characteristics of permafrost and ground ice distribution in the Northern Hemisphere *Polar Geography* **23 (2)** p 132–154

[2] Jorgenson M, Yoshikawa K, Kaveskiy M, Shur Y, Romanovsky V, Marchenko S, Grosse G, Brown J and Jones B 2008 Permafrost characteristics of Alaska *Proc. 9th International Conference on Permafrost* p 121–22

[3] Zhao L, Wu Q, Marchenko S and Sharkhuu N 2010 Thermal state of permafrost and active layer in Central Asia during the international polar year *Permafrost and Periglacial Processes*, **21 (2)** p 198–207

[4] Jorgenson M, Romanovsky V, Harden J, Shur Y, O'Donnell J, Schuur E A G and Kanevskiy M 2010 Resilience and Vulnerability of Permafrost to Climate Change *Canadian J. Forest Research* **40 (7)** p 1219–36

[5] Westermann S, Langer M, Boike J, Heikenfeld M, Peter M, Etzelmüller B and Krinner G 2016 Simulating the thermal regime and thaw processes of ice-rich permafrost ground with the land-surface model CryoGrid 3 *Geosci. Model Dev.* **9** p 523–46 doi: 10.5194/gmd-9-523-2016

[6] Knutti R and Sedlacek J 2013 Robustness and uncertainties in the new CMIP5 climate model projections *Nature Climate Change* **3** p 369–73.

[7] Anisimov O, Fitzharris B, Hagan J, Jeffries R, Marchant H, Nelson F, Prowse T and Vaughan D 2001 *The Contribution of Working Group II of the Intergovernmental Panel on Climate Change* p 801–41

[8] Yu W, Gu M, Chen L, Lai Y, Yi X and Xu L 2014 Influence of urbanization on permafrost: a case study from Mohe County, northernmost China *The Cryosphere Discuss* **8** p 4327–48 doi: 10.5194/tcd-8-4327-2014

[9] Nelson F, Anisimov O and Shiklomanov N 2002 Climate Change and Hazard Zonation in the Circum-Arctic Permafrost Regions Natural Hazards 26 p 203–25

[10] Vaganova N and Filimonov M 2015 Computer simulation of nonstationary thermal fields in design and operation of northern oil and gas fields AIP Conf. Proc. 1690 (020016) doi: 10.1063/1.4936694

[11] Zhang Y, Chen W and Cihlar J 2003 A process-based model for quantifying the impact of climate change on permafrost thermal regimes J. Geophys. Res. 108 (D22) p 4695 doi: 10.1029/2002JD003354

[12] Filimonov M and Vaganova N 2015 Simulation of Technogenic and Climatic Influences in Permafrost Lect. Notes Comput. Sci. 9045 p 178-85 doi: 10.1007/978-3-319-20239-6_44

[13] Vaganova N and Filimonov M 2016 Different shapes of constructions and their effects on permafrost AIP Conf. Proc. **1789** (020019) doi: 10.1063/1.4968440

[14] Filimonov M and Vaganova N 2016 Simulation of permafrost changes due to technogenic influences of different engineering constructions used in northern oil and gas fields *Journal of Physics: Conference Series* **754** (112004) doi: 10.1088/1742-6596/754/11/112004

[15] Patankar S 1980 *Numerical Heat Transfer and Fluid Flow* (New York etc.: Hemisphere Publication)

[16] Samarsky A and Vabishchevich P 1995 *Computational Heat Transfer Volume 2 The Finite Difference Methodology* (Chichester, Wiley) 432 p

[17] Yanenko N 1971 *The method of fractional steps (The solution of problems of mathematical physics in several variables)* (Berlin etc.: Springer-Verl.)

[18] Vaganova N 2014 Mathematical model of testing of pipeline integrity by thermal fields *AIP Conf. Proc.* **1631** p 37–41 doi: 10.1063/1.4902455

[19] Vaganova N and Filimonov M 2015 Simulation and Numerical Investigation of Temperature Fields in an Open Geothermal System *Lect. Notes Comput. Sci.* **9045** p 385–91 doi: 10.1007/978-3-319-20239-6_44

[20] Filimonov M 2000 Application of special coordinated series to the solution of nonlinear partial differential equations in bounded domains *Differential equations* **36** (11) p 1685–91 doi: 10.1007/BF02757370

[21] Bashurov V, Vaganova N, Kropotov A, Pchelintsev M, Skorkin N and Filimonov M 2012 Nonlinear model of a pipeline in a gravity field with an ideal fluid moving through it *Journal of Applied Mechanics and Technical Physics* **53** (1) p 43–48 doi:10.1134/S0021894412010063

[22] Cherepanov A, Shapeev V, Fomin V and Semin L 2006 Numerical Modeling of Thermophysical Processes in Laser-Beam Welding with Formation of a Vapor Channel *Journal of Applied Mechanics and Technical Physics* **47** (5) p 688–95

[23] Amosov A 2005 Global solvability of a nonlinear nonstationary problem with a nonlocal boundary condition of radiative heat transfer type *Differential Equations* **41** (1) p 96–109

[24] Prokhorov I 2009 Structure of the continuity of the solution of the boundary value problem for the equation of radiative transfer *Math.* **86** (2) p 256–72

# Atmospheric pollution in cities of Russia: statistics, causes and characteristics

**A Petrov**[1] **and D Petrova**[2]

[1] Tyumen Industrial University, 72 Melnikaite str., Tyumen, 625000, Russian Federation
[2] Ural Federal University named after the first President of Russia B. N. Yeltsin, 16a Chapaev str., Ekaterinburg, 620002, Russian Federation

E-mail: ArtIgPetrov@yandex.ru

**Abstract**. The article considers the issues of air pollution assessment in Russian industrial regions (2014) and cities (2012). The statistical data is presented both in terms of absolute emissions of pollutants into the atmosphere of Russian cities and relative air pollution calculated for 1 average statistical inhabitant. Classifications of the ecological state of Russian cities on the basis of specific (per inhabitant) air pollution and the air pollution by predominant type source (stationary or non-stationary) are proposed.

## 1. Introduction

A simple idea that sustainable economic development is only possible through continuous improvement of production technology of the gross domestic product closely linked to the concept of technological paradigm, i.e, aggregate of conjugate productions that have a comparable scientific and technological level and develope synchronously.

Today, in the majority of industrialized countries manufacturing industry of the fifth technological stage reached maturity phase, and an urgent task are practical problems of development and production relating to the sixth technological stage. At the same time all around the world, including Russia, there are still existing and functioning productions, based on the achievements not only of the fifth, but the second and the third and the fourth technological stages. It is mainly the steel industry, heavy machinery, petrochemistry. Typically, the geographical location of such enterprises is localized and there are whole clusters of urban areas, with a common history and similar trends and development issues. Vivid examples of such clusters in Russia may be Ural and Kuzbass.

One of the most problematic features of the people living in these regions is a bad environment [1]. Significant parameters of environmental pollution, mainly – from stationary sources [2, 3], are inherent for the entire districts of Sverdlovsk regions, Chelyabinsk regions, Kemerovo regions. Of course, in recent years much has been done to improve the environmental situation in the industrialized regions of the country. Many industries, especially the sphere of mining and processing and metallurgical production, are equipped with filters and cleaning systems of emissions. In general we can say that in recent years in Russian cities emissions of pollutants into the environment has decreased significantly. However, the problem is not completely solved. Emissions of contaminants into the atmosphere in many industrial cities remain relatively high. And in the cities, where the industrial environment has more favorable situation, today the role of the main pollutant of the atmosphere performs road transport [4, 5].

In this article issues of the industrial regions and cities of Russia ranking in terms of "Absolute values of pollutants emissions" and "Unit (per person per year) emissions of pollutants" are considered.

## 2. Basic data

In [6] shows chronologically the most actual statistics, characterizing the actual emissions of pollutants into the atmosphere in t he most problematic territories (total 34 areas) from the position of ecology in cities and industrial regions of Russia in 2014.

On the basis of the data [6] table 1 is made up. Commenting on the data of Table 1 it should be noted that for the comparative assessment of the environmental situation in the cities of the country, with different area, population and other characteristics, it is necessary to use specific indicators, such as "Specific emissions of pollutants (per pers. per year)" SEP (1):

$$SEP = Specific\ emitions\ of\ pollutiants = [\frac{Emission\ of\ pollutiants - total}{Population}], \quad (1)$$

where *Specific emissions of pollutants* – Specific emissions of pollutants, tons/pers./year;
*Emissions of pollutants – total* – Emissions of pollutants – total, thousands, tons/year;
*Population* – Population (2014), pers.

**Table 1.** The ranking of cities of the Russian Federation in terms of actual emissions of pollutants into the atmosphere (2014) [6]

| No. of ranking | Cities and industrial areas of Russian Federation | Population (2014), pers. | Emissions of pollutants – total, thousands, tons/year | Specific emissions of pollutants, tons/pers./year |
|---|---|---|---|---|
| 1 | Norilsk | 176559 | 1841.3 | 10.428 |
| 2 | Reftinskiy industrial area | ≈ 66800 | 315.8 | 4.727 |
| 3 | Cherepovets | 316758 | 313.7 | 0.990 |
| 4 | Lipetsk | 509719 | 291.1 | 0.571 |
| 5 | Usinsk | 39831 | 289.8 | 7.275 |
| 6 | Novokuznetsk industrial area | ≈ 600000 | 276.4 | 0.461 |
| 7 | Magnitogorsk industrial area | ≈ 633700 | 227.7 | 0.359 |
| 8 | Vorkuta | 61638 | 191.1 | 3.100 |
| 9 | Omsk | 1166092 | 174.3 | 0.149 |
| 10 | Ufa | 1096702 | 148.2 | 0.135 |
| 11 | Nizhny Tagil | 357280 | 145.7 | 0.407 |
| 12 | Chelyabinsk industrial district | ≈ 1250000 | 140.2 | 0.112 |
| 13 | Krasnoyarsk | 1035528 | 129.8 | 0.125 |
| 14 | Bratsk | 238825 | 119.3 | 0.499 |
| 15 | Novocherkassk | 173464 | 108.7 | 0.627 |
| 16 | Astrakhan | 530863 | 104.4 | 0.196 |
| 17 | Novosibirsk | 1547910 | 102.6 | 0.066 |
| 18 | Mezhdurechenskiy industrial area | ≈ 110000 | 100.8 | 0.916 |
| 19 | Kachkanarsky industrial area | 40260 | 77.8 | 1.932 |
| 20 | Myskoskiy industrial area | ≈ 43000 | 70.2 | 1.632 |
| 21 | Kaltanskiy industrial area | ≈ 22000 | 69.2 | 3.145 |
| 22 | Irkutsk | 612973 | 68.0 | 0.111 |

| No. of ranking | Cities and industrial areas of Russian Federation | Population (2014), pers. | Emissions of pollutants – total, thousands, tons/year | Specific emissions of pollutants, tons/pers./year |
|---|---|---|---|---|
| 23 | Novotroitsk | 93578 | 65.5 | 0.700 |
| 24 | Serov industrial area | 106572 | 65.3 | 0.612 |
| 25 | Verkhneufaley industrial area | 33366 | 64.0 | 1.918 |
| 26 | Polysayevskiy industrial area | ≈ 29500 | 63.7 | 2.159 |
| 27 | Tula | 490508 | 62.7 | 0.127 |
| 28 | Belovsky industrial area | ≈ 75000 | 62.5 | 0.833 |
| 29 | Surgut | 332313 | 57.5 | 0.173 |
| 30 | Kostomuksha industrial area | 29586 | 56.0 | 1.892 |
| 31 | Sterlitamak | 277048 | 53.8 | 0.194 |
| 32 | Troitzky industrial district | 77176 | 53.8 | 0.697 |
| 33 | Nazarovo | 51437 | 51.2 | 0.995 |
| 34 | Stary Oskol industrial area | 257128 | 51.0 | 0.198 |

More extensive information about specific (per pers. per year) emissions of air pollutants in the cities of Russia is shown in [7] as in 2012. It should be noted that this document provides information on the actual specific emissions (t/pers./year) in an atmosphere of 180 cities of the Russian Federation.

Figure 1 shows a histogram of the distribution of cities in Russia (2012) in groups depending on the value of specific total emissions of air pollutants per inhabitant per year.

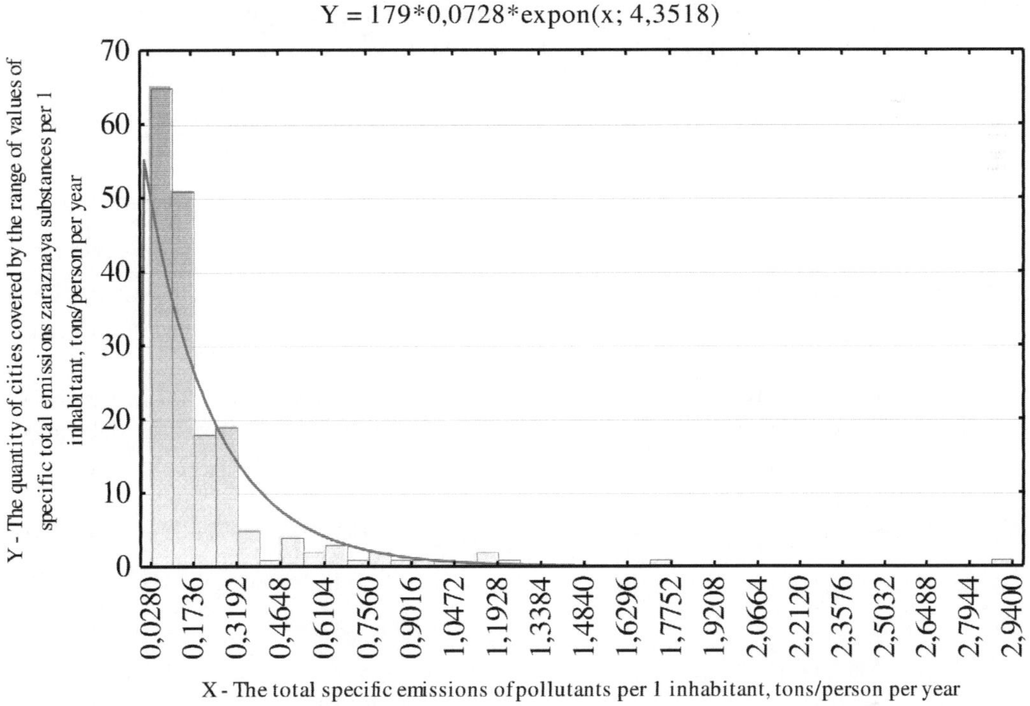


**Figure 1.** The histogram of distribution of cities in Russia (2012) in groups depending on the value of specific total emissions of air pollutants per inhabitant per year

## 3. Results and Discussion

Analysis shows that the cities can be classified into groups depending on the value of this parameter (table 2).

**Table 2.** The frequency characteristic of Russian cities classification (2012) on the basis of "Specific total emissions of air pollutants per inhabitant per year"

| Indicator | Groups of cities on the basis of "Specific total emissions of air pollutants per inhabitant per year", tons/pers./year | | | | |
|---|---|---|---|---|---|
| | I | II | III | IV | V |
| Range of values | 0.028–0.100 | 0.101–0.173 | 0.174–0.246 | 0.247–0.319 | > 0.319 |
| Quantity of cities in group | 65 | 52 | 18 | 18 | 27 |

Almost two thirds of the number of examined cities (117 of 180) refer to classification groups I and II, i.e. to relatively safe from the point of city ecology. At the same time the existence of large number of cities, where for 1 person annually 200 kg or more air pollutants are thrown is a demonstrative example of complex environmental situation.

Table 3 shows the rating of Russian cities with the most polluted atmosphere. Analysis of the table 3 data shows that in 2014 in some Russian cities some progress in reducing the emission of pollutants into the atmosphere has been achieved. Thus, positive changes were observed in almost all the cities of the Russian Federation, for which there is relevant statistics, with the exception of Vorkuta. At the same time, emissions statistics in Norilsk is so shocking that the positive developments in 2014 in comparison with 2012 seem to be very insignificant.

**Table 3.** The ranking of Russia's most polluted cities (2012/2014) in terms of "Specific total emissions of air pollutants per inhabitant per year" (Group V in 2012) [6, 7]

| No. of ranking | Russian city of group V in environmental classification | Value of indicator "Specific total emissions of air pollutants per inhabitant per year", tons/pers./year | |
|---|---|---|---|
| | | 2012 | 2014 |
| 1 | Norilsk | 11.054 | 10.428 |
| 2 | Vorkuta | 2.940 | 3.100 |
| 3 | Krasnouralsk | 1.707 | No data |
| 4 | Angarsk | 1.198 | No data |
| 5 | Cherepovets | 1.158 | 0.990 |
| 6 | Krasnoturinsk | 1.156 | No data |
| 7 | Novodvinsk | 1.042 | No data |
| 8 | Monchegorsk | 0.907 | No data |
| 9 | Rezh | 0.864 | No data |
| 10 | Serov | 0.797 | 0.612 |
| 11 | Novocherkassk | 0.774 | 0.627 |
| 12 | Zarinsk | 0.735 | No data |
| 13 | Lipetsk | 0.636 | 0.571 |
| 14 | Magnitogorsk | 0.624 | 0.359 |
| 15 | Kstovo | 0.614 | No data |
| 16 | Novokuznetsk | 0.584 | 0.461 |
| 17 | Bratsk | 0.553 | 0.499 |
| 18 | Orsk | 0.517 | No data |

| No. of ranking | Russian city of group V in environmental classification | Value of indicator "Specific total emissions of air pollutants per inhabitant per year", tons/pers./year | |
|---|---|---|---|
| | | 2012 | 2014 |
| 19 | Achinsk | 0.514 | No data |
| 20 | Okha | 0.510 | No data |
| 21 | Noyabrsk | 0.477 | No data |
| 22 | Nizhny Tagil | 0.415 | 0.407 |
| 23 | Anadyr | 0.392 | No data |
| 24 | Gubkin | 0.378 | No data |
| 25 | Stary Oskol | 0.363 | 0.198 |
| 26 | Artem | 0.337 | No data |
| 27 | Surgut | 0.331 | 0.173 |

Researcher of urban environment atmosphere is interested in the question of the source of contamination. All sources can be classified into stationary and mobile (vehicles) [8]. Information about sources of pollution in urban areas has been obtained in [7] and presented as a classification in Table 4.

Figure 2 ... 3 shows charts of values of indicator "Proportion in the total volume of pollution emissions into urban atmosphere of the internal combustion engine emissions" in the cities of I and V classification groups (Table 4). Note that cities of I group are cities where practically no industrial enterprises, implementing emissions of pollutants into the atmosphere (cities of resort specialization), or large in terms of population and the level of the city motorization, where the problem of cities' significant pollution by car takes rampant.

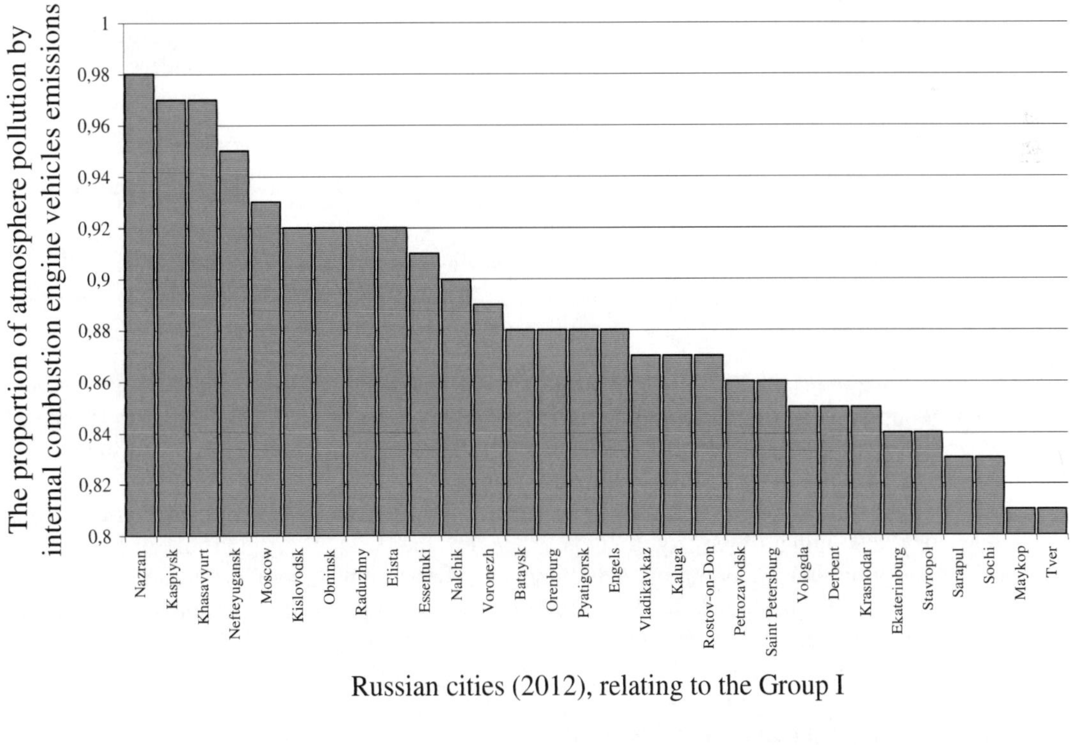


**Figure 2**. Russian cities (2012), relating to the Group I of classification on the basis of indicator "The proportion of atmosphere pollution by internal combustion engine vehicles emissions"

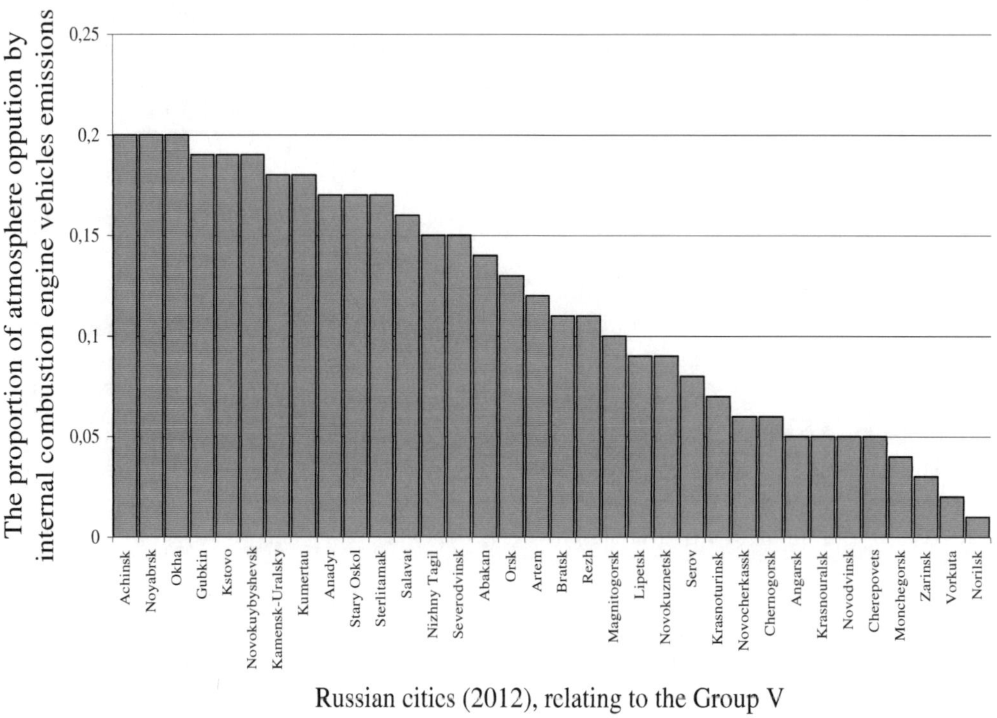


**Figure 3**. Russian cities (2012), relating to the Group V of classification on the basis of indicator "The proportion of atmosphere pollution by internal combustion engine vehicles emissions"

**Table 4.** The frequency characteristic of the classification of Russian cities on the basis of "Proportion in the total volume of pollution emissions into urban atmosphere of the internal combustion engine emissions"

| Indicator | Groups of cities on the basis of "Proportion in the total volume of pollution emissions into urban atmosphere of the internal combustion engine emissions" | | | | |
|---|---|---|---|---|---|
| | I | II | III | IV | V |
| Range of values | 0.81…1.00 | 0.61…0.80 | 0.41…0.60 | 0.21…0.40 | 0…0.20 |
| Quantity of cities in group | 30 | 47 | 37 | 32 | 34 |

Cities of V group are mainly monotown of industrial specialization, in which pollution is formed in large quantities by stationary sources. Atmospheric pollution of automobile origin in such cities cannot compete with emissions from industrial manufactures.

Cities of intermediate II, III and IV groups have mixed (in different proportions) signs of cities of I and V groups. By belonging of the city to any group on the basis of "Proportion of the internal combustion engine emissions in pollution emissions into urban atmosphere" you can make basic conclusions about the specialization of city's production and the leading role of productions of certain technological paradigm in the city's economy.

## 4. Conclusion.

In conclusion, the following points should be noted.

1. Generally in recent years in most cities of Russia a slight improvement in air pollution can be noted.

2. Air pollution of Russian cities varies in a wide range of values (from 0.03 to 11.05 tons/person/year).

3. Russian cities can be classified according to the degree of air pollution into 5 groups. 27 of the 180 examined cities in the country refers to the most problematic fifth group.

4. In different cities of the country priority sources of air pollution are cars and stationary pollution sources.

**Reference**

[1] State Report 2015 On the state of protection of population and territories of Russian Federation from natural and technogenic emergency situations in 2014 (Moscow: Emercom of Russia. Federal State Budgetary Institution) p 352
[2] Bityukova V R 2011 *Regional Research of Russia* **1 (4)** p 416–20
[3] Bityukova V and Kasimov N 2012 *Geofizika* **29** p 39
[4] Cole T B 2004 *The Journal of American Medical Association* **291 (21)** p 2531
[5] Molina M J and Molina L T 2004 *Journal of the Air & Waste Management Association* **54** p 644
[6] Main indicators of environment in 2015 Available from: http://www.gks.ru/wps/wcm/connect/rosstat_main/rosstat/ru/statistics/publications/catalog/doc_1140094699578 [Accessed 30th April 2017]
[7] Main indicators of environment in 2013 Available from: http://www.gks.ru/wps/wcm/connect/rosstat_main/rosstat/ru/statistics/publications/catalog/doc_1140094699578 [Accessed 30th April 2017]
[8] Chainikov D, Chikishev E, Anisimov I and Gavaev A 2016 *IOP Conference Series: Materials Science and Engineering* **142** 012109

IOP Conf. Series: Earth and Environmental Science **72** (2017) 012008 doi:10.1088/1755-1315/72/1/012008

# Strategy of Construction and Demolition Waste Management after Chemical Industry Facilities Removal

**I N Tashkinova, G M Batrakova and Ya I Vaisman**

Department of Environmental Protection, Perm National Research Polytechnic University, Komsomolsky Ave. 29, 614990 Perm, Russian Federation

E-mail: i.n.tashkinova@gmail.com

**Abstract.** Mixed waste products are generated in the process of irrelevant industrial projects' removal if conventional techniques of their demolition and dismantling are applied. In Russia the number of unused chemical industry facilities including structures with high rate of wear is growing. In removing industrial buildings and production shops it is used conventional techniques of demolition and dismantling in the process of which mixed waste products are generated. The presence of hazardous chemicals in these wastes makes difficulties for their use and leads to the increasing volume of unutilized residues. In the process of chemical industry facilities' removal this fact takes on special significance as a high level of hazardous chemicals in the waste composition demands for the realization of unprofitable measures aimed at ensuring environmental and industrial safety. The proposed strategy of managing waste originated from the demolition and dismantling of chemical industry facilities is based on the methodology of industrial metabolism which allows identifying separate material flows of recycled, harmful and ballast components, performing separate collection of components during removal and taking necessary preventive measures. This strategy has been tested on the aniline synthesis plant being in the process of removal. As a result, a flow of 10 wt. %, subjected to decontamination, was isolated from the total volume of construction and demolition waste (C&D waste). The considered approach allowed using the resource potential of more than 80wt. % of waste and minimizing the disposed waste volume.

## 1. Introduction

At present the acceleration of unused chemical industry facilities' removal including structures characterized by high rate of wear is observed in Russia. Conventional techniques of structures removing involve demolition works and mechanical failure resulted in generation of mixed waste containing recyclable, ballast and environmentally hazardous components [1]. Also it takes place considerable polluting powder-gas emission which is accumulated in the working space. In this case all the waste products are considered to be non-utilized and their distribution in the environment becomes economically inexpedient and environmentally dangerous as there is a strong probability of the secondary pollution of neighbouring areas and toxic components migration into the environment [2–4].

Enterprises of chemical industry founded in the 1930s – 1950s are characterized by a heavy deterioration of structures and facilities, availability of obsolete equipment and strongly contaminated sites of solid waste and liquid effluents accumulation. More than 10 thousand potentially hazardous chemical facilities belonging to different branches of industry continue to operate in Russia.

Moreover, 70 % of them are located in the cities the population of which is more than 100 thousand people [5].

Conducted research shows that removal of industrial buildings directly involved in production processes leads to considerable generation of hazardous waste. This fact defines the necessity of applying the strategies [6, 7] which includes: the separation of waste into the recyclable, hazardous and ballast flows; determination of preventive measures aimed at minimizing the volume of hazardous waste flows; identification of hazardous waste [1, 8, 9] and its preliminary removal from the waste flow; evaluation of suitable for recycling waste's resource potential, the choice of their recycling technologies [10, 11], the ways of their recycling and evaluation of competitiveness of secondary materials obtained on their basis; the choice of environmentally-friendly, economically affordable and technologically feasible method of demolishing or dismantling.

The algorithm of realizing the waste management strategy after demolition and dismantling of chemical industry facilities is given in Figure 1.

At the first stage a reconnaissance inspection of buildings and facilities is carried out. Then the construction and archival documents denoting the operating conditions of the building are analyzed. These conditions include: the period of the building use; the main parameters of engineering process; conditions of hostile industrial environment; information about emergency situations; production control data and monitoring of fruits and intermediates of production in the air of working area.

At the second stage expert-analytical evaluation of the object of removal is carried out: identification of resource potential and competiveness of materials which can be derived from waste after removal; analysis of the factors of the waste's resource potential change; selection of economically affordable, environmentally-friendly and technologically feasible ways of removal; assessment of waste's environmental safety.

According to the results of the made analysis recyclable, ballast and hazardous waste streams are separated; the amount of unutilized fractions' landfill is determined; the choice of hazardous waste neutralization techniques and methods of utilized waste flows' processing is made [12]. After that the projects of buildings' demolishing and dismantling as well as the project of further recovery of contaminated areas are developed.

The third stage consists of demolishing, dismantling or reshaping of buildings and of the territory's recovery. Refined territory is assigned for the secondary use. After completing these kinds of works, the territory is provided for the secondary use. To increase the resource potential of demolishing and dismantling waste it is necessary to neutralize the dangerous flow, to landfill the ballast flow and to recycle the utilized waste flow aimed at its further reuse.

Strategy of waste management after industrial buildings removal has been tested at one of typical industrial facilities of aniline production. In the course of work material flow of waste has been identified, their hazard has been estimated and the techniques of recycling and the waste resource potential use aimed at minimizing the volume of unutilized residue intended for safe distribution in the environment have been proposed.

The following tasks have been solved beforehand:

- The rate of wear of building structures have been analyzed on the base of information about the change of physical-mechanical properties of materials;
- The modeling of C&D waste material flows has been made for several variants of the test subject removal.
- The volume, the composition and the risk of C&D waste generated after removal have been defined;
- The guidelines for C&D waste resource potential use and reduction of waste level in the environment have been developed.

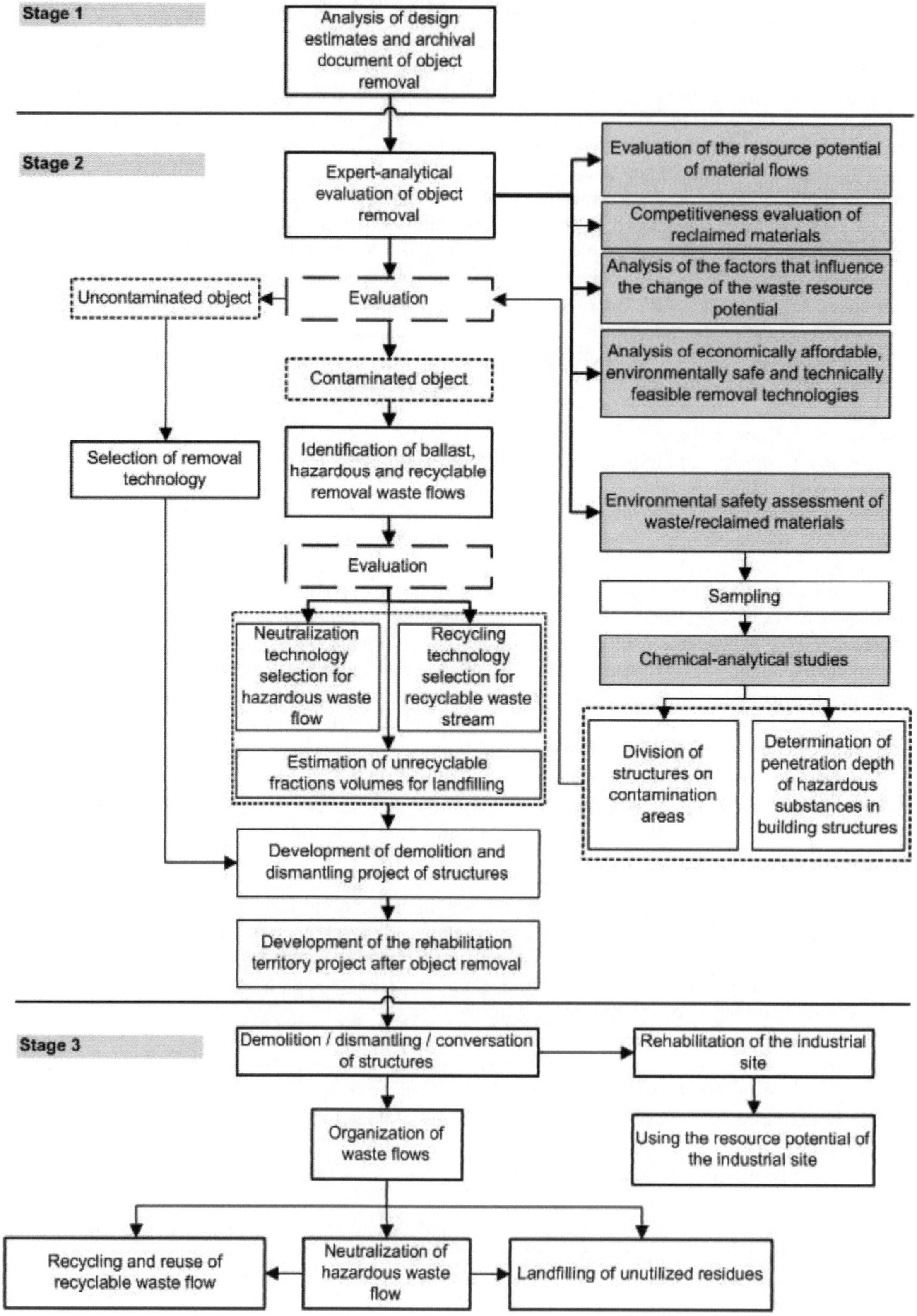


**Figure 1**. Algorithm of C&D waste strategy realization after removing chemical facilities

## 2. Materials and methods

*Analysis of the rate of wear of building structure in chemical production*
Under the influence of industrial aggressive factors the intensity of surface damage and corrosion of building structures is increased. Thus, toxic products are accumulated in building materials.

Floor and wall ceramic tile, plaster, cores of wall constructions' concrete and other materials have been sampled. It is known that the greatest amount of pollutants is accumulated in the finishing materials.

Therefore, studies of changes in the physical and mechanical characteristics of finishing agents [13] based on the ranking of the building according to the degree of contamination have been carried out (Table 1).

**Table 1**. Modification of physical and mechanical characteristics of finishing agents

| Characteristic, measure | Starting value (standard) | Area of high level of contamination | Area of average level of contamination | Area of low level of contamination |
|---|---|---|---|---|
| Residual bending strength, % | 100.0 | 63.7–87.0 | 67.0–97.8 | 75.0–94.6 |
| Water absorbing rate, % of weight | 1.80–2.00 | 4.50–7.47 | 2.57–3.20 | 1.84–2.41 |

*C&D waste hazard assessment*
High level of contamination of finishing agents remaining in the area of direct contact with the basic reactant has been identified during the examination of industrial workshops. Accumulation of aniline as the main pollutant has been found at a depth of 30–35 mm. The aniline content in the samples of building materials varies from 4.8 to 832.3 mg/kg of dry substance of the sample. Moreover, the amount of aniline migrating in the aqueous medium has been estimated in the amount of 17.8 wt. % upon the average. High level of aniline migration in aqueous medium proves the potential environmental hazard of C&D waste during their allocation in the environment and reuse without preliminary decontamination.

The obtained data and identified dependences of the major pollutant migration made possible to justify the requirements for necessary degree of wall constructions' neutralization.

We have obtained the following approximated dependencies which characterize the migration of a pollutant deep into constructions of two buildings situated in the closed industrial complex ($\Delta c$ – the concentration of water-soluble aniline in wall construction material, h – depth of sampling):

$$f_1(\Delta c) = 709.1824\, e\, -0.601243\, h, \quad (1)$$

$$f_2(\Delta c) = 669.25\, e\, -0.074\, h, \quad (2)$$

(1) – building 1, (2) – building 2.

The obtained data and identified dependences (1) and (2) allow justifying the requirement for decontamination of wall construction and the importance of making surface pre-cleaning from finishing agents which performed a barrier function in the aggressive environment of production shops.

To prevent secondary pollution of the environment during dismantling and demolition it should be neutralized chemical contamination of the finishing agents' surfaces. According to the assessment of substance migration in the construction the treatment is to be made at the depth from 5 to 30 mm.

The proposed approach aimed at preventing secondary pollution of environment can be used in the development of model schemes designed for the C&D waste treatment.

*Analysis of the component and quantitative composition of C&D waste*

There are two ways of C&D waste volume determination:

- According to the passport of the object (estimation and project construction documentation);
- According to the technical inspection of the object. Such inspection includes measuring of its parameters and calculation based on the average density of each building material defined by regulations.

In order to calculate the total volume of C&D waste it was used the first way. The analysis of the project documentation of the removed production revealed the total waste volume amounted to 25 000 tones. The calculated component composition of C&D waste is presented in Table 2. The silicate brick, concrete and ferroconcrete building structures, metal and other materials predominate in the waste composition. The amount of contaminated C&D waste which is subject to decontamination and landfill as well as nondifferentiable waste makes more than 10 wt. %.

**Table 2**. Component and quantitative composition of waste from main industrial buildings removing

| Type of waste | Amount of waste, tn. | | | |
|---|---|---|---|---|
| | Building 1 | Building 2 | Building 3 | Total |
| Concrete | 3756.29 | 4880.33 | 1992.62 | 10629.25 |
| Ferroconcrete | 2944.86 | 3505.84 | 572.06 | 7022.77 |
| Calcium-silicate brick | 2500.23 | 2746.30 | 570.01 | 5816.55 |
| Metal | 714.57 | 376.73 | 23.28 | 1114.58 |
| Soft roofing | 0.02 | 0.00 | 0.01 | 0.03 |
| Felt roofing | 0.00 | 0.01 | 0.00 | 0.01 |
| Sandstone tile | 53.03 | 55.75 | 9.19 | 117.97 |
| Glass | 6.07 | 7.30 | 1.33 | 14.70 |
| Total mass of C&D waste, tn.: 24715.84 | | | | |

*Analysis of C&D waste material flows in different ways of removal*

Finding the alternative for the method of object's removal three approaches has been considered. To compare the main approaches we have carried out the modelling of different materials and waste streams [14].

The first approach is a combined method of removal. For constructions being of high hazard and rate of wear it has been proposed the method of undermining. Mechanical demolition has become the optimal technique for constructions having low risk level and rate of wear. Materials obtained in the process of mechanical demolition should be directed to recycling while blasting waste should be directed to disposal.

The second approach is based on the method of mechanical elementwise dismantling [15] with a preliminary decontamination of finishing agents and contaminated areas (Figure 2). Decontamination is implied as removing of the layers of construction materials from the walls, floor and ceiling at the depth of contaminant's penetration.

Glass, wood and metal waste runs to processing. It has been proposed to allocate strongly hazardous waste after its neutralization in the environmentally-friendly place or to reuse it in case of competitive ability.

In the third approach material and waste streams have been estimated in the process of production complex restructuring for a different function. The process consisted in mechanical dismantling of constructions having a high rate of wear, decontamination of industrial surfaces and reconstruction of buildings. After removing the elements of structures being of high rate of wear and finishing agents which accumulated dangerous chemicals the building could be used again including further restructuring of the object.

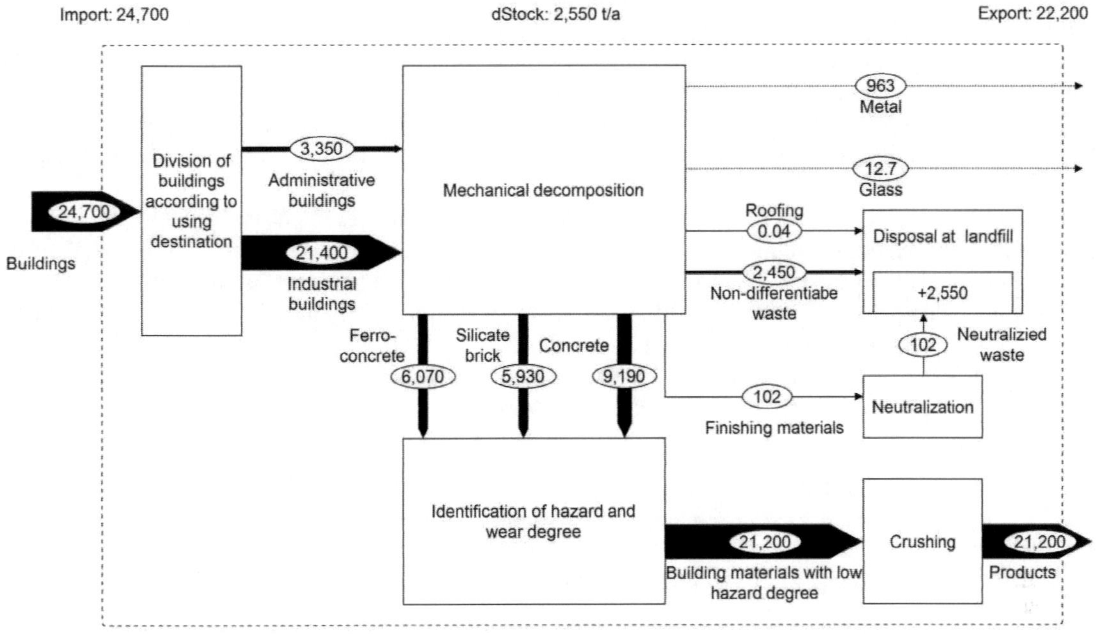


**Figure 2**. Scheme of material flows for the second approach

## 3. Results and discussion

During the comparison of approaches to removal of chemical industry facilities the following estimated figures have been considered:

- Ways of the given industrial territory use
- Technical state of constructions
- The depth and the area of mechanical pre-treatment of contaminated industrial surfaces of the object aimed at reduction of harmful effects on the environment
- The amount of waste which is subject to processing and recycle
- The forecast of the potential harmful effects on the environment
- The environmental effectiveness of the method
- Technical and economic parameters (duration of work, logistics and tooling backup, etc.).

The main possible directions of this industrial area using are the following: residential construction, agricultural use, recreation area, industrial construction, industrial restructuring and realigning into administrative objects.

The realigning of buildings into administrative objects is impermissible without preliminary decontamination of premises and hydrophobization of walls owing to the considerable accumulation of contaminants in finishing agents.

According to the existing practice of managing the closing industrial sites in Russia all the approaches can be realized with the view of further industrial development or object's realigning. Removed production is a part of a large industrial complex, and it is located near the boundaries of the enterprise. The liberated territory can be used for planting or reducing of the sanitary protection zone.

The technical state of industrial constructions characterized by the following features: minor damage, presence of cracks in some areas, partial damage of corrosion protection. The buildings can be restructured with preliminary extraction of material fractions contaminated with production chemical agents. It will be required maintenance work and local damage repair without reinforcing of constructions.

The first approach doesn't involve preliminary decontamination of surfaces which allows saving of time. However, it should be noted that waste formed by undermining without preliminary extraction of contaminated fractions becomes more hazardous for the environment. It is characterized as Hazard Class 3 and its allocation at landfills is rather unprofitable.

The preliminary decontamination of walls' surfaces is an advantage of Approach 2 and Approach 3. In both cases building structures are refined from finishing agents (tile, plaster, paint, etc.) and the layer of contaminated materials is removed.

Approach 2 is characterized by a maximal quantity of waste assigned for recycling. The first approach is recognized as the most ineffective one in terms of using resource potential of waste after production removal. Approach 3 has minimal quantity of waste directed for disposal (Table 3).

The third approach of removal simplifies the management of waste since its volume is reduced to minimum but it does not make possible to remove thoroughly C&D waste directed to disposal.

The types of removal damage effects on the environment are as follows: air pollution, a negative effect on soil-vegetable layer, pollution of hydrosphere, secondary pollution and indirect effect on the ecosystem.

**Table 3**. Management of C&D waste in different approaches of plant removal

| Removal Approach | Amount of waste for disposal, tn. | Amount of waste for recycling, tn. | Percentage of waste for disposal, % | Percentage of waste for recycling, % |
|---|---|---|---|---|
| No.1 | 20838.68 | 3876.97 | 84.31 | 15.69 |
| No.2 | 2549.15 | 22166.69 | 10.31 | 89.69 |
| No.3 | 1376.82 | 5712.30 | 5.57 | 23.11 |

The first approach includes all the listed and most hazardous processes. So, it will be the maximal environmental stress at its application. Approach 2 and Approach 3 contain two processes which have a negative impact on the environment. However, taking into account little amounts of disposal waste and partial mechanical decomposition in Approach 3 application environmental stress will be minimal. In terms of ecological effect Approach 1 does not prevent negative effect on environment. Approaches 1 and 2 have maximum ecological effect. In addition, forecasting of environmental consequences has been carried out. The consequences include: secondary emission of pollutants into water and soil, suppression of ecosystems in the area of potential impact, agreements for temporary use of land, potential harmful effect on a man (direct contact, inspiration). All these consequences are probable for the first Approach. Approach 3 has a risk of air quality problem in the working area after realigning of buildings in case of their careless clearing from the previous emissions.

In the capacity of technical-and-economical performance of each approach there were considered the following: working efficiency, the number of additional expendable materials, engineering and tooling backup, mass of waste which is subject to transportation, mass of waste which is subject to disposal, mass of waste which is subject to processing.

The final choice of removal technique has been carried out by the method of expert evaluation. According to the final comparison of different approaches of removal the optimal one are as follows: elementwise mechanical method of dismantling with preliminary decontamination of surfaces (Approach 2) and the method of restructuring (Approach 3).

Results of the evaluation allowed making a conclusion that in terms of environmental safety of the last life cycle stage the Approach 3 is the most favourable. Approach 3 has such peculiarities as neutralization, reconstruction and repair of buildings with further restructuring. This Approach provides a minimal amount of waste destined for disposal being the urgent task of environmental safety. Large duration of work, i.e. time required for decontamination and for works aimed at renovation of industrial facility could be recognized as the main disadvantage of Approach 3.

Approach 2 makes possible to achieve environmental safety of the removal process and to minimize the mass of waste destined for disposal for the less period of time. Moreover, during its application is possible to process maximum amount of building materials.

## 4. Conclusion

It has been proposed the management strategy for construction and demolition waste of chemical industry as well as an algorithm of its realization which allows reducing the risk of generated waste to an acceptable level, minimizing the volumes of unutilized residues and making the most out of the waste resource potential use.

Proposed strategy has been tested at one of the typical objects of organic synthesis – the aniline production. On the base of the analysis of three approaches to the removal of aniline synthesis structures it has been selected the optimum alternative which includes the elementwise mechanical dismantling with preliminary mechanical removal of the most contaminated layers of materials. The use of the utilized waste stream resource potential and decontaminated dangerous stream enables to minimize the amount of landfills.

## References

[1] Hennebert P, Papin A, Padox J-M and Hazebrouck B 2013 The evaluation of an analitical protocol for the determination of substances in wastefor hazard classification *Waste Management* **33 (7)** p 1577–88

[2] Beyer C, Konrad W, Rügner H, Bauer S, Liedl R and Grathwohl P 2009 Model-based prediction of long-term leaching of contaminants from secondary materials in road constructions and noise protection dams *Waste Management* **29 (2)** p 839–50

[3] Chowdhury R, Apul DS, Fry T 2010 A Life cycle based environmental impacts assessment of construction materials used in road construction *Resources Conservation and Recycling* **54 (4)** p 250–55

[4] Eikelboom R T, Ruwiel E and Gounmans J J J M 2001 The building materials decree: an example of a Dutch regulation based on the potential impact of materials on the environment *Waste Managemen* **21 (3)** p 295–302

[5] 2008 The national system of chemical and biological safety of the Russian Federation to 2009–2013 *Passport of the Federal Target Program*

[6] Abhijith H, Sreejith M H and Priya A J 2014 Minimizing construction wastes by efficient site practices *International J of Education and Appl. Research* **4** p 46

[7] Bhander G S, Christensen T H, Hauschild M Z 2010 EASEWASTE-life cycle modeling capabilities for waste management technologies *InternationalJ of Life Cycle Assessment* **15** p 403

[8] Hennebert P, Hans A van der Sloot, Rebischung F, Weltens R, Geerts L and Hjelmar O 2014 Hazard property classification of waste according to the recent propositions of the EC using different methods *Waste Management* **34 (10)** p 1739–51

[9] Weltens R, Vanermen G, Tirez K, Robbens J, Deprez K and Michiels L 2012 Screening tests for hazard classification of complex waste materials – selection of methods *Waste Management* **32 (12)** p 2208–17

[10] Hendriks Ch F, Janssen G M T 2001 Construction and demolition waste: general process aspects *Heron* **46** p 79

[11] Carpenter A C, Gardner K H, Fopiano J, Benson C H and Edil T B 2007 Life Cycle Based Risk Assessment of Recycled Materials in Roadway Construction *Waste Management* **27 (10)** p 1458–64

[12] Hemalatha B R, Prasad Nagendra and Venkata Subramanya B V 2008 Construction and demolition waste recycling for sustainable growth and development *J of Environmental Research and Development* **2** p 759–65

[13] Korth D, Lippok J Bauing and Dexheimer R 1981 *Abbruch-arbeiten. Vorbereitung und durchführung* (Berlin:VEB Verlag für Bauwesen)
[14] Brunner P H and Rechberger H 2004 *Practical handbook of material flow analysis* ed P H Brunner (USA: CRC Press LLC).
[15] Shvetcova I N, Batrakova G M and Shirinkina E S 2012 Investigation of mechanical properties and hazards of building materials after use in aggressive industrial environment *Building Materials* **8** p 46

IOP Conf. Series: Earth and Environmental Science **72** (2017) 012009 doi:10.1088/1755-1315/72/1/012009

# Study of surface mud sediment in an urban environment

**A Seleznev, A Savastianova and I Yarmoshenko**

Institute of Industrial Ecology, Ural Branch of Russian Academy of Sciences, S. Kovalevskoy str., 20, 620219, Ekaterinburg, Russia

E-mail: seleznev@ecko.uran.ru

**Abstract**. Surface mud sediment is a media integrating pollution over space and time. Sampling of the mud sediment allows obtaining additional information about environmental state. The results of the study of surface mud sediment in Ekaterinburg city (Russia) are represented in the paper. Particle-size composition of the sediment is primarily represented by dust and fine sand. Study of the sediment allows ranking the territories over pollution degree with heavy metals, identifying technogenic and typomorphic geochemical associations of the elements in environmental compartments.

## 1. Introduction

The process of forming fine grained particles and dust of various origins occurs in an urban environment permanently. Natural and anthropogenic processes (primarily landscaping and planning) cause the constant redistribution and accumulation of fine grained particles and dust at surfaces and local depressed areas of micro landscape, thus forming the recent anthropogenic sediments. The loose sedimentary material is a product of soil erosion, decomposition and abrasion of various surfaces and materials [1–7]. The sediment may be composed of mineral constituents, natural biogenic materials, anthropogenic inorganic/organic matters. Sediment substance accumulates pollution (pathogens, industrial chemicals and metals) and provides a potential source for a present-day secondary and nonpoint source of pollution in urban environment. Some researchers note that the transported sediment is the main source of particle matters and pollutants of surface water environment [7, 8]. Nature of the transfer of sediments and pollution makes obtaining reliable information on the dominant sources of sediments and the association between the sediment substance and contaminants within urban landscape difficult. Thus the US EPA recommends removing the sedimentary material from the surfaces in an urban area due to its toxicity [9–14].

The sediment characterizes the environment where it is formed. Surfaces participating in the accumulation of the sediment are constructed in various time periods. The sediment accumulates pollution over time from the moment of forming the landscape and over its area. Thus, the sediment is a media integrating pollution over space and time. Sampling of the mud sediment allows obtaining additional information about environmental state and geochemical behavior of pollutants in an urban environment [1, 3–5, 15–17].

The current paper represents the generic and recently obtained results of the study of surface mud sediment in an urban environment (on the example of Ekaterinburg city, Russia).

## 2. Object and methods of the study

### *2.1. Object of the research*

The object of the research is the mud sediment deposited on the surfaces and in local surface depressed zones of landscape (in other words puddle sediment, PS) in residential districts with multistory buildings in the city. Surface mud sediment can be classified as specific surface facie of the recent anthropogenic sediment in urban environment. Studied sediments participate in migration and accumulation of pollutants in an urban environment [18, 19].

### *2.2. Methods of the research*

Concentrations of metals (typical pollutants and elements of the parent rock), pH level were measured in 210 PS samples collected in 2007–2010. The 'total' concentrations of Pb, Zn, Cu, Mn, Fe and Al in PS samples were determined with ICP-MS Perkin Elmer ELAN 9000 (Perkin Elmer Inc., USA). The pH level the aqueous extract solution of the PS samples was determined with ionomeric method.

Particle-size composition was determined in 30 samples of mud sediment collected in 2016. To conduct particle-size composition analysis the sampling of the sediment was conducted at the 6 sampling sites (residential blocks located in different geographical districts of the city) according to the following scheme. The sampling of the sediment was conducted inside and outside the residential block. Samples were collected from asphalt, pavement, green zone and parking (authorized and unauthorized). For particle-size separation the representative subsample with mass approx 200–300 g was taken from the sample of the sediment. The subsample was crushed and homogenized with rubber pounder. Than the particle-size fraction > 1 mm was separated with dry sieving the subsample. The remaining substance was added with distilled water. After that, the particle-size fractions 0.05–0.1 mm, 0.1–0.25 mm and 0.25–1 mm were separated with wet sieving of the remaining material of the sediment subsample. Particle-size fractions 0.002–0.01 mm and 0.01–0.05 mm were obtained from the solution after its decantation and vacuum filtering with membrane filters.

## 3. Results of the study

### *3.1. Distribution of heavy metal concentrations in the mud sediment*

The analysis of spatial redistribution of concentrations of pollutants in PS samples was conducted by ten enlarged residential districts in Ekaterinburg. The districts can be ranked by pollution degree. The distribution of concentrations of Pb, Zn and Cu in Ekaterinburg is shown in Table 1. City districts where the average concentrations are above the average for the city are highlighted in bold.

**Table 1.** Distribution of concentrations of Pb, Zn and Cu in puddle sediment by the city districts

| Disrtict | Geometric mean | | |
|---|---|---|---|
| | Cu, mg/kg | Zn, mg/kg | Pb, mg/kg |
| East | **127** | **473** | **104** |
| West | **104** | **380** | **90** |
| North | 93 | 277 | 71 |
| North-West | **96** | 237 | **75** |
| Center | **103** | **433** | **91** |
| South | **104** | **509** | **101** |
| South-East | 77 | 230 | 57 |

| Disrtict | Geometric mean | | |
|---|---|---|---|
| | Cu, mg/kg | Zn, mg/kg | Pb, mg/kg |
| South-Eastern suburbs | 82 | 281 | 52 |
| South-West | 91 | **399** | **82** |
| Southern suburbs | 79 | 285 | 46 |

### *3.2. Correlation between the heavy metal concentrations in the sediment*

The Spearman's correlation coefficients were assessed for metals Al, Mn, Fe, Cu, Zn и Pb in PS samples in Ekaterinburg (Table 2).

**Table 2.** The Spearman's correlation coefficients for metals Al, Mn, Fe, Cu, Zn и Pb in PS samples in Ekaterinburg

| **Metal** | **Metal** | | |
|---|---|---|---|
| | **Cu** | **Zn** | **Pb** |
| **Al** | -0,11 | -0,27* | -0,23* |
| **Mn** | 0,33* | 0,05 | -0,04 |
| **Fe** | 0,31* | 0,09 | 0,04 |

* – correlation is significant under $p < 0.05$

For the studied metals Spearman's correlation coefficient does not show the strong statistically significant correlation between the elements, though it is significant under $p < 0.05$. Such correlation may correspond to the existence of different sources of the elements.

The analysis of variances showed two geochemical associations of metals in PS: typomorphic (Mn, Fe and Al) and anthropogenic (Cu, Zn and Pb).

### *3.3. Geochemical associations of metals in mud sediment and bottom sediment of water bodies*

Comparative analysis of heavy metal content in PS and bottom sediments of anthropogenic water bodies was conducted for Ekaterinburg territory. Studied anthropogenic water bodies are the former open-cut mines flooded with atmospheric precipitations and fire ponds in the suburbs of the city. The composition of sediments of the water bodies is mainly represented by pulp. The characteristic geochemical association for heavy metals Mn-Zn-Cu-Pb is the same for PS and bottom sediments of anthropogenic water bodies in the city.

### *3.4. Association between metals concentration and pH level in mud sediment*

The pH level of PS samples ranges from 6.13 to 8.78 with a mean of 7.66 and median value of 7.68. Portion of samples with a weak-acid medium (pH = 6–7) is 12 %, with an alkalescent (pH = 7–9) – 88 %. The statistically significant correlation between concentrations of Pb, Zn and Cu vs pH was not found. It means that pH level doesn't determine the significant impact to the variability of content of Pb, Zn and Cu in PS.

### *3.5. Particle-size composition of mud sediment*

Each sample of surface mud sediment was divided into six particle-size fractions. The 180 granulometric subsamples were obtained for 30 collected samples. The results of particle-size composition analysis are represented on Figure 1.

The surfaces were divided into two types: ground and asphalt located outside and inside the block. Thus, the sediment is primarily formed with dust and fine sand redistributed within the territory of residential block.

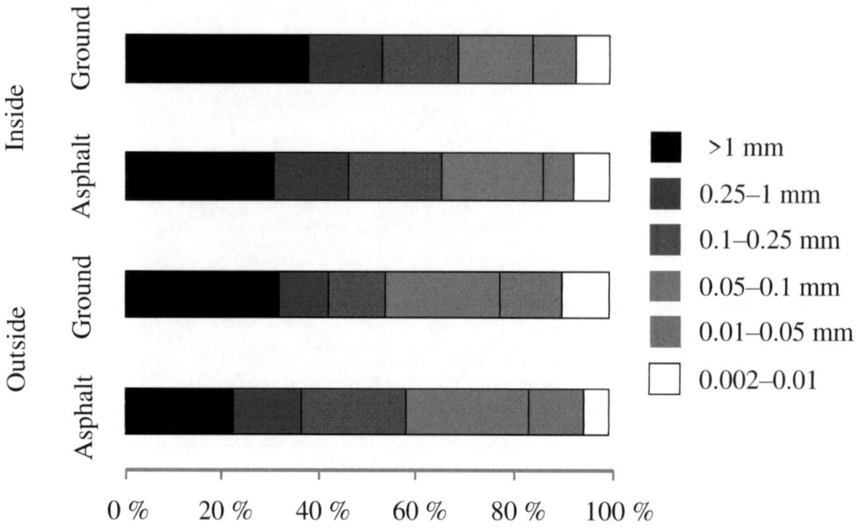


**Figure 1.** Results of particle-size composition analysis of the mud sediment

## 4. Conclusion

The studied surface mud sediment represents the environmental compartment that spatially and temporally integrates surface migration fluxes of pollutants in an urban environment. Study of the sediment as environmental indicator allows assessing the degree of pollution of residential areas, ranking the city territories over pollution degree and identifying genesis of pollutants in urban environment. The results of the current study allowed elaborating the methodology of environmental assessment based on the sampling of the surface mud sediment. The methodology was implemented in Ekaterinburg and (Russia).

**Acknowledgement**

The reported study was funded by RFBR, according to the research project No. 16-35-60044 mol_a_dk.

## References

[1] Wong C S C, Li X and Thornton I 2006 *Env. Pollution* **142(1)** p 1–16
[2] Jartun M, Ottesen R T, Steinnes E and Volden T 2008 *Sci. of the Tot. Env.* **396(2–3)** p 147–63
[3] Selbig W R, Bannerman R and Corsi S R 2013 *Sci. of the Tot. Env.* **444** p 381–91
[4] Wei B and Yang L 2010 *Microchemical J.* **94(2)** p 99–107
[5] Apeagyei E, Bank M S and Spengler J D 2011 *Atmospheric Env.* **45(13)** p 2310–23
[6] Sabin L D, Lim J H, Stolzenbach K D and Schiff K C 2005 *Water Research* **39(16)** p 3929–37
[7] Hilliges R A, Schriewer A and Helmreich B 2013 *J. of Env. Management* **128** p 306–12
[8] Shi G, Chen Z, Bi C, Li Y, Teng J, Wang L and Xu S 2010 *Env. Pollution* **158** p 694–703
[9] US EPA 2005 National Management Measures to Control Nonpoint Source Pollution from Urban Areas (Washington, DC: United States Environmental Protection Agency Office of Water) EPA-841-B-05-004
[10] US EPA 2001 Stormwater to Street Trees. Engineering Urban Forests for Stormwater Management (Washington, DC: United States Environmental Protection Agency) EPA-841-B-13-001

[11] US EPA 1991 Construction Site Stormwater Discharge Control. An Inventory of Current Practices (Washington, DC: United States Environmental Protection Agency Office of Water) EPA 833-D-91-100
[12] US EPA 2009 Protecting Water Quality with Smart Growth Strategies and Natural Stormwater Management in Sussex County, Delaware. EPA-NOAA Smart Growth Implementation Assistance for Coastal Communities for Sussex County, Delaware (Washington, DC: United States Environmental Protection Agency)
[13] US EPA 2008 Managing Wet Weather with Green Infrastructure. Municipal Handbook. Green Streets (Washington, DC: United States Environmental Protection Agency) EPA-833-F-08-009
[14] US EPA 2009 Developing Your Stormwater Pollution Prevention Plan: A Guide for Industrial Operators (Washington, DC: United States Environmental Protection Agency) EPA 833-B-09-002
[15] Nolde E 2007 *Desalination* **215(1-3)** 1–11
[16] Kose T, Yamamoto T, Anegawa A, Mohri S and Ono Y 2008 *Desalination* **226(1–3)** 151–159
[17] Duzgoren-Aydin N S 2007 *Sci. of the Tot. Env.* **385(1–3)** 182–195
[18] Seleznev A A, Yarmoshenko I V and Sergeev AP 2015 *J. of Env. Radioactivity* **142** 9–13
[19] Seleznev A A and Yarmoshenko I V 2014 *Env. Technology & Innovation* **1–2** 1–7

# Entropy modeling of sustainable development of megacities

**A N Tyrsin[1] and G G Gevorgyan[2]**

Ural Federal University, 620002, Ekaterinburg, Russia

[1] E-mail: at2001@yandex.ru
[2] E-mail: garnik.ggg@gmail.com

**Abstract**. The entropy approach of modeling multidimensional stochastic systems is described. It is based on the system representation as a multidimensional random vector and on the use of its differential entropy as a mathematical model. The possibilities of using this entropy model are considered for problems of monitoring the state of complex systems, including megacities, regions and critical infrastructure. Examples of practical implementation of the model are presented for the study of the sustainable development of megacities and regional environmental protection systems.

## 1. Introduction

The city is a complex ecological-socio-economic system consisting of a huge number of interacting elements. Large industrial centers are characterized by the enormous diversity and complexity of factors, elements of infrastructures, and the connections between them that affect the quality of its population life [1]. It should be noted that the quality of population life is a joint indicator reflecting the state of environmental protection, ecology, economy, education, health, housing conditions, culture, etc. These characteristics make it vital using a system approach for the sustainable development of these megacities, ensuring the growth of the quality of the population life [2]. A system approach involves the integration of the entire urban community into a system consisting of interrelated elements of infrastructures (environmental protection, ecology, economics, education, health, culture, politics, etc.), functioning of which is aimed to achieve the global goal of improving the quality of life of everyone [3].

However, despite the generally understandable main goal, the development of universal formal indicators characterizing the sustainable development of megacities is a complex problem. The existing diversity of various indicators that characterize various urban infrastructures and industries does not allow us to characterize the state of a megacity as a system.

Along with multidimensionality, another feature of the city as a system is its stochastic behavior. The stochasticity of the system is comes out in its probabilistic (unpredictable) nature of behavior, depending on random factors that can cause instability of individual parameters of the system as a whole. Peculiarity of such systems is the presence of a set of elements that are interconnected in a complex, probabilistic way. In this kind of situations, a multidimensional stochastic system is often modeled as a random vector.

One of the perspectives of modeling complex stochastic systems is based on the application of entropy. It is known that entropy is a fundamental property of all systems with an ambiguous or probabilistic behavior [4]. The concept of entropy is rather flexible and it can be clearly interpreted in terms of that specific area, where it is applied. It is being widely used in modern science to describe

IOP Conf. Series: Earth and Environmental Science **72** (2017) 012010 doi:10.1088/1755-1315/72/1/012010

the structural organization and disorganization, the degree of destruction of the connections between the elements of the system [5–8], including city systems [9, 10]. Therefore, it seems that the entropy could act as a universal parameter and it is ideal for solving the considered problems of the behavior of complex stochastic systems.

However, despite the frequent use of this term, the use of entropy for the modeling of open systems, unlike thermodynamics, is not formalized enough and is mainly of a qualitative and particular nature, there are no sufficiently simple and adequate mathematical models to link entropy with actual characteristics of states of multidimensional stochastic systems.

In these works the common thing is the use of information entropy:

$H(S) = -\sum_{i=1}^{M} p_i \ln p_i$, where $p_1,\dots,p_M$ the probability of the system S taking a finite number of corresponding values. The use of information entropy for the modeling of multidimensional open stochastic systems faces several serious difficulties. Let's list them.

1) The calculation of the information entropy requires an estimate of the pi probabilities of the elementary states of the system. This requires large samples to ensure sufficient accuracy of entropy calculation.
2) Often there are difficulties, both with an unambiguous identification of a fixed finite set of states in a complex system, and with the fact that some states may not be known in advance.
3) Information entropy is not designed for the case of multidimensional systems. Therefore, it is difficult to model the relationships between the system elements.
4) The change in the variance of the studied process is not considered.

Therefore, the existing adequate entropy models of real objects are designed to solve particular problems.

The aim of the article is to consider another approach that allows one to universally construct and interpret entropy models of such complex stochastic systems as megacities.

## 2. The entropy model of a multidimensional stochastic system

The proposed approach is based on using the differential entropy [11]

$$H(\mathbf{Y}) = -\int_{-\infty}^{+\infty}\dots\int_{-\infty}^{+\infty} p_{\mathbf{Y}}(x_1, x_2, \dots, x_m)\ln p_{\mathbf{Y}}(\mathbf{x})dx_1dx_2\dots dx_m \quad (1)$$

of the random variable $\mathbf{Y} = (Y_1, Y_2, .., Y_m)$ with the $p_Y(x_1, \dots, x_m)$ with the joint probability density function (PDF) without knowing its distribution type.

Let's represent the stochastic system S as a random variable $\mathbf{Y} = (Y_1, Y_2, .., Y_m)$. Each $Y_i$ element of the vector $\mathbf{Y}$ is a one-dimensional random variable which is characterizing the functioning of the particular element of the system under study.

The entropy (1), being a functional from the set of probability densities of random variable $\mathbf{Y}$, is a number. And therefore it cannot be counted as an adequate mathematical model of a multidimensional system. In [12] it is proved that if all the components of $Y_i$ have variances of $\sigma^2_{Y_i}$, then the $H(\mathbf{Y})$ differential entropy (the entropy, from now on) of the random vector $\mathbf{Y}$ is equal to:

$$H(\mathbf{Y}) = H(\mathbf{Y})_V + H(\mathbf{Y})_R = \sum_{i=1}^{m} \ln\sigma_{Y_i} + \sum_{i=1}^{m} \kappa_i + \frac{1}{2}\sum_{k=2}^{m} \ln(1 - R^2_{Y_k/Y_1Y_2\dots Y_{k-1}}), \quad (2)$$

where $\kappa_i = H(Y_i/\sigma_{Y_i})$ is the entropy indicator of the distribution type of the random variable $Y_i$, $R^2_{Y_k/Y_1Y_2\dots Y_{k-1}}$ the coefficients of determination of regression dependencies.

The first two terms $H(\mathbf{Y})_V = \sum_{i=1}^{m} H(Y_i) = \sum_{i=1}^{m} \ln\sigma_{Y_i} + \sum_{i=1}^{m} \kappa_i$ are the entropy of a random vector with mutually independent components and are called the randomness entropy, the third one – $H(\mathbf{Y})_R = \frac{1}{2}\sum_{k=2}^{m} \ln(1 - R^2_{Y_k/Y_1Y_2\dots Y_{k-1}})$ – self-organization entropy, which characterizes the tightness of the joint correlation between the components of $Y_i$.

According to (2), the $H(\mathbf{Y})$ entropy possesses threealism. There are three reasons for its change: the change of the level of scattering of its components, the change of the distribution type of its components, and the change of the strength of correlations between its components.

The advantage of formula (2) is that the entropy modeling of multidimensional stochastic systems on its basis doesn't require any definition of the distribution type of the multidimensional stochastic variable $\mathbf{Y}$, which is practically unfeasible in real problems. Herewith, unlike the methods of multivariate statistical analysis, the formal rigor and correspondence of model (2) to real experimental data is not lost. This allows us to use formula (2) to model and study real multidimensional stochastic systems and processes on limited experimental data set.

For multidimensional normally distributed random variable $\mathbf{Y}$ we have [12]:

$$H(\mathbf{Y})_V = \sum_{i=1}^{m} \ln\sigma_{Y_i} + m\ln\sqrt{2\pi e},\ H(\mathbf{Y})_R = \frac{1}{2}\ln(|\mathbf{R}|), \quad (3)$$

where $\mathbf{R}$ is the correlation matrix of the random variable $\mathbf{Y}$.

According to (2), the parameters of the entropy model are:

- $\sigma_{Y_i}$ standard deviations of $Y_i$ components,
- entropy indicators $\kappa_i$ of distributions, $i = 1, 2, \ldots, m$
- $R^2_{Y_k/Y_1Y_2\ldots Y_{k-1}}$ coefficient of determination of regression dependencies of the components of random vector $\mathbf{Y}$, $k = 2, 3, \ldots, m$.

The diagnostic model should explain the changes that occur in the studied object during functioning, in dynamics. Let's now consider the entropy of random vector from this point (2).

Let the stochastic system be presented as a random vector $\mathbf{Y}$. Then, based on the model (1) it is possible to monitor the state of the stochastic system by analyzing the change in its entropy. This can be done as follows. Let's assume that two random vectors and describe the previous and current periods of functioning of the system. We assume that the dispersions of all the components of random vector are finite.

To diagnose the state of a multidimensional stochastic system we'll adhere to the following steps [12]:

1) system behavior definition (stable/unstable), search for time dependencies of system behavior, and critical values;
2) detecting changes in the nature of the system ("randomness" or "self-organization") in critical periods;
3) analysis based on the detected reason: which element of the system turned to be the cause of the change in its state;
4) drawing conclusions about the impact of changes on the system, considering the identified critical points and their causes.

**Case 1:** First consider the case when distributions of all relevant components of $Y_i^{(1)}$, $Y_i^{(2)}$ $(i = 1, 2, \ldots, m)$ are described by the same type of distribution with some location and scale parameters. This means that $\forall i\ \kappa_i^{(1)} = \kappa_i^{(2)}$. Then the difference of entropies $\Delta H(\mathbf{Y}) = H(\mathbf{Y}^{(2)}) - H(\mathbf{Y}^{(1)})$ is defined as:

$$\Delta H(\mathbf{Y}) = \sum_{i=1}^{m} \ln\frac{\sigma_{Y_i}^{(2)}}{\sigma_{Y_i}^{(1)}} + \frac{1}{2}\sum_{k=2}^{m} \ln\frac{1-R^2_{Y_k^{(2)}/Y_1^{(2)}\ldots Y_{k-1}^{(2)}}}{1-R^2_{Y_k^{(1)}/Y_1^{(1)}\ldots Y_{k-1}^{(1)}}}.$$

Having pair of components $Y_i^{(1)}$, $Y_i^{(2)}$ with the same types, instead of threealism we'll have a particular case of dualism, i.e. the entropy can only change either due to the change of variance of the components of random vector or due to the change of the tightness of the correlation between these components.

The contribution of any $l$-th component in the change of the entropy of randomness:

$$\Delta H(\mathbf{Y})_{V,l} = \ln\frac{\sigma_{Y_l^{(2)}}}{\sigma_{Y_l^{(1)}}}, l = 1, 2, \ldots, m.$$

Since $R^2_{Y_m/Y_1Y_2\ldots Y_{m-1}} \geq R^2_{Y_m/Y_1Y_2\ldots Y_{m-2}} \geq \ldots \geq R^2_{Y_m/Y_1}$, the ultimate coefficient of determination $R^2_{Y_m/Y_1Y_2\ldots Y_{m-1}}$ most accurately describes the dependence of the $Y_m$ component from the rest $(m-1)$ components. Therefore, it is expedient to assess the contribution of the *l*-th component in the change of the self-organization entropy through the ultimate values of coefficients of determination.

$$\Delta H(\mathbf{Y})_{R,l} = \frac{1}{2}\ln\frac{1-R^2_{Y_l^{(2)}/Y_1^{(2)}\ldots Y_{l-1}^{(2)}Y_{l+1}^{(2)}\ldots Y_m^{(2)}}}{1-R^2_{Y_l^{(1)}/Y_1^{(1)}\ldots Y_{l-1}^{(1)}Y_{l+1}^{(1)}\ldots Y_m^{(1)}}}, l = 1, 2, \ldots, m.$$

The total contribution of the *l*-th component in the change of entropy of the random vector is defined as $\Delta H(\mathbf{Y})_l = \Delta H(\mathbf{Y})_{V,l} + \Delta H(\mathbf{Y})_{R,l}$.

**Case 2:** Consider the general case when at least one pair of components $Y_i^{(1)}$, $Y_i^{(2)}$are not described by the same type of distributions. Then the difference of entropies $\Delta H(\mathbf{Y}) = H(Y^{(2)}) - H(Y^{(1)})$ is defined as:

$$H(\mathbf{Y}) = \sum_{i=1}^{m}\ln\frac{\sigma_{Y_i^{(2)}}}{\sigma_{Y_i^{(1)}}} + \sum_{i=1}^{m}(\kappa_i^{(2)} - \kappa_i^{(1)}) + \frac{1}{2}\sum_{k=2}^{m}\ln\frac{1-R^2_{Y_k^{(2)}/Y_1^{(2)}\ldots Y_{k-1}^{(2)}}}{1-R^2_{Y_k^{(1)}/Y_1^{(1)}\ldots Y_{k-1}^{(1)}}}.$$

Since in this case $\kappa_i^{(2)} - \kappa_i^{(1)} \neq 0$, there is now third factor of the entropy change caused by the distribution type change.

Thus, the case of the conservation of the distribution types of the random vector components is easier to implement and does not require the definition of the entropy indicators of the components. But the violation of this condition can cause significant errors in the assessment of the entropy dynamics, and, consequently, a decrease in the reliability of the diagnosis of the system state.

By monitoring the change of the $\Delta H(\mathbf{Y})$ entropy as a whole and of its components, we can conclude about the state of the studied stochastic system and detect emerging trends in its changes. Analysis of changes in each component of the random vector **Y** will reveal those of them having the greatest impact on change in entropy, and hence on change in the state of multidimensional stochastic system.

Since the city is an open system, its entropy can both increase and decrease. Systems with different randomness $H(\mathbf{Y})_V$ and self-organization $H(\mathbf{Y})_R$ entropies can have the same entropy values $H(\mathbf{Y})$. Therefore, it is necessary to consider the entropy (2) in vector form as:

$$h(\mathbf{Y}) = (h_V;\ h_R) = (H(\mathbf{Y})_V;\ H(\mathbf{Y})_R). \qquad (4)$$

## 3. Practical use of the entropy approach in modeling the sustainable development of megacities

The sustainable development of a megacity as a complex system according to the proposed entropy approach consists in the simultaneous growth of diversity, opportunities for all elements of this system and the existence of a close relationship between these elements. This comes out with the fact that with the development of the megacity, its randomness entropy should gradually increase, and the self-organization entropy should decrease.

On the example of the analysis of the main indicators characterizing the state of critical infrastructures (environment, ecology, economy, education, healthcare, culture, etc.), we will consider the possibilities of practical use of the vector form (4) of the entropic model for studying the stability of the development of megacities and regions.

*3.1. Comparative entropy analysis for the dynamics of Moscow, St. Petersburg and Ekaterinburg development in 1992–2015*

The analysis will be carried out on the statistical dataset of Federal State Statistics Service [13]. From the large set of basic socio-economic indicators of cities, a system of 12 features is formed, characterizing all the main aspects of the city infrastructure:

1) The natural increase in population per 1000 people;
2) Employment-to-population ratio (in organizations), %;
3) Accrued nominal monthly average wages, thousand rubles;
4) The proportion of pensioners listed by social protection bodies, %;
5) The total area of residential premises, an average per an urban resident (at the end of the year), $m^2$;
6) Number of pupils in pre-school educational organizations, thousand people;
7) Number of doctors per 1000 people;
8) The number of registered crimes per 1000 people;
9) Volume of industrial production, thousand rubles per capita;
10) The amount of work performed under construction contracts, thousand rubles per capita;
11) Retail trade turnover, thousand rubles per capita;
12) Investments in fixed assets, million rubles.

This system of indicators was formed according to the following rules. Pair correlations were considered between all the available in [13] main socio-economic indicators of cities, which are about twenty. It turned out that a number of indicators are very strongly correlated (coefficients of pair correlation exceeded 0.9). One indicator of these mutually correlated indicator groups was kept. The indicator, which characterized the city as a socio-economic system more meaningfully, was prioritized.

The inflation was counted by recalculation in the prices of 2015 based on consumer price indexes, different population of cities was taken into account on the transition to relative indicators per capita. The small sample size makes it possible to use the system as a Gaussian random vector.

For the entropy calculations the assessments were done for periods of 13 years. This period proved to be optimal by mean of statistical smoothing, on one hand, and considering the dynamics of entropy change, on the other. The entropy was assessed in vector form (4).

Since the sample was quite small, the deviations of the empirical distributions of the features considered from the normal distribution cannot be practically established. Therefore, when calculating the randomness and self-organization entropies, we'll use the formulas (3).

In Figure 1–3 there are the plots of changes in the randomness and self-organization entropies in Ekaterinburg, Moscow and St. Petersburg, respectively.

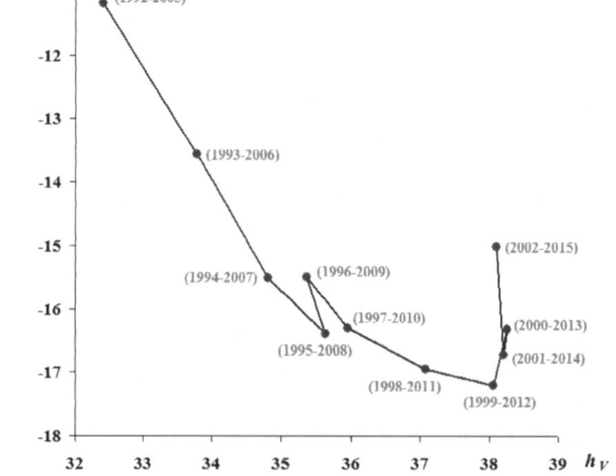


**Figure 1.** The randomness and self-organization entropy changing in Ekaterinburg

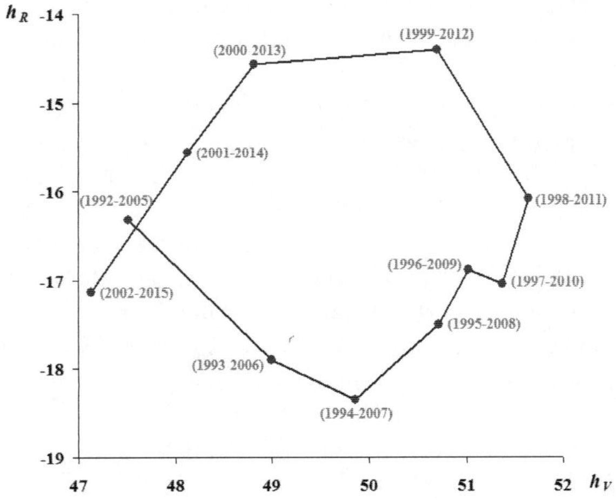


**Figure 2.** The randomness and self-organization entropy changing in Moscow

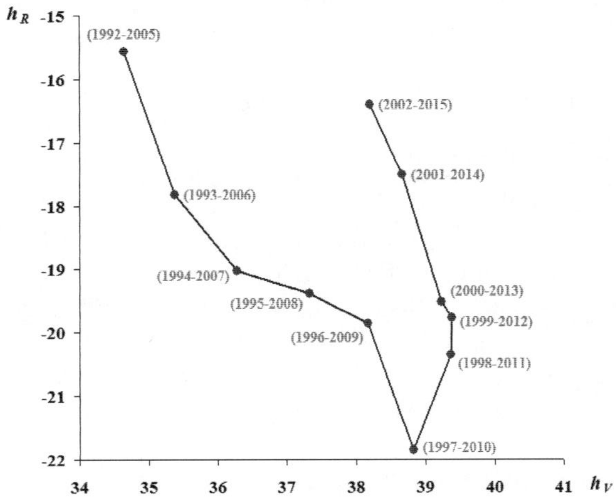


**Figure 3.** The randomness and self-organization entropy changing in St. Petersburg

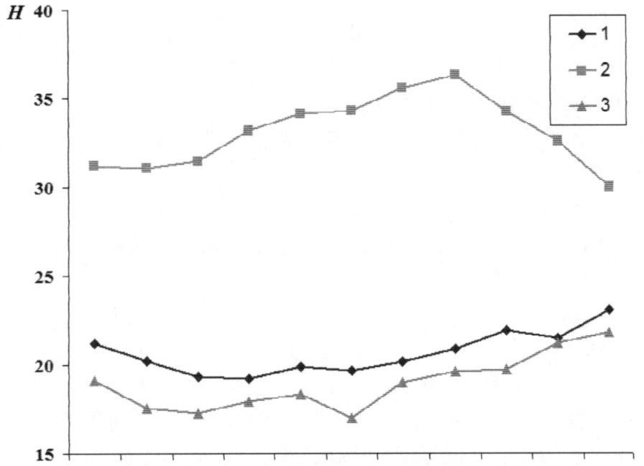


**Figure 4.** Dynamics of entropy in: 1 – Ekaterinburg, 2 – Moscow, 3 – St. Petersburg

Figure 4 shows the dynamics of entropy (2) for all three megacities.

The analysis of Figure 1–4 plots according to the entropy model allows us to draw the following conclusions.

1) Initially, Moscow was at a higher level of development, because it had a greater variety of infrastructures functioning (the greatest value of randomness entropy) and, at the same time, demonstrated a higher level of interaction between infrastructures (minimum value of self-organization entropy). St. Petersburg was the second, both in terms of the diversity of the infrastructures functioning, and of the interaction between them.
2) Ekaterinburg was the third place in the initial period. But during the analyzed period, Ekaterinburg showed the best dynamics of development and almost equaled with St. Petersburg in levels of randomness and self-organization entropies. Moscow as a multidimensional stochastic system showed no development. The plot of Figure 2 shows a cycle, i.e. the system returned to its original state.
3) The graphs clearly show the specific areas of instability, corresponding to the global financial crisis of 2008–2009, the period of slower growth in gross domestic product in 2012 and the period of sanctions, since 2014.
4) From the graphs in Figure 4 we see that the total entropy in all three megacities has practically not changed during the examined period. This speaks about the need of considering the entropy of a multidimensional stochastic system, not in scalar, but in vector form.

### *3.2. Comparative entropy analysis for the environmental protection situation of the regions of Ural Federal District*

Entropy analysis can be carried out both at the level of a megacity or region, and at the level of their subsystems (elements), for example, critical infrastructures.

The problem of environmental protection is now becoming more acute. Therefore, as an example of the possibilities of entropy analysis for infrastructures, we can consider the environment. In the statistical reporting of the Federal State Statistics Service, the most complete data are presented for regions. We will perform a comparative entropy analysis of the state of environmental protection for four regions of the Ural Federal District – Kurgan, Sverdlovsk, Tyumen and Chelyabinsk.

The system of attributes will be formed and entropy calculation will be carried out the same way as in example 3.1.

The analysis will be carried out on the statistical data of Federal State Statistics Service for 2005–2015 [14].

From the variety of the indicators that characterize the environmental protection a system of 9 features is formed:

1) Investments in fixed assets aimed to the environmental protection, rubles per capita;
2) Current costs of environmental protection, rubles per capita;
3) Emitted pollutants of the atmosphere, kg per capita;
4) Water abstraction from natural reservoirs, $m^3$ per capita;
5) Loss of water during transportation, cc
6) Sewerage discharge/ contamination of surface water, $m^3$ per capita;
7) Area of urban settlements, that are cleaned by mechanized method, $m^2$ per capita;
8) The total area of the forest funds and other categories of land where forests are located, ha per capita;
9) The area of dead forest stands, ha per capita.

Just as in the example 1, the inflation was counted by recalculation in the prices of 2015 based on consumer price indexes; the different population of regions was taken into account by the transition to relative indicators per capita. The small sample size makes it possible to use the system as a Gaussian random vector.

In Figure 5 we have the randomness and self-organization entropies of the environmental protection system of the Kurgan, Sverdlovsk, Tyumen and Chelyabinsk regions in 2005–2015. The results of entropy analysis speak about the following.

The most developed is the environmental protection system of the Tyumen region (the largest value of the randomness entropy and the lowest value of the self-organization entropy), the next one is the Sverdlovsk region, and the Kurgan and Chelyabinsk regions are the third and fourth places, respectively.

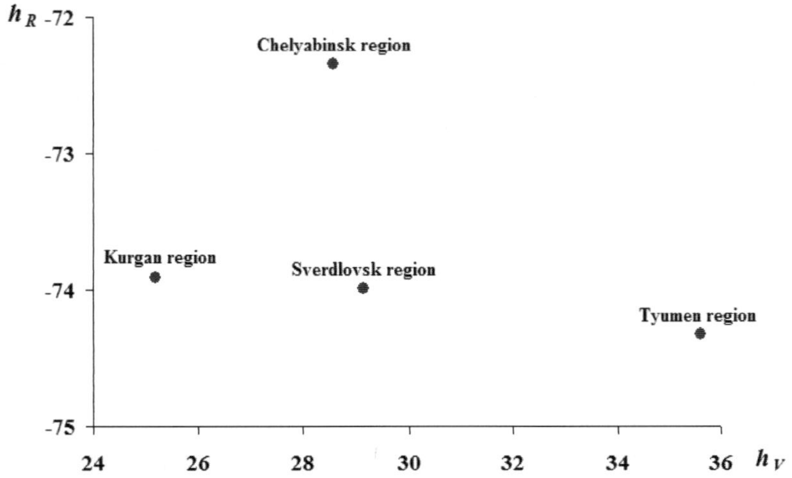


**Figure 5.** The randomness and self-organization entropies of the environmental protection system of Kurgan, Sverdlovsk, Tyumen and Chelyabinsk regions in 2005–2015 years

## 4. Conclusion

1. Hereby it is considered the possibilities of using the entropy modeling for the problems of sustainable development of large cities and regions, as well as the critical infrastructures on the example of environmental protection.

2. It is shown that the vector representation of the system entropy allows us to reveal the main trends in the development of multidimensional stochastic systems.

**Acknowledgements**

The reported study was carried out by financial support of the Russian Fund of Fundamental Research grant No. 17-01-00315 “Development of models and methods of monitoring, forecasting and control optimization of multidimensional regional socio-economic systems based on entropy and minimax approaches”.

## References

[1] O'Sullivan A 2012 *Urban Economics* 8th edition (McGraw-Hill/Irwin)
[2] Satzinger J W, Jackson R B and Burd S D 2008 *Systems Analysis and Design in a Changing World* 5th edition (Course Technology)
[3] Popkov Yu S 2014 *Mathematical Demoeconomics: Integrating Demographic and Economic Approach* (Berlin: De Gruyter)
[4] Nicolis G and Prigogine I 1977 *Self-Organization in Nonequilibrium Systems: From Dissipative Structures to Order through Fluctuations* (New York: Wiley)
[5] Lindblad G 1983 *Non-Equilibrum Entropy and Irreversibility* (Dordrecht: Delta Reidel)
[6] Haken H 2000 *Information and Self-Organization: A Macroscopic Approach to Complex Systems* (Springer)
[7] Ellis R S 2006 *Entropy, Large Deviations, and Statistical Mechanics* (Springer)

[8] Timashev S A and Tyrsin A N 2011 Entropy Approach to Risk-Analysis of Critical Infrastructures Systems *Proc. First Int. Conf. on Vulnerability and Risk Analysis and Management and Fourth Int. Symposium on Uncertainty Modeling and Analysis* (Hyattsville, Maryland: ASCE) p 147–54
[9] Wilson A G 1970 *Entropy in Urban and Regional Modelling* (London: Pion)
[10] Webber M J 1979 *Information Theory and Urban Special Structures* (London: Croom Helm)
[11] Shannon C E 1948 A Mathematical Theory of Communication *The Bell System Technical Journal* **27** pp 379–423, 623–56
[12] Tyrsin A N 2016 *Entropy modeling of multidimensional stochastic systems* (Voronezh: Scientific Book)
[13] *Regions of Russia The main social-economic indicators of cities* Federal State Statistics Service Available from: http://www.gks.ru/wps/wcm/connect/rosstat_main/rosstat/en/statistics/publications/catalog/
[14] *Environmental protection in Russia* Federal State Statistics Service Available from: http://www.gks.ru/wps/wcm/connect/rosstat_main/rosstat/en/statistics/publications/catalog/

# Strategic guidelines of a megalopolis's development: new industrialization and ecological tension

**Yulia Lavrikova, Victoria Akberdina and Elena Mezentseva**

The Institute of Economics, the Ural Branch of Russian Academy of Sciences, 620014, Ekaterinburg, Russia

E-mail: akb_vic@mail.ru

**Abstract**. The article is devoted to the integration of environmental concerns in the development strategy of the metropolis. The authors substantiate the relationship of the new industrialization and reduce the burden on the environment. For example, a large city in Russia – Ekaterinburg – projections and strategic directions of ecological development of the city for the period up to 2035. The basis for the forecast were the methods of aggregation of economic sectors, functional relationship and forecasting algorithms. The article describes three scenarios for the development of Ekaterinburg, the results of calculations by the author's method. The authors have shown the relationship of industrial development of the metropolis and the anthropogenic load, which is assessed using such indicators as emissions of harmful substances into the atmosphere, discharges of sewage, green spaces per inhabitant. The authors note that the ecological security of the residents is directly related to the improvement of controllability of the municipal economy, the increased control in the field of the environment, reduction of environmental burden on humans and the environment, zoning of the city with a goal differential application of indicators of the quality of the environment.

## 1. Introduction

The future of million-plus cities in Russia in a medium-term horizon is approximately the same. It aims to create a quality of urban environment, not inferior to the world cities, the most accessible of which is Moscow. In fact, we are talking about the construction of a "post-industrial city", with developed economy of services, modern trade, entertainment and public spaces. At the same time, industrial megacities, which historically developed as centers of industrial territories, faced a key contradiction – how to assess an objectively occurring deindustrialization process. On the one hand, a growth of trade, public catering and services sector certainly changes a face of a city, attracts investors and population, raises a quality of life level, forming a "city for a person". However, on the other hand, a decline in an industry share in the city's economy structure may be disastrous for the city's economy, since an industrial complex, being a material basis where innovations are realized, provides high-performance and highly paid jobs, each of which places a demand for highly skilled professionals.

Considering an issue of de-industrialization, it is widely believed that an industrial segment squeezing will solve environmental problems. Socio-ecological tension indeed characterizes a relationship between a city development and a real anthropogenic load on the natural environment. A need to increase an economic growth rate of a city reflects an universal (for post-industrial society) setting for a growth of material well-being of population. According to German philosopher W. Hesle, with a transformation of "liberal legal states into social services states" a maximum satisfaction of people material demands became the main goal of an existing socio-economic system [1]. Formed on a basis of consumer culture, a lifestyle of megacity population aggravates a relationship between a society and the environment, stimulating an economic development. To meet new needs, a society accelerates a scientific and technical progress, which, in accordance with prevailing attitudes in public consciousness, is also perceived as "a foundation of material prosperity and as a foundation for conquering nature" [2].

However, at present in the post-industrial countries the orientation towards ecology becomes a significant factor that intensifies economic development. Consistent implementation of environmental activities leads to accelerated economic progress, since both environmental protection and economic growth have a common basis – a technological progress. Environmental programs stimulate economic development, an implementation of structural changes, new jobs creation and further progress in science and technology.

Thus, a new industrialization goes hand in hand with a new environmental imperative at a level of the urban environment. An industrial sector is qualitatively changing in terms of a set of technologies used. It ensures a slowdown in a growth rate of emissions into the environment. Reindustrialization, namely, a strengthening of an industry share in city economy, is associated with a development of new sectors of industry that were previously absent, a growth of a high value-added industry share, an increase in a production in more complex, convergent activities, a replacement of sectors going obsolete and sectors with low productivity by modern industry with the philosophy of Industry 4.0.

Considering Ekaterinburg as an example of one of the largest megacities of Russia – we will trace basic tendencies of industrial development and ecology.

## 2. Methods

In our study, we adhere to the national school of spatial development (N.N. Kolosovsky [3], G.M. Lappo [4], P.M. Polyan [4] and others [5–8]) whose representatives study large cities with a high concentration of industrial production, complex production and technological and labor connections, a concentration of a cultural, information and educational potential.

The Institute of Economics of the Ural Branch of RAS has developed a megacity model with a system of interrelations (Figure 1) and a methodology for forecasting socio-economic indicators. The model formed a basis of a software complex which was used to develop a long-term socio-economic forecast for Ekaterinburg for a period up to 2035. The methodology includes a logical scheme of aggregation of economic sectors, functional relationships, prediction algorithms, formulas for calculating indicators, links to statistical compilations. The methodology provides an opportunity to get a systematically coordinated forecast of structural changes in the economy for a long term, as well as to take into account changes in an ecological situation of a city.

## 3. Object of study

An object of our research is Ekaterinburg. Among Russian million-plus cities, Ekaterinburg has a leading position after Moscow and St. Petersburg. According to Russian Statistical Office, Ekaterinburg has a combined rating equal 1. This high rating is due primarily to high rates of wages, a trade, services and housing construction development. An Eurostat data analysis allowed to single out European million-plus cities with the same structure of a city's economy as in Ekaterinburg. Currently, a share of Ekaterinburg population working in industry, is 19.4 %. Industrialized European million-plus cities with the closest values for this indicator are: Greater Milan (19.3 %), Greater Naples (15.7 %), Lyon (11.9 %), Munich (11.3 %), Greater Manchester (10.7 %), Budapest (10.7 %), Hamburg (10.3 %) and Cologne (10.2 %). Comparison of Ekaterinburg with these cities in terms of level and quality of life, as well as environmental friendliness, will help identify targets for the city development in a planning horizon until 2035.

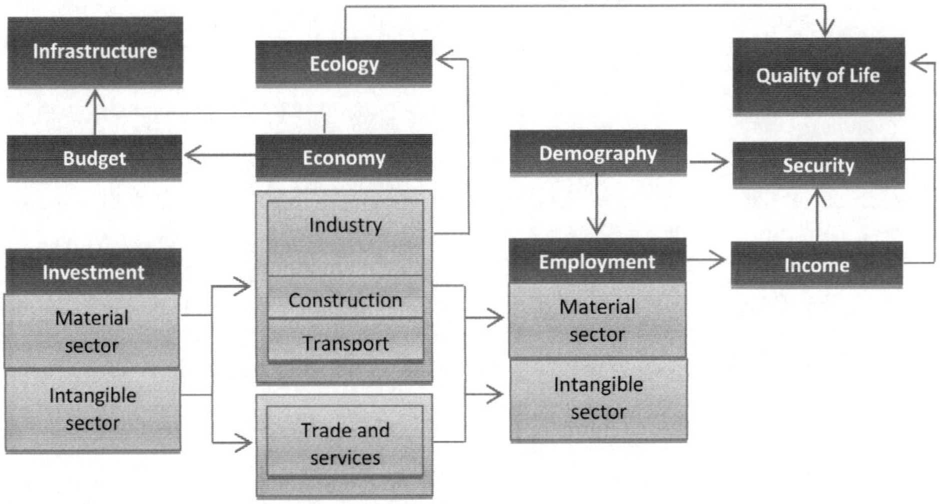


**Figure 1.** Main interrelations of a city development according to the methodology of the Institute of Economics, UB RAS

## 4. Development analysis

Ekaterinburg has historically been the capital of an industrial region – Sverdlovskaya Oblast', and therefore, an industry share in a organization turnover structure is significant – over the past decades about one-third of an turnover on average. However, a share of industry is declining, driven out by accelerated growth in trade. While in 2006 the share of an industry was higher than the share of an trade and service sector (39.1 % vs. 31.6 %, respectively), then in 2015 a trade and service sector had already a larger share than the industrial sector (40.5 % against 37.4 %). The share of an industrial sector in a number of employed at the end of 2015 was 20.5 %, of which 16.3 % were employed in processing industries.

Throughout the analyzed period, an industry share in a turnover structure reached its maximum in 2010 (43.2 %), and then the slowdown in an industrial growth began, which led to a loss of industrial sector positions. As a result, we can talk about an emerging *trend of de-industrialization of Ekaterinburg city.*

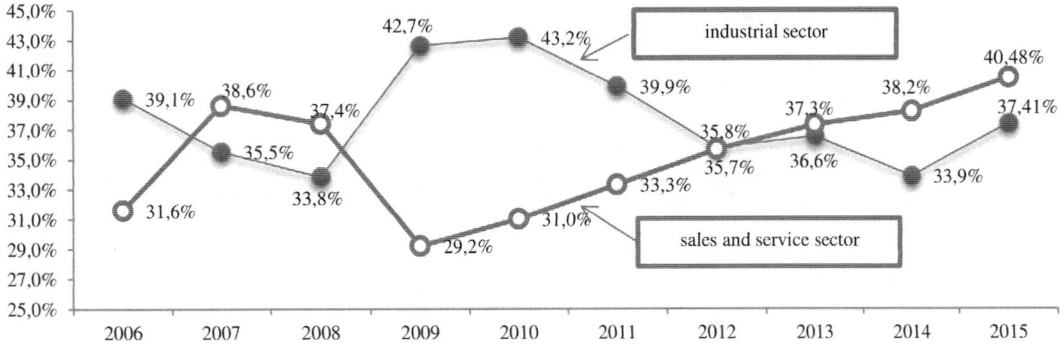


**Figure 2.** Dynamics of shares of industrial and trade-service sectors for a period 2006–2015, %

*An index of industrial production* in Ekaterinburg for a period 2000–2015 amounted to 107.7 % annually, however there were periods of a significant decline in 2008–2009 and 2014–2015. A fairly high average annual growth rate over a 15-year period is provided by a high rate of recovery growth in 2010–2011.

IOP Conf. Series: Earth and Environmental Science **72** (2017) 012011 doi:10.1088/1755-1315/72/1/012011

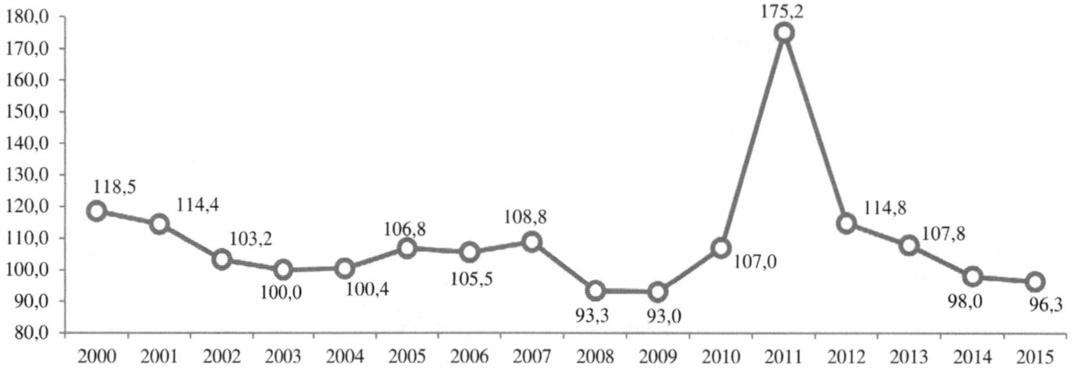


**Figure 3.** An index of industrial production for a period 2000–2015, %

Thus, *today Ekaterinburg is at a fork* in whether it will remain the capital of an industrial region or whether a trend of a de-industrialization will continue. Despite an inevitable accelerated development of a tertiary sector, an industrial sector of Ekaterinburg has a significant innovative potential. This is due to the fact that a core of a city industrial sector is made up of military-industrial complex enterprises, where high-tech productions of new technological structures are developing. Non-alternative processes of new industrialization will form prerequisites for building up an innovative potential of city industrial enterprises. An implementation of these opportunities will become available in a formation of a sound industrial policy in the city.

An ecological situation of Ekaterinburg as a whole is characteristic for any megacity – a volume of an anthropogenic pressure on environment increases every year. So, an average annual growth rate of emissions of harmful substances (both from stationary and mobile sources) for a period 2000–2015 was 103.2 %. *A situation is positive only in a part of waste water* – an average annual decrease in a discharge volume during an analyzed period was 1.8 %. This fact is related to activities of a strategic project "Water for Life" aimed to improve a quality of water in surface water bodies, to reduce a risk to health of a younger generation associated with a quality of drinking water from water supply networks, to observe sanitary requirements for coastal areas, to monitor a water quality of water bodies. One of components which improve these indicators is an increase in a number of enterprises that have received permits for a right to use water bodies up to 94.2 %.

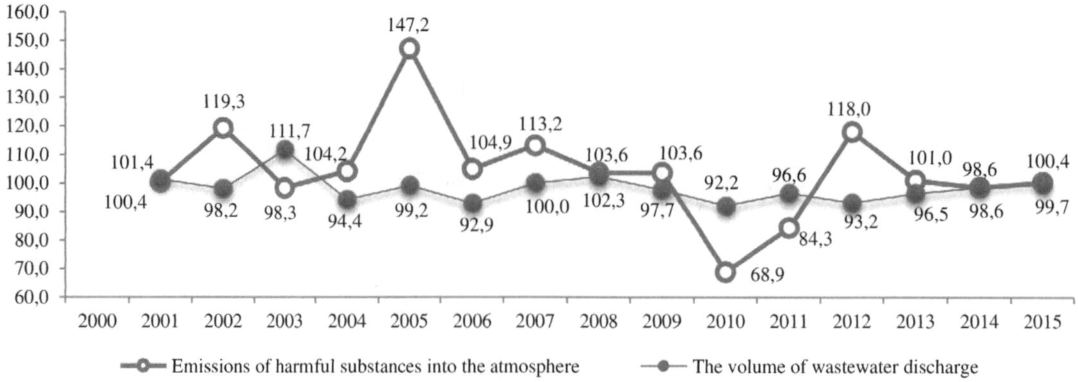


**Figure 4.** Average annual growth rates of the main indicators of anthropogenic load on environment for a period 2000–2015, %

Today, a specific volume of emissions of pollutants per an inhabitant in Ekaterinburg equals 0.148 tons. However, an emissions share of stationary sources (industrial enterprises) is only 25.6 %. The main polluter is a road transport and city buses. A development of any megalopolis is inevitably associated with an increase in a *level of motorization* – in Ekaterinburg only in the last 10 years this figure has grown by 1.4 times, amounting to 410 cars per 1000 inhabitants – in fact, every second resident of Ekaterinburg has a car.

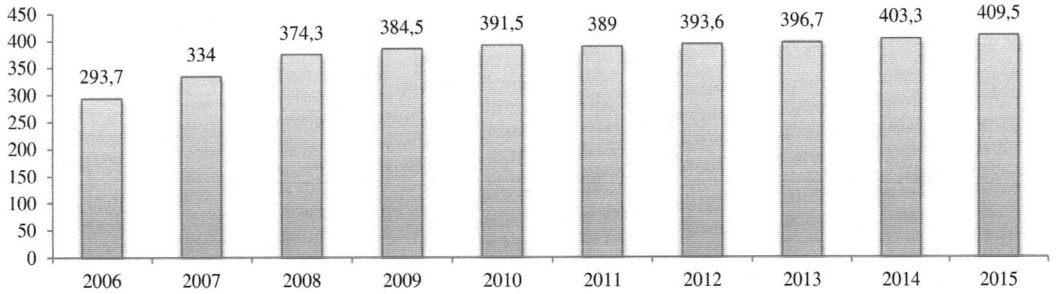


**Figure 5.** A level of motorization for a period 2006–2015, cars per 1000 inhabitants

## 5. Forecast of industrial development and ecological situation

In a horizon of strategic forecasting up to 2035, one can single out a set of contradictions that allow to structure a scenario choice, to identify possible strategies for a city and substantiate purposes of its development in new conditions.

A key contradiction in a development of a historically industrial Ekaterinburg is the contradiction in a structure of the city's economy. This is what determines a quality of life in future. Since 2000, an indicator characterizing a ratio of a turnover volume of a trade and service sector[1] to that of an industrial sector[2] of an economy is constantly increasing. This is due to high growth rates of a trade and service sector of economy in comparison with an industrial sector. So, for 15 years an average annual growth rate of a trade and service sector was 12 %, while an industrial sector – 8 %. A sphere of production and a sphere of circulation come into conflict, while a financial sector and a social sector (education, health) act as sectors serving a development of main sectors.

Ultimately, when we were forming scenarios for a development of Ekaterinburg, we used *three scenario forks*, each fork gave two alternative scenarios. A main fork of scenarios – structural – defines two vectors of development: an industrial and a trade-service. For a forecast calculations, one scenario was chosen within a framework of an industrial vector and two scenarios within a trade-service vector. A scenario in a framework of an industrial vector was called "innovative", or "*a new industrialization in conditions of an accelerated investment and an intensified migration*". A preservation of existing trends of an uncontrolled trade sector growth is fully visible in a inertial scenario, which can lead to complete deindustrialization of the city and loss of its status as the "capital of an industrial region". And, finally, a base scenario will be implemented within a trade vector, but with a preservation of an industrial component of the city's economy.

A forecast of city economy structure both in terms of a turnover and a number of population is a starting point for predicting a quality of life indicators. *A loss of the city industrial status is clearly visible in an inertial scenario.* By the end of 2035 a share of an industrial sector will not exceed 30 %, while a share of a trade and service sector will account for a half of a city organizations turnover. *An innovative scenario will allow talking about a parity of two the most important sectors for the city's economy* – their shares will amount to 40 %, which will allow to use a potential of a trade and service sector for investment development of an economy's real sector. A basic scenario assumes an intermediate version of a turnover structure.

---

[1] A retail and service sector of an economy includes retail trade, catering, paid services to the population, household services to the population.

[2] An industrial sector of an economy includes mining, processing industries, production and distribution of electricity, gas and water.

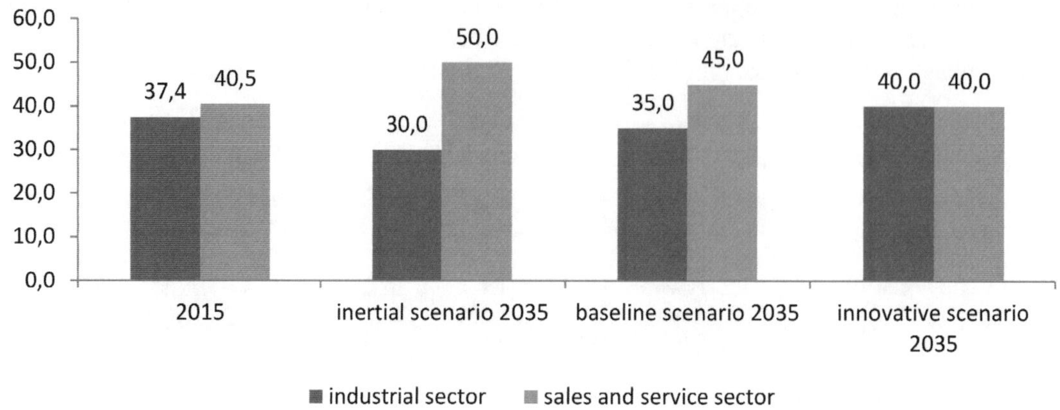


**Figure 6.** Structure of an organization turnover in 2035, %

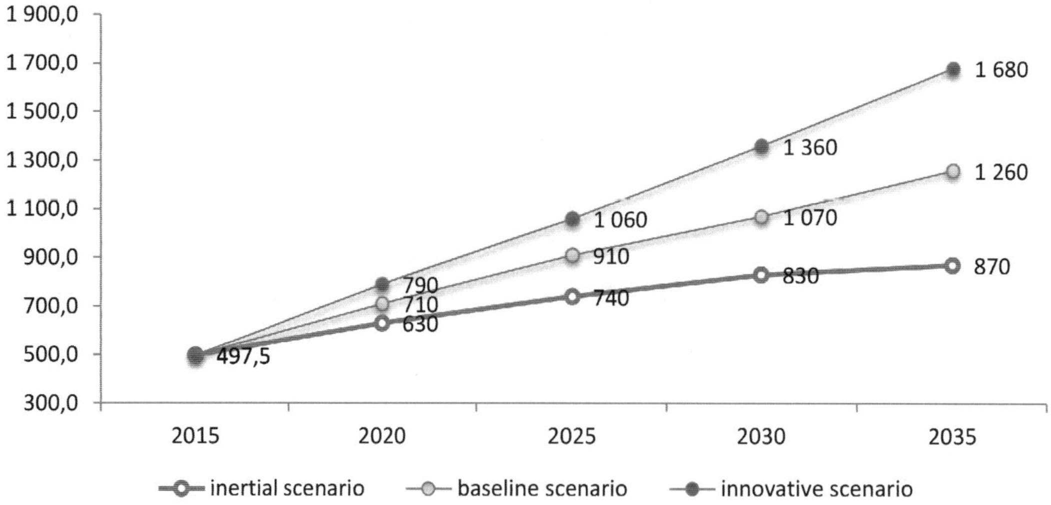


**Figure 7.** Dynamics of an industrial sector turnover, billion rubles

Without going into details of all socio-economic indicators of the forecast, let us dwell on indicators of an environmental block.

*An amount of air emissions* will continue to grow objectively, due to the fact that the main air polluter is road transport. Following an increase in a level of motorization, air emissions from mobile sources will also increase. At the same time, it should be noted that the smallest growth will be based on an innovative scenario. An amount of emissions from stationary sources due to emission control and implementation of corresponding measures at industrial enterprises of the city will decrease according to a basic and innovative scenario. A discharge of waste water will significantly improve a situation, especially in an innovative scenario.

The main goal of an overall strategy is an environmentally sound sustainable development of Ekaterinburg and to create a favorable habitat and comfortable conditions for life and reproduction of population, to ensure a protection of natural resources and biodiversity, to prevent man-made accidents and catastrophes, and to ensure a social and economic growth and people's well-being in present and future. The main indicators of sustainable development and, accordingly, the main priority in environmental policy at all levels are: health of population; a rational use of natural resources; high quality of the natural environment which is formed from a quality of specific ecosystems (an atmosphere, a water basin, flora, fauna, etc.); economic well-being of population.

**Table 1.** Forecast of ecological indicators of Ekaterinburg up to 2035

| Indicator | 2016 | 2035 | | |
|---|---|---|---|---|
| | | Inertial scenario | Base scenario | Innovative scenario |
| **Emissions of harmful substances into an atmosphere, total, thousand tons** | **218.9** | **235.0** | **222.0** | **215.0** |
| Average annual growth rate, % | | 100.4 | 100.1 | 99.9 |
| Growth rate by 2016, % | | 107.4 | 101.4 | 98.2 |
| **Stationary sources, thousand tons** | **24.2** | **25.0** | **22.0** | **20.0** |
| Average annual growth rate, % | - | 100.2 | 99.5 | 99.0 |
| Growth rate by 2016, % | - | 103.3 | 90.9 | 82.6 |
| Mobile sources, thousand tons | **194.7** | **210.0** | **200.0** | **195.0** |
| Average annual growth rate, % | - | 100.4 | 100.1 | 100.0 |
| Growth rate by 2016, % | - | 107.9 | 102.7 | 100.2 |
| **Volume of wastewater discharge into water bodies, million cubic meters** | **159.5** | **155.0** | **150.0** | **140.0** |
| Average annual growth rate, % | - | 99.8 | 99.7 | 99.3 |
| Growth rate by 2016, % | - | 100.0 | 94.0 | 87.8 |

To provide an environmental safety for city residents it is necessary to improve a manageability of urban life, to strengthen control in a field of environmental protection, to reduce and bring to safe levels a technogenic stress to human and the environment, to zone a city territory to differentiate a quality of living environment.

An important element of a long-term strategy is a formation of an ecological outlook of city residents through environmental education, upbringing and education, as well as an ecological and urban development of Ekaterinburg.

## 6. Conclusion

A development of any industrial megalopolis in its development approaches a point of bifurcation, after which a path of a development can change irrevocably: a megalopolis can lose its industrial status for good and became a large shopping center in a center of an industrial region, or it can go on a path of a radical modernization of an industry based on principles of the fourth industrial revolution.

A renewal of an industry on a new technological platform will lead to a gradual stabilization of an ecological situation and an achievement of environmental pressure regulatory indicators – a decline in a level of technogenic air basin pollution by reducing road and industrial emissions, satisfying population drinking water requirements of standard quality, etc. A simultaneous rationalization of nature use and protection of natural resources for each of the components of the natural environment will ensure environmental reproduction and environmentally oriented city development including a rationalization of a structure and level of fuel and energy resources consumption, an implementation of new low-waste technologies, as well as an optimization and dislocation of industrial production.

## Acknowledgement

The article has been prepared under the support of the The Russian Foundation for Basic Research 15-06-08375.

## References

[1] Hesle V 2013 On the relationship of morality and politics Part I *Political Studie* **4** p 45–61
[2] Elyanov A 2004 Globalization and catch-up development *World Economy and International Relationships* **1** p 3–16

[3] Kolosovskiy N N1958 *Economic zoning* (Moscow: Gospolitizdat) 200 p
[4] Lappo G, Polyan P and Selivanova T *Russian agglomeration in XXI century* Available from: http://www.frrio.ru/uploads_files/Lappo.pdf (date of access: 07 May 2017)
[5] Biyakov O A 2004 *A theory of economic space: methodological and regional aspects* (Tomsk: Tomsk University Publishing) 152 p
[6] Matafonov M E, Verbitskaya N O, Fedorov V A and Kuchinskaya V Ye 2003 *Human resources of the biggest city: models, prediction, management* (Ekaterinburg: Pravoved–2001) 148 p
[7] Nosonov A M *Theories of space development in socio-economic geography. Theoretical issues of regionology* Available from: http://prj-pgpu.narod.ru/prj_11_1.pdf (date of access: 07 May 2017)
[8] 2010 *Spatial organization of socio-labor systems: genesis and development problem* ed O A Kozlova (Ekaterinburg: Institute of Economics, Ural branch of RAS) 206 p

# Change in the quality of atmospheric air under the influence of chemical enterprises in the Sverdlovsk Region cities

**Maria N Ivantsova, Irina S Selezneva and Maxim A Bezmaternykh**

Ural Federal University, Mira str., 28, Yekaterinburg, Russia

E-mail: mivantsova@mail.ru

**Abstract**. In this article, we investigated the gaseous emission dynamics of Sverdlovsk region chemical complex from 2011 to 2015. We have analyzed the dynamics of changes in the atmospheric air state in cities where the big chemical enterprises are located. Our main attention is paid to gases emissions such as carbon monoxide, nitrogen oxides and sulfur dioxide. We also looked at the volume of purified gas emissions in all gaseous emissions from the chemical industry enterprises. The contribution of emissions from chemical industry is relatively low and accounts for approximately 4 % of emissions from all manufacturing industries, including metallurgical complexes. There is a general tendency to reduce emissions of pollutants in connection with the implementation of environmental measures. Reducing the gaseous substances emission can be achieved both by improving the production technology, and by installing new more sophisticated equipment which catches harmful substances emissions.

## 1. Introduction

At the beginning of the 20$^{th}$ century man's anthropogenic influence was not as high as at the beginning of the 21$^{st}$ century. During one hundred years the economic activity of man has so strongly influenced on the state of the environment that it can lead to irreversible changes of the future destiny of humankind. Excessive emissions into the atmosphere, dumping into the hydrosphere, solid waste disposal in the lithosphere are resulted in such negative consequences as: greenhouse effect, smog, acid rains, significant excess of TLV (threshold limit value) of many dangerous substances in places of people's lives, etc. At the present, human society is trying to solve environmental problems that have created for ourselves, in order to live and not survive on the Earth [1–3].

## 2. Methods

One of the main and acute problems of humankind is the air pollution with gaseous emissions. The largest emissions into the air environment are emissions of $SO_2$, CO, $NO_x$ and solids (dust) [4–6]. They have a detrimental effect on the health of animals and humans. Sulfur oxide (IV) causes a runny nose, cough, hoarseness, and swelling in the throat. Large concentrations of $SO_2$ lead to suffocation, speech disorders, difficulty swallowing, vomiting, and possible acute edema of the lungs. Carbon oxide (II) prevents oxygen absorption by blood hemoglobin, which weakens cognitive abilities, slows down reflexes, causes drowsiness and can cause loss of consciousness and death. Nitrogen oxides cause a weakening of smell, reduce the ability of the eye to adapt to darkness, increase the effort expended on breathing, make a person more susceptible to diseases of the respiratory system. Dust leads to irreversible changes in lung tissue. In the zones of intense dust pollution, a number of specific diseases arises, for example, silicosis and asbestosis [7–9].

Undoubtedly, the main industrial sources of harmful emissions are thermal power, ferrous and non-ferrous metallurgy. However, the contribution of the chemical and oil refining industry is also high (Figure 1).

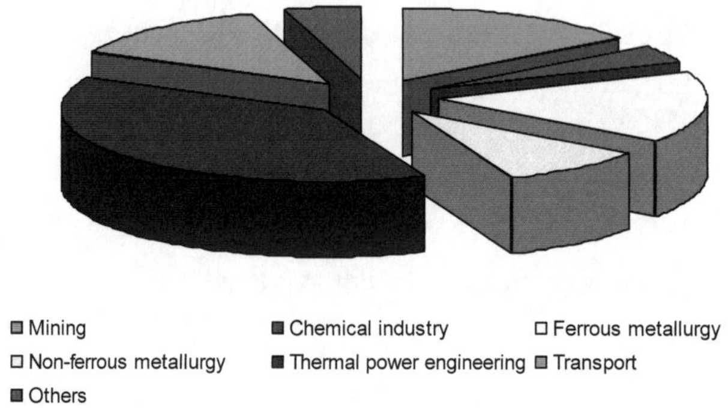


**Figure 1**. The main emission sources into the atmosphere

## 3. Results and discussion

### *3.1. The atmospheric air quality*

The state observation organizations carry out data acquisition on environmental pollution during the monitoring of pollutions [10, 11]. We have analyzed the data of state reports on air pollution in those Sverdlovsk region cities where chemical enterprises are located.

We found that for the period from 2011–2015 the number of cities with a very high level of atmospheric pollution has not changed (the high level of atmospheric pollution is remained at the same level in Nizhny Tagil and Krasnoturinsk). In Pervouralsk and Kamensk-Uralsky, the high level of pollution has decreased to an increased level. However, in 2015 the pollution level rose to a high level in Kamensk-Uralsky. The pollution level in Yekaterinburg is remained at the same level namely high during the past 2 years (Figure 2) [12, 13].

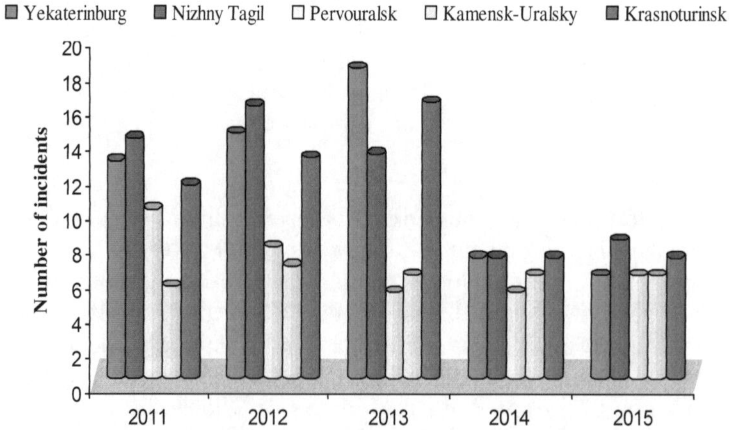


**Figure 2**. The atmospheric air quality in the Sverdlovsk region cities based on the values of the complex index (IPA) of atmospheric pollution in 2011–2015

0–4 – low level of pollution 5–6 – increased level of pollution
7–13 – high level of pollution 14 and above – very high level of pollution

In 2015, the values of the standard index (SI) which characterize the maximum value of the concentration in fractions of the corresponding TLV were very high in Yekaterinburg and Nizhny Tagil and were determined by the maximum daily average concentrations of ethylbenzene in Ekaterinburg and the maximum average monthly concentrations of benzopyrene in Nizhny Tagil (Figure 3).

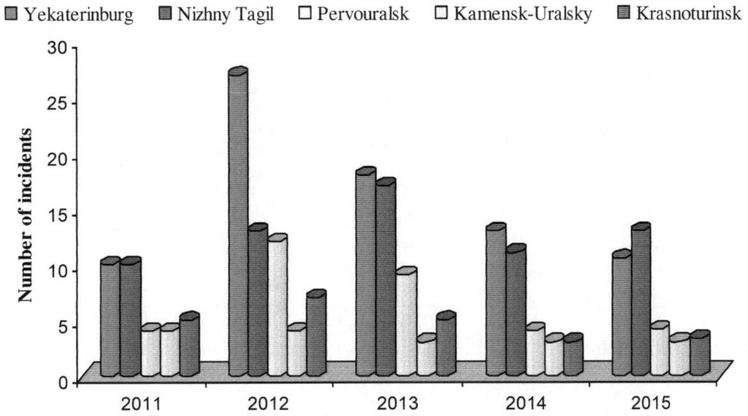


**Figure 3**. The atmospheric air quality in the Sverdlovsk region cities based on the SI parameter values in 2011–2015

0–1 – low level of pollution 2–4 – increased level of pollution
5–10 – high level of pollution 11 and above – very high level of pollution

As we can see from the diagram the SI values were elevated and were determined by the maximum monthly average concentrations of benzopyrene in Pervouralsk and Krasnoturinsk. The SI values were determined by maximum single concentrations of solid fluorides in Kamensk-Uralsky.

The values of the highest repeatability of exceeding TLV (HR parameter) were elevated in all Sverdlovsk region cities in 2011–2015. The exception was Kamensk-Uralsky, where the HR level remained at a high level. In 2015, the HR values were determined by the highest repeatability of TLV excesses, for example suspended substances in Yekaterinburg, hydrogen fluoride in Krasnoturinsk and Pervouralsk, solid fluorides in Kamensk-Uralsky, and formaldehyde in Nizhny Tagil (Figure 4).

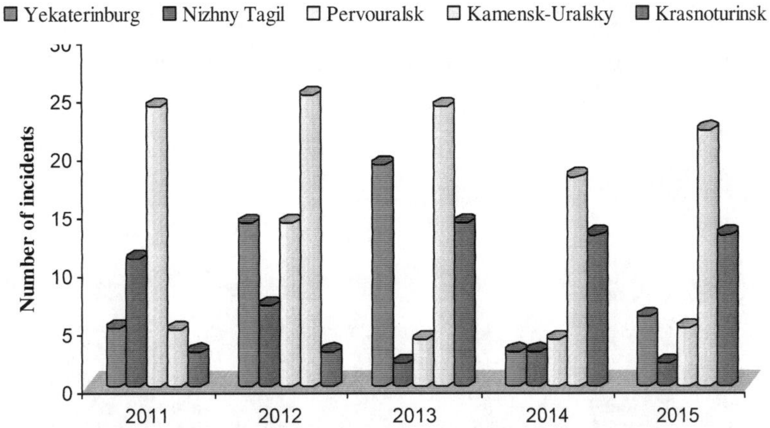


**Figure 4**. The atmospheric air quality in the Sverdlovsk region cities of the based on the HR parameter values in 2011-2015 (excluding data on benzopyrene)

0 – low level of pollution 1–19 – increased level of pollution
20–49 – high level of pollution 50 and above – very high level of pollution

### *3.2. Estimation of atmospheric air*

This article represents the study results of the gaseous emissions dynamics of the chemical complex enterprises in the Sverdlovsk Region for the period from 2011 to 2015 [12, 13].

In 2015 year, in compare to 2014, emissions of pollutants into atmospheric air from stationary sources were decreased by 37.3 thousand tons (by 3.7 %). Decrease in the total amount of emissions into atmospheric air was mainly in communication with the reduction of electricity generation, fuel consumption, a decrease in production at a number of enterprises, a reduction in the volume of repair work, and the implementation of environmental measures. Pollutants emissions into the air in the region were decreased by 119.2 thousand tons (by 10.8 %) in compare to 2011.

Volumes of pollutants emissions into the air from chemical industries, industries of rubber and plastic products, by types of economic activity are shown at figure 5 during the period 2011–2015.

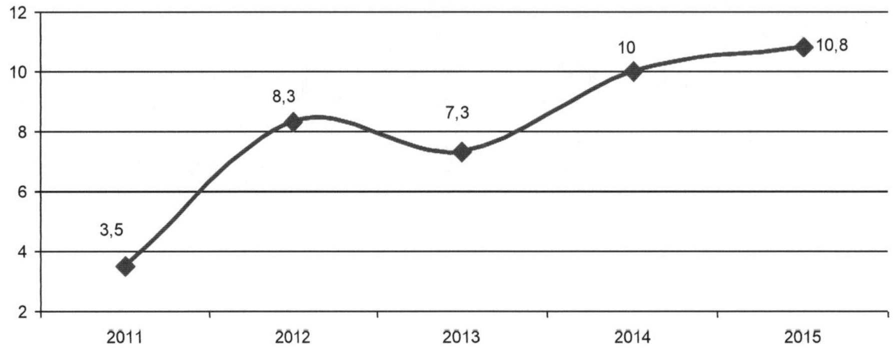


**Figure 5**. Harmful (polluting) substances emissions volumes into atmospheric air from chemical industries, thousand tons

The contribution of emissions from chemical industry is relatively low and accounts for approximately 4 % of emissions from all manufacturing industries, including metallurgical complexes.

In the Sverdlovsk region, the largest chemical enterprises are Uralchimplast, Uralplastic, Ural RTI, Ural Tire Plant and Verkhnesinyachikhinsky Wood Chemical Plant which are the main sources of atmospheric air pollution to total emissions of pollutants into the atmosphere from stationary sources in the Sverdlovsk region. Assessment of their contribution was carried out on the basis of information provided by these enterprises and is shown in figure 6 for the period 2011–2015.

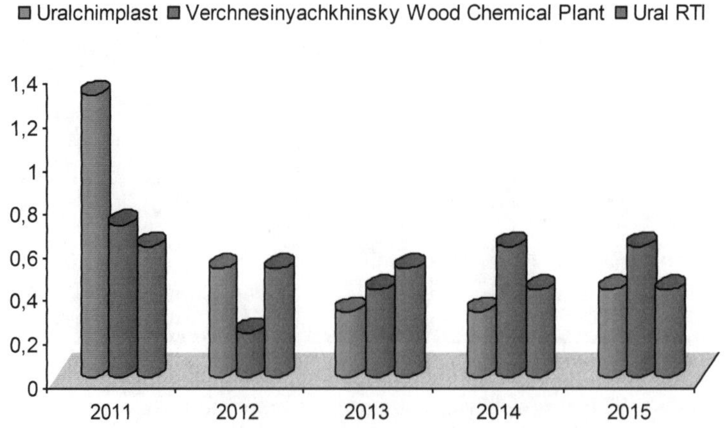


**Figure 6**. Dynamics of changes in emissions of pollutants into the atmosphere from major chemical enterprises, thousand tons

As we can see from the diagram there is a general tendency to reduce emissions of pollutants in connection with the implementation of environmental measures. However, in 2015 Uralchimplast have increased the amount of emissions by 0.1 thousand tons (by 33.3 %), which was associated with an increase in output.

### *3.3. Influence of chemical enterprises on atmospheric air*

In addition, we have analyzed the dynamics of changes in the atmospheric air state in cities where the big chemical enterprises are located (figures 7–9). The main air pollutants which enterprises emit into the atmosphere are sulfur dioxide, carbon monoxide and nitrogen oxides.

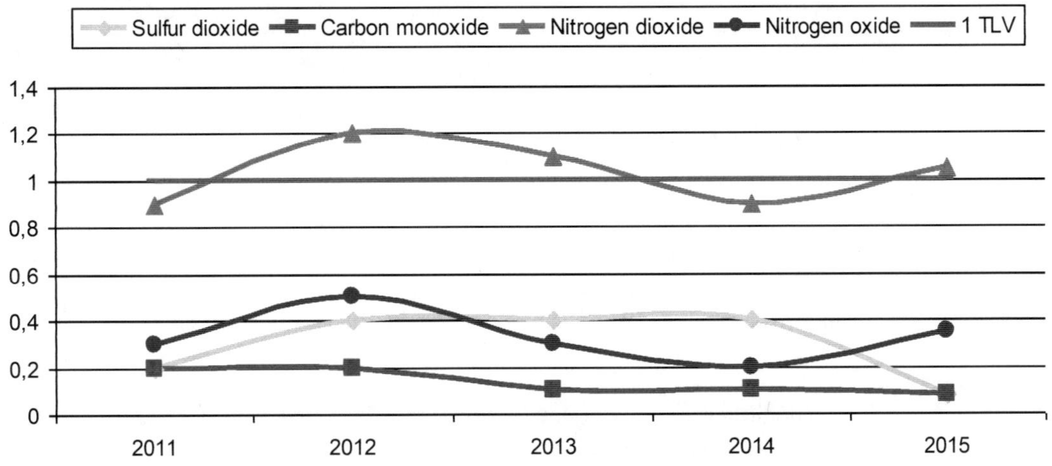


**Figure 7**. Average annual pollutant concentrations in the atmospheric air of Yekaterinburg in 2011–2015

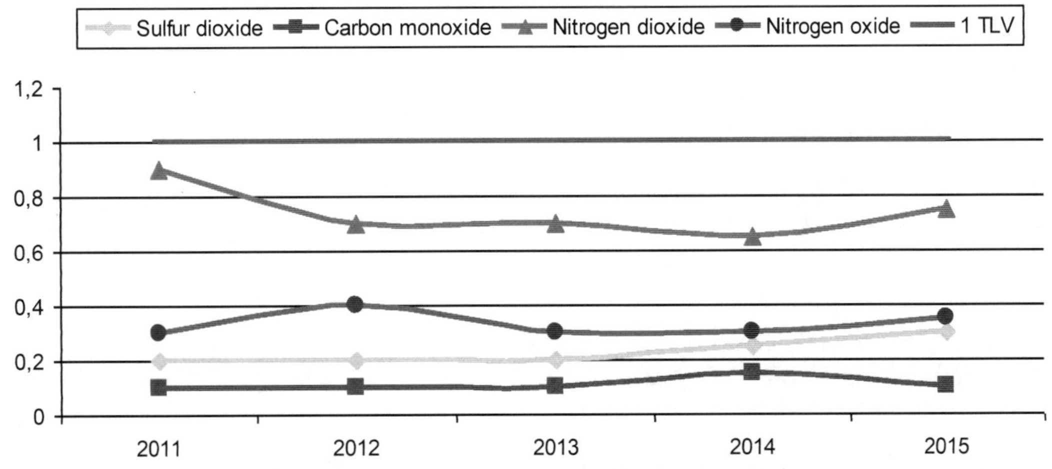


**Figure 8**. Average annual pollutant concentrations in the atmospheric air of Nizhny Tagil in 2011–2015

We found that there is a general tendency to reduce gaseous polluting emissions of $SO_2$, CO, $NO_2$ and NO in all cities with the big chemical enterprises. It should be noted that the dynamics of pollutant emissions into the atmosphere from stationary sources have an oscillatory nature. Probably, this is due to the unsustainable activity of enterprises in a market economy (alternation of recessions and upsurge in production, changes in the nomenclature of produced products, the output volume and the used raw materials quality). Reducing the gaseous substances emission can be achieved both by improving the production technology, and by installing new more sophisticated equipment which catches harmful substances emissions.

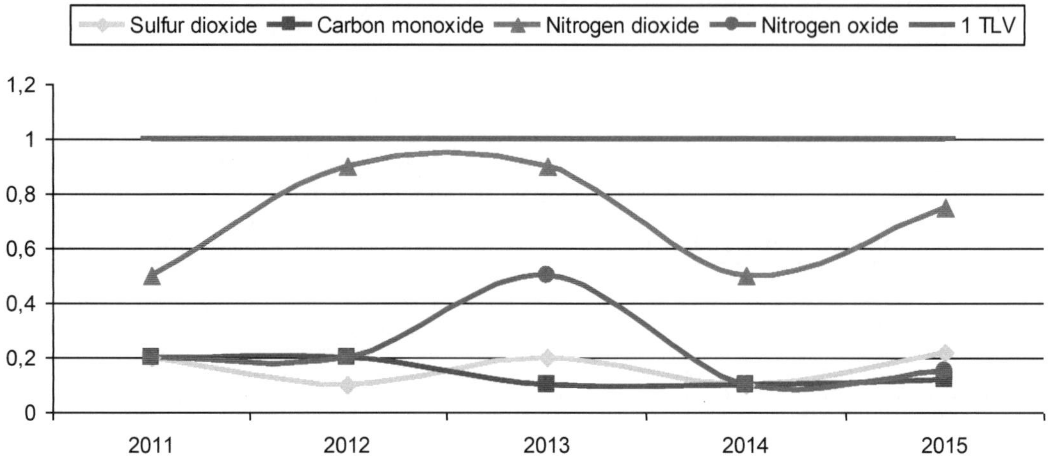


**Figure 9**. Average annual pollutant concentrations in the atmospheric air of Kamensk-Uralsky in 2011–2015

## 4. Conclusion

To sum up, it should be noted that the total pollutant emissions into the air environment by chemical enterprises are decreasing in the Sverdlovsk region. However, only 20 % pollutant emission reductions ($SO_2$, CO, $NO_2$ and NO) are associated with measures which reduce anthropogenic impact on the environment. The main reason (about 50 %) is the decline in production due to the unstable economic situation.

The total amount of purified emissions is increased due to scientific and technological development and improvement of modern industry. For example, Uralchimplast invests in modernizing its production facilities, carries out work on reduction all types industrial waste [14]. Uralchimplast operates an independent Eco-Service that is accredited up technical competence in the field of eco-analytical control and registered in the State Register of the Russian Federation. At this enterprise there are 40 dust and gas cleaning installations, thus, emissions do not exceed the established maximum permissible norms of emissions for all ingredients.

This has a favorable effect on the environmental situation in the Urals region, and in particular in the Sverdlovsk region.

## References

[1] Federal Law on Environmental Protection No. 7-FZ of January 10, 2002

[2] Federal Law No. 219-FZ of June 21, 2014 On Amendments to the Federal Law On Environmental Protection and Certain Legislative Acts of the Russian Federation

[3] Scientific and practical portal "Industrial ecology" Available from: http://www.ecoindustry.ru/global/control.html [Accessed 20th March 2017]

[4] 1979 GOST 17.2.3.02-78 Nature Atmosphere *The rules establish allowable emissions by industry* (Moscow, Publishing Standards)

[5] 2000 Russian Federation SanPiN 2.1.6.983-00 Atmospheric Air and Air in Enclosed Indoor Spaces, Sanitary Protection of Air. *Public Health Requirements with Respect to the Quality of Atmospheric Air in Populated Areas* (Moscow: Ministry of Health of Russia)

[6] 2010 *List and codes of substances polluting the atmosphere* (Saint Peteburg) Available from: http://www.complexdoc.ru/ntdtext/532540 [Accessed 20th March 2017]

IOP Conf. Series: Earth and Environmental Science **72** (2017) 012012 doi:10.1088/1755-1315/72/1/012012

[7] Rakhmanin Yu, Novikov S and Ivanov S 2005 Modern scientific problems of improving the methodology for assessing the risk to public health *Journal of Hygiene and Sanitation* **2** p 7–10

[8] Revich B 2001 *Pollution of the environment and public health. Introduction to Environmental Epidemiology*: a training manual (Moscow) 264 p

[9] 2004 Human Health Risk Assessment fron Environmental Chemicls *Manual P.2.1.10.1920-04* (Moscow) 144 p

[10] 2011 The state of works on the forecast of air pollution in the cities of the Russian Federation *News bulletin for 2010* (Saint Peteburg) 68 p Available from: http://elib.rshu.ru/files_books/pdf/img-42823.pdf [Accessed 14th April 2017]

[11] Official website of Federal State Statistic Service Available from: http://www.gks.ru/ [Accessed 14th April 2017]

[12] Official website of Ministry of Natural Resources and Environment of the Sverdlovsk Region Available from: http://mprso.midural.ru/article/show/id/1084 [Accessed 14th April 2017]

[13] Social services Official statistics *Official site of management of Federal State Statistics Service of the Sverdlovsk region and the Kurgan region* Available from: http://sverdl.gks.ru/wps/wcm/connect/rosstat_ts/sverdl/ru/statistics/sphere [Accessed 14th April 2017]

[14] Uralchimplast *Official website of the company* Available from: http://www.ucp.ru/ru/politics/ecology/ [Accessed 14th April 2017]

# Estimation of CO2-Equivalent Emission under the Copper Fire Refining Process

**Yu N Chesnokov**[1], **V G Lisienko**[1], **S I Holod**[2], **V P Anufriev**[3] **and A V Lapteva**[1]

[1] Ural Federal University, Institute of Radio Electronics and Information Technologies, Department Information Technologies and Automatics, 620002, Russia, Ekaterinburg, Mira str. 32
[2] Ural Federal University, Institute of new materials and technologies, Metallurgy of Heavy Non-ferrous Metals department, 620002, Russia, Ekaterinburg, Mira str. 17
[3] Ural Federal University, Higher School of Economics and Management, Department Economy of Production and Power Systems, 620002, Russia, Ekaterinburg, Mira str. 19

E-mail: A.V. Lapteva: annalapteva@mail.ru

**Abstract.** Non-ferrous metallurgy is one of the most energy-consuming and carbon-emissive sectors of industry. This is due to the fact that the volume of greenhouse gas (GHG) emissions is stipulated by energy consumption. Uralelectromed is a city-forming enterprise of the Verkhnyaya Pyshma. The situation is similar other cities of the old industrial regions of the Russian Federation (Krasnouralsk, Verkhnaya Salda, Karabash, etc.) Verkhnyaya Pyshma has many characteristics of "a clever city". It can be compared to Hamburg where blister copper is being produced at the center of the city at a copper smelting plant Aurubis. Following the example of such ecologically clean country as Germany and in order to assess how modern energy-efficient low-carbon technologies can provide a favorable habitat, and an acceptable level of carbon footprint, the authors estimated the level of greenhouse gas, i.e., carbon dioxide emission produced by the Uralelectromed. The emission of greenhouse gas – carbon dioxide in the process of fire refining of blister copper has been calculated. The anode melting process consists of several stages where the most important ones are melting of charge, oxidation, and copper melt reduction. Calculations are based on taking into account the mass of burnt carbon of natural gas and the thermal dissociation of fuel oil. It implies that a complete combustion of carbon takes place. The specific value of carbon dioxide emission of the copper refining process is averaged 181 kg $CO_2$ per 1 ton of anode copper.

## Introduction

It is considered to be not typical for the so-called "clever" cities to have non-ferrous metallurgy enterprises in the city. However, since non-ferrous metallurgy has often a city-forming significance, enterprises are located within the city, for example, at the center. Thereby, it is desirable to estimate the emission of carbon dioxide from the refining process at Uralelectromed in Verkhnyaya Pyshma, which is a part of the Ural Mining and Metallurgical Company (UMMC) occupying the first place in Russia and is the eleventh largest producer of refined copper in the world. The article uses an array of random sample data for the period from 2016 to 2017 for seven variants of anode melting.

The copper refining enterprise was chosen for the following reasons:

- Russia has a significant raw material base and ranks third place in the world for copper ore reserves (about 90 million tons) after Chile and Peru;
- Copper has valuable mechanical (ductility and toughness) and corrosion properties and has high thermal, and electrical conductivity;
- 85 % of copper in the world is produced using pyrometallurgical method of processing the raw copper. This process consists of several operations: extraction, and beneficiation of ore to produce concentrate; production of blister copper; production of refined copper for providing the anodic and cathodic copper (Figure 1);
- Copper is currently the only alternative material in computational digital electronics (primarily computers) in various fields of electrical engineering, such as robotics, automation, measuring instruments, radio and telecommunication devices and many other digital devices.

The diagram in Figure 1 shows the steps of removal of iron and sulfur: in three stages (roasting, melting and converting), in two stages (melting, converting), or in one stage [1]. With the exception of the latter option including the direct melting of concentrates to produce a blister copper, the technology for its production is multistage.

The most common technology includes the obligatory use of the following metallurgical processes:

- matte smelting,
- matte copper converting,
- fire and electrolytic copper refining.

Refined copper is the main raw material for production of semi-finished copper goods. Nowadays 91 % of the UMMC production consists of copper rod, of non-ferrous rolled products 8 %, of copper powder 1 %.

At the center of any metallurgical process the principle lies of transferring the processed raw materials to a heterogeneous system. The process consists of several phases that differ from each other in composition and physical properties. During this process one of the phases can be enriched by the recovered metal and impoverished by the contaminants, while other phases, on the contrary, become impoverished by the main component.

One of the phases is the gas phase, which is subjected to high environmental requirements for non-ferrous metallurgy enterprises. Polluting emissions of copper melting enterprises mostly consist of such elements as

- sulfur dioxide,
- dust,
- nitrogen dioxide,
- carbon oxide and dioxide,
- metals and their compounds (depending on the composition of the feedstock, there are primarily cadmium, copper, arsenic, mercury, lead, etc.),
- volatile organic compounds (organic carbon, etc.).

The problem of capturing the sulfur dioxide SO2 (accounting for 75–80 % of the total amount of pollutants in the waste gases) is being effectively solved at copper melting plants, for example, in the EU. Now, an average of 98.9 % of sulfur is extracted from emissions and used for production of sulfuric acid and liquid sulfur dioxide. Meanwhile, the issue of controlling and capturing other harmful components in the waste gases emitted by non-ferrous metallurgy enterprises remains quite acute. One of these undesirable components is the carbon dioxide CO2.

Carbon dioxide CO2 is the basic greenhouse gas (GHG). Presence of such gases in the atmosphere of our planet leads to the appearance of a greenhouse effect, namely to increasing the surface temperature and to climate change [2–6]. According to the IPCC Source Category, cast iron and steel belong to the category 1A2a, while nonferrous metals to 1A2b [4]. Generally, the most efficient practice is to use higher-level methods for the key categories, such as cast iron, steel, and non-ferrous metals. Decision-making schemes for each category help the GHG inventory compilers to follow the guidelines and select the appropriate multi-level methodology suitable for their conditions.

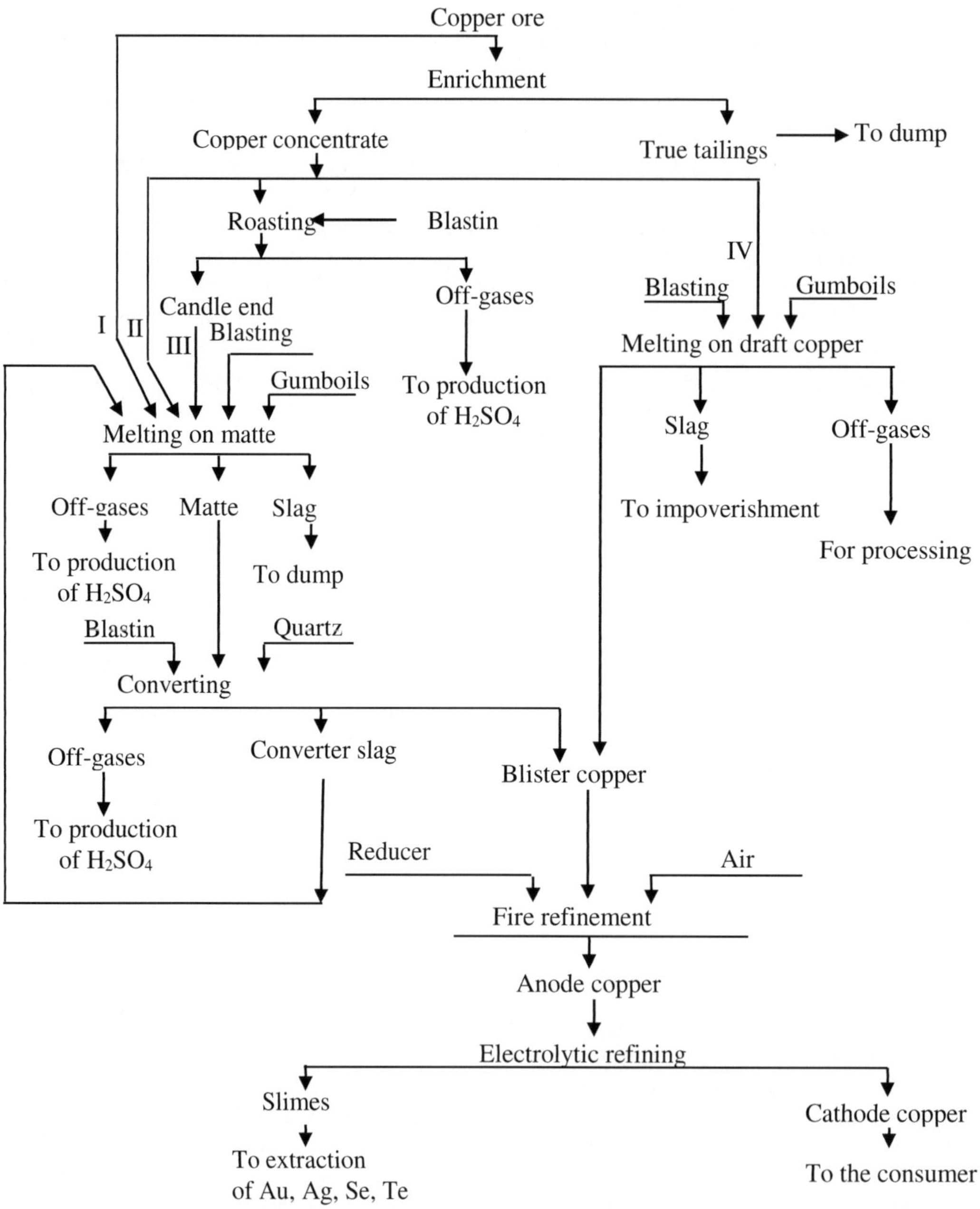


**Figure 1.** Principal technological scheme of pyrometallurgical copper production from sulphide ores; numericals indicate the possible options for processing the raw materials into thr blister copper

The widespread notion of the carbon footprint is reduced to the concept of end-to-end emission of the carbon dioxide $M_C$, This value is the sum of carbon dioxide emissions that consistently occur in all processes of the technological chain starting with the extraction of raw materials and ending with the product. Moreover, we will distinguish between the emissions of the process $M_P$ and the transit $M_T$ due to the fractions of the total mass of analyzed carbon dioxide emissions generated in previous processes. The end-to-end integrated emission of the carbon dioxide $M_C$ is determined by the sum of emissions of carbon dioxide of the process and transit emissions

$$M_C = M_P + M_T. \quad (1)$$

**Chemical Relations of the Process**

Consider in details the methods for reducing the impact of production facilities onto the environment. These methods include ones integrated into the production process: the methods associated with preventing or reducing emissions from activities such as storage, chemical reactions, separation, and purification of various materials and substances [7]. As it was mentioned above, here, we take refining enterprise Uralelectromed (Verkhnyaya Pyshma) as an example. This plant receives supplies of blister copper from various producing factories and transforms it into the charge for anodic copper smelting. The full load of a single furnace is 350 tons.

The averaged chemical composition indicators of blister copper charge are the following:

- 99.35 % Cu,
- 0.17 % Ni,
- 0.14 % Sb,
- 0.13 % Pb,
- 0.11 % As,
- 0.052% Sn,
- 0.00121 % Fe,
- 0.0001 % Bi.

The process of the copper fire refining in terms of greenhouse gas emissions was considered in [8, 9].

The refining process is based on the following regularities:

- High solubility of copper oxide (I) $Cu_2O$ in liquid copper;
- Oxidation of the most contaminants by copper oxide (I) $Cu_2O$;
- Insolubility of contaminant oxides in copper (MeO);
- The reducibility of copper oxide (I) $Cu_2O$ by various hydrocarbons after slag removal;
- The possibility of desorption of sulfur oxide (IV) $SO_2$, which is dissolved in liquid copper while mixing it with gaseous products of dry distillation of wood.

The main chemical reactions in this case are [1]

$$2[Cu] + 1/2O_2 \leftrightarrow (Cu_2O), \quad (2)$$

$$[Me] + (Cu_2O) \leftrightarrow (MeO) + 2[Cu], \quad (3)$$

$$(Cu_2O) \leftrightarrow 2[Cu] + [O], \quad (4)$$

$$[Me] + [O] \leftrightarrow (MeO), \quad (5)$$

$$Cu_2O + H_2 = 2Cu + H_2O_{steam}, \quad (6)$$

$$Cu_2O + CO = 2Cu + CO_2, \quad (7)$$

$$4Cu_2O + CH_4 = 8Cu + 2H_2O + CO_2. \quad (8)$$

As mentioned above, during this process, the carbon dioxide $CO_2$ is being produced. In practice, the temperature of the furnace working zone is maintained in the range from 1100 to 1200 °C. The main condition for burning of natural gas is the presence of oxygen and, consequently, the presence of air which leads to a chemical reaction combining oxygen from the air with carbon and hydrogen in the fuel. The reaction occurs with the release of heat, light, as well as the carbon dioxide, and water vapor.

The amount of air depends on the calorific value of natural gas: the higher it is the more air is needed.

Regardless of the aggregate state of the substance, the amount of air necessary for complete combustion is determined by equation of the combustion chemical reaction compiled on the basis of the chemical composition of the combustible substance.

In order to calculate the value of fuel combustion, we calculate the material balance of the combustion process on the basis of relative molecular masses of the main substances involved in combustion, as well as, on the basis of composition (in percentage), and density ρ of natural gas (volume %), temperature $T$, and air pressure $P$.

The calculation is carried out at $T = + 150$ °C, $P = 760$ mm Hg, $P = 0.717$ kg/m$^3$.

The equations of methane combustion reactions is

$$CH_4 + 2O_2 = CO_2 + 2H_2O. \tag{9}$$

Substitute the molecular masses of the elements in this equation.

Calculations are performed with the data for the methane combustion equation

$$\begin{matrix} 16 \text{ kg/mol} & & 64 \text{ kg/mol} & & 44 \text{ kg/mol} & & 36 \text{ kg/mol} \\ CH_4 & + & 2\,O_2 & = & CO_2 & + & 2\,H_2O. \end{matrix}$$

From the methane combustion formula, find the mass of oxygen $m_O$ necessary for the complete combustion of 1 kg of the component

$m_O = 64/16 = 4$ kg.

Find the volume $V_O$ of oxygen required for the complete combustion of 1 m$^3$ of the component by the formula

$$V = \frac{m_o}{\rho_o} \tag{10}$$

where $\rho_O$ is the oxygen density.

Combustion of methane requires the following volume of oxygen:

$$V_{O_2} = \frac{4}{1.429} = 2.799 \text{ m}^3.$$

Define the required volume of air for methane combustion under the normal physical conditions, taking into account that concentration of oxygen in the air is 21 %:

$$\text{Vacn} = 2.799/0.21 = 13.33 \text{ m}^3/\text{m}^3.$$

Calculate the volume of air under the given conditions ($T = + 150$C, $P = 760$ mm Hg):

$$\text{Vac} = 13.33 \cdot (273 + 15)/273 = 14.06 \text{ m}^3/\text{m}^3.$$

Thus, complete combustion of 1 m$^3$ of natural gas under the taken conditions requires 14.06 m$^3$ of air. One kilogram of methane has a volume of 1/0.717 = 1.395 m$^3$ provided the methane density is ρ = 0.717 kg/m$^3$. Therefore, in order to burn 1 kg of methane, 14.06/1.395 = 10.08 m$^3$ of air is needed. The amount of air determined in this way is called theoretically necessary. In practice, the combustion process in burners of various types has a much greater actual consumption.

Completeness of combustion is determined by the amount of air involved in the combustion process. With a sufficient supply of air, complete combustion of natural gas takes place and, as a result, are the following its combustion products are the following non-flammable gases:

- carbon dioxide $CO_2$,
- nitrogen $N_2$,
- $H_2O$ water vapor.

If the amount of air is insufficient then the combustion of natural gas is incomplete or chemical undercooking of combustible constituents occurs. So, the following flammable gases appear in combustion products:

- carbon monoxide CO,
- methane $CH_4$,
- hydrogen $H_2$.

Thus, the lack of air leads to incomplete combustion of natural gas, while the excess of air leads to excessive cooling of the flame temperature. To provide complete combustion of fuel, it is necessary to supply air to the furnace somewhat more than required by the calculations, which is regulated by the air excess coefficient.

Besides emissions from burnt natural gas necessary to maintain the melt temperature, appearance of significant concentrations of the carbon monoxide is possible in the recovering atmosphere of the furnace. Removal of the carbon monoxide CO is carried out by burning furnace gases with the consumption of additional fuel; so, the amount of the carbon dioxide $CO_2$ in the plant emissions increases.

Provided that all carbon is oxidized to the carbon dioxide $CO_2$ and with availability of data from the plant's practice, calculations of carbon dioxide emissions can be simplified without using formulas (2)–(8). Articles [10, 11] give us the formula from the industrial method for determining carbon dioxide emission $E_{CO2}$

$$E_{CO2} = (M \cdot C_M - P \cdot Cp) \cdot K_u. \quad (11)$$

Here, $M$ is consumption of carbon-containing materials in thousand tons; $C_M$ is the carbon content in the material, t/t; $P$ is output of carbon-containing products, t/t; $Cp$ is the carbon content of the product, t/t; $K_u$ is the coefficient of conversion of the carbon C mass into the mass of the carbon dioxide $CO_2$, equal to $44/12 \approx 3.667$. In paper [12], we find calculations of carbon dioxide emissions based on generalized coefficients made by enterprises. However, there is no data for non-ferrous metallurgy in this study. Let us use formula (9), taking into account that the output of carbon-containing products is absent in non-ferrous metallurgy.

**Calculations of emission of the carbon dioxide**

On elaboration of blasting 1 $m^3$ of air it is needed about 0.43 kWh of electric power that leads to emission of the carbon dioxide 1.086 kg/$m^3$ [13].

Taking it into account, calculation of values of emissions of the carbon dioxide is carried out by the following formulas:

$$E_{ghg} = V_{ghg} \cdot \rho_{ghg} \cdot C_{ghg} \cdot 44/12, \text{ kg } CO_2, \quad (12)$$

$$E_{fo} = m_{fo} \cdot C_{fo} \cdot 44/12, \text{ kg } CO_2, \quad (13)$$

$$E_a = 1.086 \cdot V_{ghg} \cdot V_a, \text{ kg } CO_2. \quad (14)$$

where $V_{ghg}$ is the volume of natural gas, $m^3$; $\rho_{ghg}$ is density of natural gas, kg/$m^3$; $C_{ghg}$ is the content of carbon in natural gas, %; $m_{fo}$ is the mass of fuel oil, kg; $C_{fo}$ is the content of carbon in fuel oil, %; $V_{ghg}$ is a specific consumption of air for burning of 1 $m^3$ of natural gas; 44 and 12 are the molar massed of the carbon dioxide.

According to practice of Uralelectromed for chosen the furnace charge at fire refinement of copper in the reflective furnace, usually it is spent of 28 855 $m^3$ or $m_{ghg} = 28\,855 \cdot 0.717 = 20\,689$ kg of natural gas of density $\rho_{ghg} = 0.717$ kg/$m^3$ with content of the carbon $C_{ghg} = 75$ %. Combustion of natural gas requires $V_a = 275\,784$ $m^3$ or $m_a = 337\,835$ kg of air.

Resultant emission of process of the carbon dioxide (if in it there is no transit emission of $CO_2$ of power plant) at the rate of full load of the furnace (350 t) of the considered process is

$$E_{pr} = E_{ghg} + E_{fo}, \text{ kg } CO_2, \quad (15)$$

Taking into account emission of the carbon dioxide at power generation the through emission of $CO_2$ at process of manufacturing the draft copper at anode will be

$$E_{ti} = E_{pr} + E_a + E_{fo}, \text{ kg } CO_2, \quad (16)$$

From the loaded furnace charge $m_{ac}$ of tons of anode copper be received. Specific emission of the carbon dioxide of process of fire refinement then will be

$$E_{scfr} = E_{da}/m_{ac}, \text{ kg } CO_2, \tag{17}$$

As mentioned above, the basic data are taken from practice of the plant, which were used for calculations for formulas (12)–(17) and are given in Table 1. Existence of data for these concrete swimming trunks significantly increases the accuracy of calculations. The last line represents average values of basic initial and calculated data.

**Table 1.** Basic and calculated data of emissions of the carbon dioxide

| $m_{cop}$, t | $V_{ghg}$, $m^3$ | $m_{ghg}$, kg | $E_{ghg}$, kg $CO_2$ | $V_a$, $m^3$ | $E_a$, kg $CO_2$ | $m_{fo}$, t | $E_{fo}$, kg $CO_2$ | $E_{ghg}$, kg $CO_2$ | $E_{ti}$, kg $CO_2$ | $E_{scfr}$, kg $CO_2$ |
|---|---|---|---|---|---|---|---|---|---|---|
| 342 | 28442 | 20393 | 56081 | 398188 | 432432 | 1.22 | 3802 | 59883 | 492315 | 175 |
| 341 | 29411 | 21088 | 57991 | 411754 | 447165 | 1.34 | 4173 | 62164 | 509329 | 183 |
| 338 | 28605 | 20510 | 56402 | 400470 | 434910 | 1.22 | 3815 | 60217 | 495127 | 178 |
| 332 | 29867 | 21415 | 58890 | 418138 | 454098 | 2.03 | 6336 | 65226 | 519324 | 197 |
| 353 | 31517 | 22598 | 62144 | 441238 | 479184 | 1.08 | 3350 | 65494 | 544679 | 185 |
| 346 | 27186 | 19492 | 53604 | 380604 | 413336 | 1.56 | 4865 | 58469 | 471805 | 169 |
| 344 | 28855 | 20689 | 56895 | 403970 | 438711 | 1.28 | 3989 | 60884 | 499596 | 177 |
| 342 | 29126 | 20883 | 57429 | 407766 | 442834 | 1.39 | 4333 | 61763 | 504596 | 181 |

Emission of the carbon dioxide of $E_{ti}$ is not a carbon trace of copper production since here the transit emissions of the carbon dioxide passing from processes of receiving draft copper and electrolytic refinement of anode copper are not considered. Formation of a carbon trace of copper production illustrates columns of emissions of the carbon dioxide (Figure 2) made for one chain of possible chains of production of copper production. The notations $TE_i$ correspond to specific through emission of the carbon dioxide in this or other process. The symbol $\Psi_j$ has designated expenses (consumption) of this or other resource. The through emission at each top point is calculated as sum of emission corresponding to this point with the through emissions of joined top points multiplied by values of consumptions of spent resources.

For example, through emission (a carbon trace) of fire refinement of copper is defined by the sum $TE_5 = (TE_{ghg} \cdot \Psi_{2ghg} + TE_4 \cdot \Psi_4 + TE_a \cdot \Psi_{3a} + TE_{fo} \cdot \Psi_{fo})/342 + 181$ kg $CO_2$/t of anode copper,
where

$$TEa \cdot \Psi 3a = (TEI \cdot \Psi 2I/342) \cdot \Psi 3a = 1\ 474 \text{ kg CO2/t of anode copper.}$$

Division by 342 serves for reduction of total emissions to specific ones. Value of a through emission of the carbon dioxide is at production and transportation of $TE_{ghg} =$ 0.234 kg of $CO_2/m^3$ of fuel oil, $TE_{fo} = 233$ kg of $CO_2$/t [14, 15]. Values of the sizes $TE_4$, $\Psi_{2ghg}$, $\Psi_4$, $\Psi_{fo}$ are unknown. For this reason, it is only possible to claim that through emission or a carbon trace of process of fire refinement of copper exceeds the value

$$181 + 504\ 596/342 = 1\ 655 \text{ kg CO2/t of anode copper.}$$

**Appendices**

1. Specific emission of the carbon dioxide in fire refinement of draft copper process makes 181 kg of $CO_2$ on ton of anode copper.

2. Annual emission of the carbon dioxide on the Uralelectromed is no more than 0.12 % of the total volume of emissions of greenhouse gases by all enterprises of Sverdlovsk region.

3. Calculations are carried out according to the concrete swimming trunks determined by method of random selection.

4. Results of article confirm that work of the refining shops placed in the urban environment does not lead to significant growth in a carbon trace and critical level of ecological damage to the environment and the population Verkhnyaya Pyshma. It is one of the main indicators, on which Verkhnyaya Pyshma can be referred in the short term to category of "the clever cities".

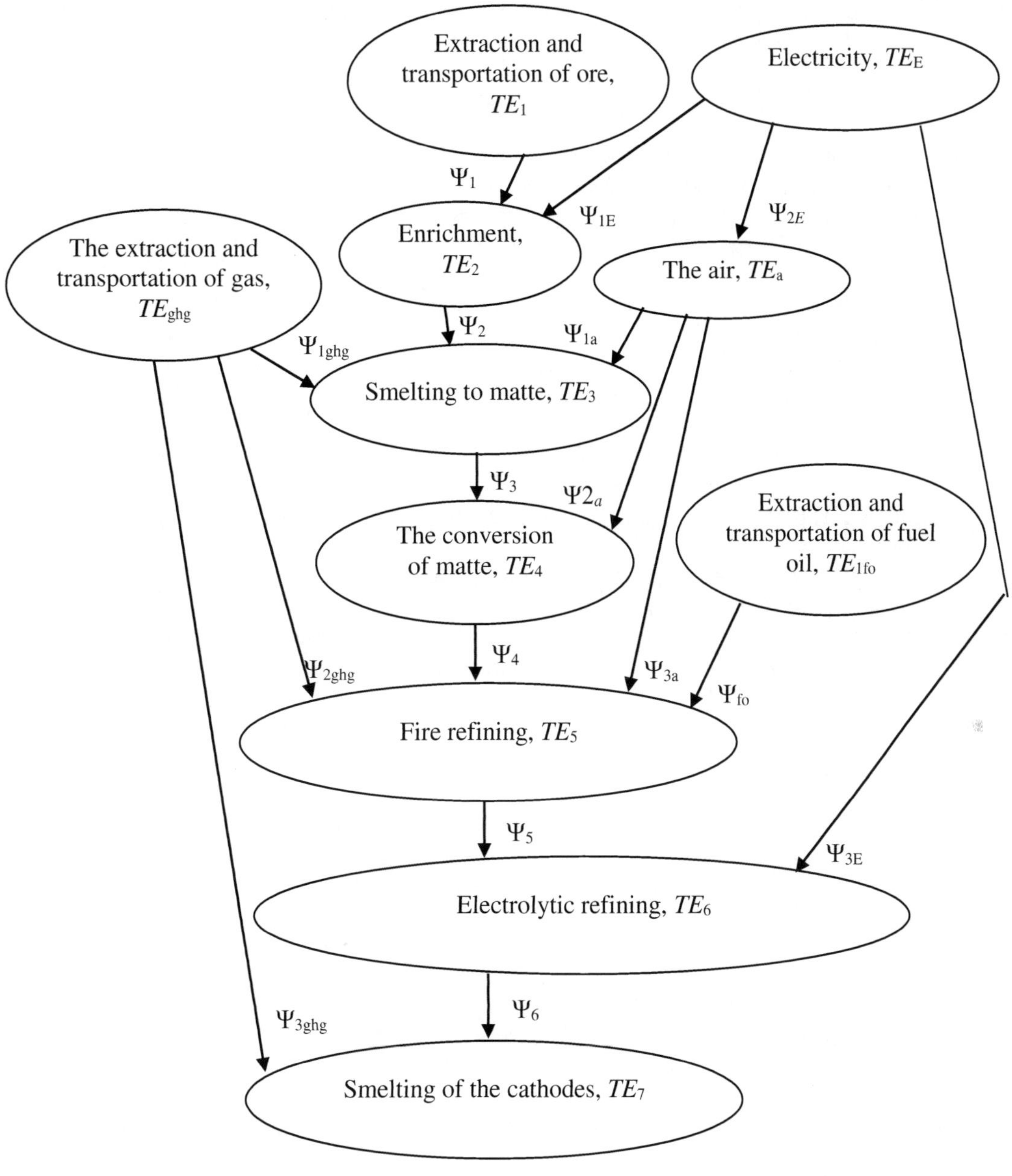


**Figure 2.** Chart of emissions of the carbon dioxide of process of manufacturing of copper production

## 1. References

[1] Zhukov V P, Skopov G V and Holod S I 2016 *Pirometallurgiya of copper* (Ekaterinburg: UO of RAS) 640 p

[2] The Kyoto puncture to the Framework convention of the United Nations on climate change Available from: http://unfccc.int/kyoto_protocol/items/2830.php [Accessed 18th May 2017]

[3] 2010 GOST P ISO 14064-1-2007 Greenhouse gases P 1 Requirements and the guide to quantitative definition and the reporting on emissions and removal of greenhouse gases at the level of the organization (Moscow: Standartinform) 23 p
[4] 2010 *The guidelines of national inventories of greenhouse gases of IPCC of 2006* (edition of June 24)
[5] *Analysis of possible means to reach emission reduction targets and of relevant methodological issues* Technical paper FCCC/TP/2008/2
[6] Anufriyev V P and Chazov A V 2006 *Energy efficiency and problem of climate change* (Moscow) 192 p
[7] 2015 *The reference book on the best available technologies "Production of Copper"* is approved by the order of Rosstandart of December 15, 2015 No. 1573
[8] Kupryakov Yu P 1976 *Reflective melting of copper concentrates* (Moscow) 352 p
[9] Voskoboynikov V G, Kudrin V A and Yakushev A M 1998 *General metallurgy* (Moscow) 768 p
[10] Shevelyov L N 2007 Methodical bases of inventory of greenhouse gases in ferrous metallurgy of Russia *Steel* **4** p 97–102
[11] Shevelyov L N 2008 An assessment of emissions of greenhouse gases in ferrous metallurgy of Russia OJSC Chermetinformation *Bulletin of scientific, technical, and economic information Ferrous metallurgy* **8 (1304)** p 3–8
[12] Kalensky I V 2007 Recommendations about the accounting of emissions of CO2 at the enterprises of ferrous metallurgy *Steel* **5** p 121–29
[13] Rozengart Yu I 1985 *Power system of steel works studies for higher education institutions* (Moscow) 303 p
[14] Chesnokov Yu N, Lisienko V G and Lapteva A V 2013Assessment of a carbon trace when smelting of steel in the arc furnace *Metallurgist* **9** p 23–26
[15] Chesnokov Yu N, Lisienko V G and Lapteva A V 2011 Mathematical models of indirect estimates of emission of CO2 in some metallurgical processes *Steel* **8** p 74–77

# Urbanization and Greenhouse Gas Emissions from Industry

**N I Didenko[1], D F Skripnuk[2] and O V Mirolyubova[3]**

Peter the Great St. Petersburg Polytechnic University, St. Petersburg, Russia

[1] e-mail: didenko.nikolay@mail.ru
[2] e-mail: djamilyas@mail.ru
[3] e-mail: olga.v.mir@gmail.com

**Abstract**. This article analyses the global environment. The article describes processes that characterize the global environment, specific indicators are suggested, that can be used to measure the change in the global environment. It is said that cities and all urbanized territories have a negative effect on the global environment. Originally, the authors wanted to call the article «City as a source of destruction of the global environment». But taking into account the fact that urbanization contributes to improving the economic efficiency of the state, cities are the centers of the economic, cultural and informational potential that provide a «breakthrough» into the development of the economy. The article assesses the impact of urbanization on the global environment. For the analysis of the impact of urbanization on the natural habitat, the autoregressive distributed lags (ADL-model) are chosen.
**Key words:** global environment; city; impact of urbanization; Greenhouse Gas Emissions from Industry; ADL-model.

## 1. Introduction

The global space surrounding modern man, includes the natural habitat, artificial habitat, created by a man and social habitat. City is an artificial environment created by a man.

Global space is currently characterized with a set of processes, capable of rending direct or indirect, immediate or remote impact on human activity, its health and future generations. A person always has a need for space, clean air, water, and a presence of some wild forest plants.

In a large modern urbanized areas it is hard to talk about the presence of sufficient volume and quality of a space enough to satisfy a person's needs minimum human requirements. A lot of different products were created, and at the same time a pollution levels were also rapidly growth. As a industrialization and urbanization result, a global space surrounding a man gradually became very «aggressive» for mans senses, senses that were evolutionary adapted for the congenital space through a millions years.

A concentration of harmful substances in the air inside is becoming much greater than in the outdoors in many cases.

With its activity a man initiates many processes that affect a global space, for example: disasters associated with climate change, growth of CO2 proportion in the air, increase in greenhouse gases in the atmosphere, increase global temperature, reduction of freshwater resources on the planet, reduction of the forest area as a result of cutting down The main negative contribution to the characteristics of environmental processes is provided by the artificial habitat, created by a man – a cities.. All cities have a great influence on the growth and development of modern economy. More than half the world's population lives in cities since 2015. Six hundred of the world's largest cities will be producing sixty percent of world GDP by 2025 according to the research organization Mc Kinsey Global Institute (MGI).

Mc Kinsey Global Institute recommends companies to switch their strategy from the largest cities and include 577 rapidly developing cities of the medium size. It is expected that their contribution will cap at 35 % of the worlds' GDP [1]. The modern vector of urbanization is the growth of the industries in the cities, transformation of the rural villages into urbanized areas, motorization and therefore pollution of the atmosphere [2]. The air pollution and indirectly the water pollution consists of greenhouse gases [3], that have a large contribution into the global warming, emission of the nitrogen gasses, dust and aerosols, carbon monoxide, volatile organic substances, that affect mans health in an extremely negative way [4–6].

A problem of global space pollution is that the quality of atmospheric air is continuously deteriorating, and this causes serious concern. To measure the level of the atmospheric pollution in urbanized areas, was created a network of observation posts service and control of atmospheric pollution as part of the global space [7, 8]. The network determines the content in the atmosphere of different harmful substances, coming from anthropogenic sources of emissions.

It is important in the conditions of constant pollution of the global environment to be able to predict the impact of the city as a source of pollution on variables that assess the state of the global environment.

The aim of this article is to demonstrate the basis for methods for predicting the pollution of the global environment, by sources of an artificial environment – a city.

## 2. Estimation of the natural conditions state

Characterizing a current state of the natural conditions as critical, we distinguish several following causes, which lead to the destruction of the natural conditions.

First, climate disasters (CD), estimated by economic damage from natural disasters caused by processes of long-lived / mesomacroscale processes (in the range from seasonal to multi-decadal climate variability).

Secondly, it is pollution, poisoning of the natural habitat of man, greenhouse gas emissions(GGE) including CO2 emissions(COE), oxygen depletion, ozone holes, etc.

Thirdly, this is a change in the global temperature (CGT) on the planet because of the greenhouse effect and the resulting increase in temperature and climate change.

Fourthly, it is predatory use of available natural resources, traditional sources of energy and sources of raw materials, reduction of fresh water supplies (RFW), and reduction of forest area (RFA).

The fifth reason is a rapid, unregulated population growth. According to the forecast for 2040, the world population will reach 13 billion people.

It is clear that with such a rapid increase of the population of the Earth, the negative consequences of anthropogenic activity will sharply increase. The negative results are not only the growth of the population itself, which increases the burden on nature beyond the permissible limits, but also the need to further accelerate the extraction of the raw materials of nature, with the consequent increase in their processing, the accumulation of industrial waste and the detrimental impact on the biosphere.

The authors suggest the status of the environment to be assessed by the following seven indicators (Endogenous indicators):

1) Climatological Disasters (CD), USD million; Climatological Disasters: refers to the value of economic damage from climatological disasters, i.e. events caused by long-lived/meso to macro scale processes (in the spectrum from intraseasonal to multidecadal climate variability).
2) CO2 Emissions per Unit of Output(COE), gr./USD (grams per USD in constant prices);
3) Greenhouse Gas Emissions from Industry(GGEI), tonnes of CO2 equivalent;
4) Greenhouse Gas Emissions from Agriculture(GGEA), tonnes of CO2 equivalent;
5) Mean Temperature, Year, Growth (%), global temperature change (GTC), %C;

6) Fresh Surface Water Withdrawal (FWW), million cu m, reduction of the fresh water resources, mil.m$^3$;
7) Forest Land, reduction in the area of the forests and forest vegetation in the process of urbanization (RFA), sq km.

Using this indicator as an example – Greenhouse Gas Emissions from Industry (GGEI) the article shows the creation of the basis for methods predicting the pollution of the global environment by sources of an artificial environment – urbanized areas.

## 3. Peculiarities of the process of negative impact of human activities on the ecology

To show the impact of the human activity on the ecology, we use the example of the change of GGEI in the life environment. The process of the change of GGEI is characterized by the following peculiarities:
a) Different dynamic factors of human activity influence the current assessment of the process of the change of GGEI in the life environment;
b) other processes occurring in the environment have an impact on the current assessment of the process of the change of GGEI in the life environment;
c) the pre-history of the process of the change of GGEI in the life environment affects the current assessment of the process of the change of GGEI in the life environment;

The influence of the factors of human activity on the process of the change in GGEI in the environment. Factors of human influence on the environment are characterized by dual nature, although the consequences are the same and have a negative impact on the environment. On the one hand, humankind needs more and more sources for nutrition, and because of that, they consume more and more lands of the nature. On the other hand, humanity has a constant need to increase energy.

The influence of the humankind on the environment always leads to the planned or forecasted consequences. And one of the most dangerous of them is the global warming of the climate. Forest destruction during the economic activity and GGEI during the fuel combustion lead to a decrease of oxygen in the atmosphere, and the invention of aerosols cause ozone depletion. This way the Earth's shell began to accumulate more heat. As a result, it can lead to the accelerated melting of glaciers, an increase in the ocean level and the sinking of land. Man's productive activity constantly exerts an increasing negative influence on the quality of the environment. All the growing scales of the man's productive activity typical for the period of the scientific and technological revolution, has violated the natural cycle of substances and energy in nature.

The evaluation of the factors of the human activity impact on the environment was fulfilled by a number of quantitative and qualitative economic indicators, which are:

GDP, World, US$ Per Capita; Employed Population, World, Unit; Economically Active Population, World, Unit; Exports (fob) by Commodity + Imports (cif) by Commodity, World, USD million; Energy, Utilities and Recycling: Production (turnover) MSP, USD million, USD million; Primary Materials Forestry- Production (turnover) MSP, USD million; and others.

By using of the qualitative and statistical methods of analysis it was found that there are six indicators that have an impact on the current assessment of the process of the change of GGEI in the environment:

$X_{1t}^3$ – Energy, Utilities and Recycling: Production (turnover) MSP, USD million, USD million;

$X_{2t}^3$ – Exports (fob) by Commodity + Imports (cif) by Commodity, World, USD million;

$X_{3t}^3$ – Road Freight Traffic (volume of transported goods). World, Million net tonne-kilometres;

$X_{4t}^3$ – Transport and Communications: Production (turnover) MSP, USD million, USD million;

$X_{5t}^3$ – Economically Active Population, World, Unit;

$X_{6t}^3$ – Material Resource Productivity, USD per kg in constant prices.

The influence of other processes occurring in the environment on the change in GGEI. The change of GGEI in the environment is influenced by other processes occurring there, which are:

a) Climatological Disasters(CD); b) CO2 Emissions per Unit of Output (COE), g / USD (grams per USD in constant prices); c) Greenhouse Gas Emissions from Agriculture(GGEA); d) Mean Temperature, Year(GTC), Growth (%); e) Fresh Surface Water Withdrawal(FWW), million cu m; f) Forest Land(RFA), sq km.

Using qualitative and statistical methods of analysis, it is indicated that the current assessment of the process of change of GGEI is affected by three processes estimated by the indicators:

$y_t^2$ –COE, grams / USD (grams per USD in constant prices);

$y_t^4$ – GGEA , tonnes of CO2 equivalent;

$y_t^7$ – RFA, sq km.

The influence of pre-history of the course of GGEI in the environment change process on the change of GGEI.

The main greenhouse gases, placed in order of the estimated impact on the heat balance of the Earth, are water vapor, carbon dioxide, methane and ozone. The planet is heating up. Scientists have practically the same view on this: the cause of warming is the industrial activity of man. The result of this activity is an increased allocation of the so-called greenhouse gases. Concentration of one of them, CO2, is not only growing, it is increasing exponentially.

Exponential growth is an increase of the value when the growth speed is proportional to the value itself. This fact is reflected in the influence of previous values of GGEI on the current value of GGEI.

## 4. The empirical models

As a theoretical model for analyzing the impact of urbanization on the environment a autoregressive distributed lags (ADL-model) is choosen, in which the current values of the time series depend on both on the past values of this series, and on the current and past values of other time series [9]. The model is generalized in the case of having several exogenous variables. In general we can assume that all exogenous variables are included in the model with the same number of lags, it is possible to exclude some lag of some variables

Additive form of the ADL model:

$$y_t = a_0 + \sum_{i=1}^{n} a_i y_{t-i} + \sum_{j=0}^{q^1} b_j x_{t-i}^1 + \ldots + \sum_{j=0}^{q^k} b_j x_{t-i}^k + E_t \qquad (1)$$

where $k$ – number of exogenous variables;

$q_i$ – number of lags in exogenous variable $(i = 1, 2, \ldots k)$;

$n$ – the depth of delays in the endogenous variable;

$E_t$ –remnants forming the process of white noise.

This model suggests that if at some time $t$ the independent variable $X$ changes, then this change will affect the values of the variable $y_t$ during the future moments of time.

## 5. Date and analysis of the primary data

Working with the input data consisted of two steps: a) data collection for the selected variables; b) the primary processing of endogenous and exogenous variables that correspond to the analyzed process and reflect the essence of the problem.

Primary data collection. The primary data was taken from the following sources: Euromonitor Passport Database (http://www.euromonitor.com/), World Bank Open Data (http://data.worldbank.org/).

The primary statistical information on endogenous and exogenous variables is collected for the countries of the world for the period from 1998 to 2015. Endogenous and exogenous parameters for use in the model were determined on the basis of the primary country statistical information as the mean geometric indicators of the countries of the world in the year $t$.

## 6. The procedure for finding of the model coefficients

A methodology for empirical verification of the model is developed [10].The methodology includes the following main steps: checking time series of variables for stationarity, using Dickey-Fuller test; checking exogenous variables for multicollinearity; selection of endogenous variable lags that have a strong correlation with the value of the variable in the last period and verification of the significance of the autocorrelation coefficients using Ljung-Box Q-test; compilation of the structural form of the model; determining the coefficients of the model using regression analysis; checking the significance of the regression equation and finding the coefficients of the regression equation..

Dickey-Fuller testis a finding of the coefficient in an autoregressive equation of following form:

$$y_t = a\, y_{t-1} + \varepsilon_t \tag{2}$$

where $y_t$ – time series, and $\varepsilon_t$ – error.

If $|a| < 1$, then the time series is stationary. If $|a| = 1$, then the process has a single root, in this case the time series is not stationary, it is an integrated time series of the first order.

The coefficient $a$ for all time series is in the range $0.47480575 \le a \le 0.718665$, which according to the Dickey-Fuller test indicates that all time series are stationary.

Checking exogenous variables for multicollinearity consisted in finding the coefficients of pair correlation between exogenous variables and with the value of the coefficient of pair correlation greater than | 0.7 | it was necessary to exclude one variable from the pair from the further analysis. As a result of this check, the variables $X^3_{1t}$ and $X^3_{3t}$ were removed and as having high closeness with variables $X^3_{2t}, X^3_{4t}, X^3_{5t}, X^3_{6t}$, at the significance level α = 0,05, the number of degrees of freedom n-2 = 17 of the Student's t-criterion is greater than $t_{table} = 2{,}11$.

The choice of the endogenous variable lags, which have a strong correlation with the value of the variable in the last period, is realized using the autocorrelation coefficients. The verification of the significance of the autocorrelation coefficients was confirmed by the Ljung-Box test. The analysis showed that the endogenous variable $y^3_t$ depends on a given level only on the value of one past period $y^3_{t-j}$. The coefficient of autocorrelation of the endogenous variable $y^3_t$ to $y^3_{t-1}$ is 0.8055. The Q-statistics of Ljunga-Box is 11,182 at $p = 0.000827$.

Taking into account the results of previous analyzes, the ADL-model takes the following structural form:

$$y^3_t = a_0 + a_1\, y^3_{t-1} + a_2\, y^2_t + a_3\, y^4_t + a_4\, y^7_t + a_5\, y^7_{t-1} + a_6\, x^3_{2t} + a_7\, x^3_{4t} + a_8 x^3_{5t} + a_9 x^3_{6t} \tag{3}$$

The coefficients of the model were found using the regression analysis module from SPSS 17.0. According to the results of the coefficients found, the equation is written in the following form:

$$y^3_t = 32515.056 + 0.705 y^3_{t-1} + 3.806 y^2_t + 1.081 y^4_t - 0.206 y^7_t - 0.905 y^7_{t-1} - 0.041 x^3_{2t} +$$
$$0.002 x^3_{4t} - 0.001 x^3_{5t} + 3947.703 x^3_{6t}$$

The accuracy of the regression equation is estimated using the coefficient of determination, which equals 0.97765481 for $F_{calculate} = 62{,}50337804 \ge F_{table} = 2{,}51$.

The statistical significance of the coefficients of the regression equation was verified on the basis of $t$ statistics. The calculated value of $t_p$ with a confidence probability $p$=0.9 for all coefficients of the higher level than the table value with the exception of $y_t^7$ and $y_{t-1}^7$. The statistical significance of the coefficients of the regression equation for $y_t^7$ and $y_{t-1}^7$ is visible with a confidence probability of $p = 0.6$. In such cases, the calculation should be repeated, deleting the variables $y_t^7$ and $y_{t-1}^7$. But we did not do this at this stage of the analysis, planning at the next stage, to introduce other additional exogenous variables estimating the urbanization.

## 7. Consideration

The article shows the influence of human activity on the environment by the example of the change of GGEI in the environment. The calculated model of the influence of exogenous variables on the endogenous variable (GGEI) shows the dependence of the endogenous variable of the current period on the variable of the past period with a lag of one year. We revealed the dependence of the endogenous variable (GGEI) on the following processes. Firstly, it is a dependence on other processes occurring in the environment that are: CO2 Emissions per Unit of Output and a reduction in the area of forests and forests in the process of urbanization.

Secondly, it is the dependence on macroeconomic indicators: an indicator of foreign economic activity, calculated as a foreign trade turnover between countries; Indicator of economic activity, calculated as an indicator of the productivity of material resources, as an indicator of the number of the employed in the sectors of the economy, capable of participating in the production of material values and the provision of services; And also as an indicator of the quantity of production in the industry of transport and communications.

We revealed the increase of the processes of pollution of the global environment by sources of an artificial environment, urbanized areas. The emerging markets, most affected by climate change and deforestation, including China, Indonesia and Brazil, represent the greatest concern about environmental damage.

The authors believe that the term of "environmentalism", as an isolated concept in the context of growing foreign trade processes between countries, under conditions of intensive economic development, is not relevant anymore.

In case when a modern urbanized areas becomes a source of destruction of the global ecology, and at the same time, the urbanization processes contribute to increasing efficiency of the economy, it is necessary to realize that the "green economy" should become a part of our life. In the life of companies, in the lives of individuals, "green economy" should maintain in different spheres of life – both culture and in production.

## Acknowledgements

This paper is based on research carried out with the financial support of the grant of the Russian Science Foundation (Project No. 14-38-00009). Peter the Great St. Petersburg Polytechnic University.

## References

[1] Dobbs R et al 2011 *Urban world: Mapping the economic power of cities* (McKinsey Global Institute) Available from: http://www.mckinsey.com/global-themes/urbanization/urban-world-mapping-the-economic-power-of-cities [Accessed 10th May 2017]

[2] Gong X et al 2016 Validity of the Emission Weighted Proximity Model in estimating air pollution exposure intensities in large geographic areas *Science of The Total Environment* **563–564** p 478–85

[3] Jones C and Kammen D M 2014 Spatial Distribution of U.S. Household Carbon Footprints Reveals Suburbanization Undermines Greenhouse Gas Benefits of Urban Population Density *Environ. Sci. Technol.* **48 (2)** p 895–902

[4] Magaril E, Abrzhina L and Belyaeva M 2014 Environmental damage from the combustion of fuels: Challenges and methods of economic assessment *WIT Transactions on Ecology and the Environment* **190** p 1105–15

[5] Leclercq B et al 2016 Differential responses of healthy and chronic obstructive pulmonary diseased human bronchial epithelial cells repeatedly exposed to air pollution-derived PM4 *Environmental Pollution* **218** p 1074–88
[6] Lazić L et al 2016 Traffic contribution to air pollution in urban street canyons: Integrated application of the OSPM, moss biomonitoring and spectral analysis *Atmospheric Environment* **141** p 347–60
[7] Patton A P et al 2016 Scripted drives: A robust protocol for generating exposures to traffic-related air pollution *Atmospheric Environment* **143** p 290–99
[8] Laña I et al 2016 The role of local urban traffic and meteorological conditions in air pollution: A data-based case study in Madrid, Spain *Atmospheric Environment* **145** p 424–38
[9] Didenko N and Kunze K 2014 Relationship between Energy Policies and Quality of Life in Russia *WIT Transactions on Ecology and the Environment* **186** (WIT Press)
[10] Rudenko D, Didenko N and Skripnuk D 2015 Environmental security issues in the Russian Arctic *15th International Multidisciplinary scientific Geoconference (SGEM 2015) Ecology, Economics, Education and legislation. Conference proceedings* **III** (Bulgaria)

IOP Conf. Series: Earth and Environmental Science **72** (2017) 012015 doi:10.1088/1755-1315/72/1/012015

# Modern technologies of processing municipal solid waste: investing in the future

**A Rumyantseva[1], M Berezyuk[1], N Savchenko[2] and E Rumyantseva[2]**

[1] Department of Environmental Economics, Ural Federal University named after the first President of Russia B.N. Yeltsin, 19 Mira str., Ekaterinburg, Russia
2 Department of Finance and tax management, Ural Federal University named after the first President of Russia B.N. Yeltsin, 19 Mira str., Ekaterinburg, Russia

E-mail: alenarum@mail.ru

**Abstract.** The problem of effective municipal solid waste (MSW) management is known to all the municipal entities of the Russian Federation. The problem is multifaceted and complex. The article analyzes the dynamics of municipal solid waste formation and its utilization within the territory of the EU and Russia. The authors of the paper suggest a project of a plant for processing municipal solid waste into a combustible gas with the help of high temperature pyrolysis. The main indicators of economic efficiency are calculated.

## 1. Introduction

The organization of effective waste production and consumption management is one of the most pressing environmental problems of the Russian Federation. The problem is connected with the organization of waste collection, generation, transportation, processing, recycling, waste disposal and neutralization. Among the aspects to be discussed are the following:

1) the need to solve the problems of dual nature related both with environmental protection and resource conservation issues, recycling and reuse of waste;

2) the steady increase of waste disposal, i.e. the need for constant removal of new land for waste storage and disposal, as well as the reclamation of disturbed (polluted, degraded) lands within the territories of closing landfills and waste dumps;

3) territorial specificity of the problem connected with the fact that the bulk of industrial waste generates in a relatively small number of regions, that is, within the locations of chemical, petrochemical, steel, pulp and paper and other industries and areas of mining. However, municipal solid waste [1], is formed everywhere, but at the same time it is utilized within the territory of the Russian Federation in a very uneven manner.

## 2. Analysis of the world and domestic trends in municipal solid waste treatment

The overall amount of production and consumption of waste, which has been collected and recorded within the territory of the Russian Federation by the end of 2015, reached the number of 31.5 billion tones [2]. The annual volume of production and consumption waste reached 5060.3 million tons in 2015. The share of used and neutralized waste was 53 %, which made 2685.2 million tons (Figure 1). In contrast, the volume of municipal solid waste reached 48.6 million tons and the volume of used and neutralized waste was only 7 % or 3.4 million tons. The dynamics of formation and usage of production and consumption waste within the territory of the Russian Federation can be seen in Figures 1 and 2 [2–4].

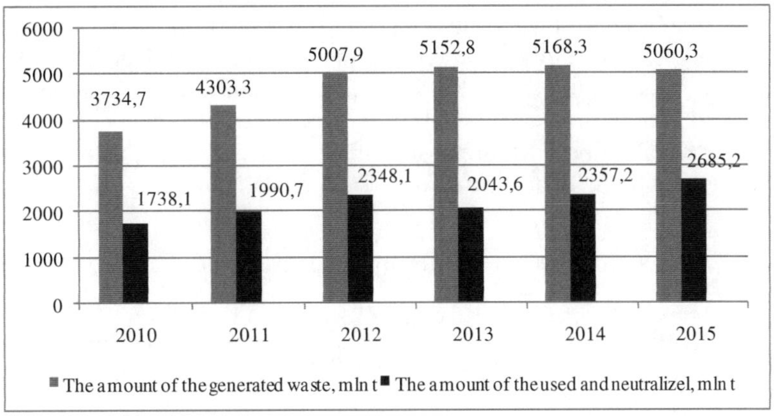


**Figures 1.** The dynamics of formation and usage of production and consumption of waste within the territory of the Russian Federation

The above statistics, to some extent, are estimates because it is hard to take into consideration all the waste generated decades ago; it is also difficult to identify all the previously accumulated waste from the point of its decomposition, dilution, weathering, corrosion, overgrown vegetation, and other processes.

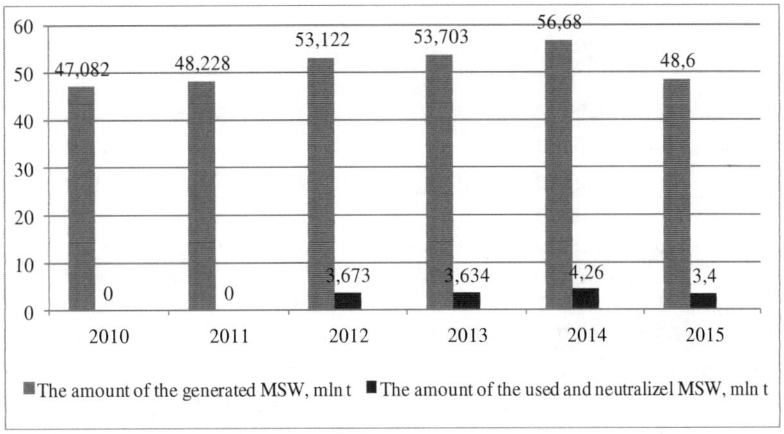


**Figures 2.** The dynamics of formation and usage of production and consumption waste

Only a relatively small part of overall solid waste undergoes waste disposal, processing and incineration within the special facilities. Thus, in 2015 waste processing facilities received about 21 million $m^3$ of municipal solid waste (7 % of the total waste disposal volume), while waste incinerating facilities received less than 7 million $m^3$ of solid waste (about 2 %). It is important to remember that besides the municipal solid waste taken from the residential areas, there is waste liquid (42 million $m^3$) and snow (62 million $m^3$), which is contaminated with various harmful impurities trapped in during the winter period [2].

The comparison with some EU countries shows that the share of incinerated solid waste may reach up to 70 %. For example, in Switzerland it reaches 79 %, in Japan it makes up to 72 % (Table 1) [2]. The EU countries use the concept of municipal waste (MW), which makes it possible to pursue the objective international comparisons.

In many cases, the processes of solid waste removal and recycling have their own specifics in many large cities of Russia. There was a significant increase in the volume of solid waste removed in 2001–2015 period in most cities. However, the increase rates vary.

In 2015, several subjects of the Russian Federation started working with landfills of solid waste and waste-processing complex. Among them are the Republic of Adygea, Bashkortostan, Buryatia, Ingushetia, Karachay-Cherkessia, Komi, Mari El, Udmurtia and Chuvash Republic; Krasnodar, Perm, Stavropol and Khabarovsk territories; Belgorod, Bryansk, Vladimir, Vologda, Ivanovo, Kirov, Kostroma, Kurgan, Kursk, Lipetsk, Murmansk, Omsk, Oryol, Rostov, Samara, Saratov and Sakhalin, Smolensk, Tambov, Tula and Tyumen regions; St. Petersburg, Khanty-Mansi Autonomous District of Yugra [2].

**Table 1.** The Proportion of Incineration of Municipal Solid Waste in Developed Countries

| Country | Population, mln. | Municipal Solid Waste, mln. tons | The Number of Waste Burning Plants | The share of Burning Waste, % |
|---|---|---|---|---|
| Switzerland | 7 | 2,9 | 29 | 79 |
| Japan | 123 | 44.5 | 1900 | 72 |
| Denmark | 5 | 2.6 | 32 | 65 |
| Sweden | 9 | 2.7 | 21 | 59 |
| France | 56 | 18.5 | 100 | 41 |
| Holland | 15 | 71 | 9 | 39 |
| Germany | 61 | 40.5 | 51 | 33 |
| Italy | 58 | 15.6 | 51 | 17 |
| USA | 248 | 180 | 168 | 16 |
| Spain | 38 | 11.8 | 21 | 6 |
| England | 57 | 35 | 7 | 5 |
| Russia | 145 | - | 8 | - |

Table 2 contains the data (in the per capita of the country) on the dynamics of municipal waste formation and disposal in the landfills of the EU countries and Russia [2]. If to talk about the relevant trends in the United States, it should said that its "trash market" is on the rise: there are about 550 waste recycling plants and about 56 thousand of waste collection and disposal enterprises there. According to the forecasts, the amount of solid waste sent for recycling can reach 30–55 % in this country in the coming years.

The analysis of EUROSTAT data [5] for the period from 2000 to 2014 shows that in many countries the volume of municipal waste dumping largely decreased – both in terms of the overall tonnage and per an inhabitant. In such countries as Belgium, Germany, the Netherlands and Sweden, the corresponding figures went down from 10 to 30 times. At the same time, there was no waste dumpling carried out in Switzerland at all.

The USA reduced the waste of dumping in 2000–2012 (about 3–4 % percent).

**Table 2.** The Dynamics of Municipal Waste Formation and Disposal within the Landfills of the EU and the Russian Federation (1 kg per inhabitant)

| Country | 2005 | 2010 | 2012 | 2013 | 2014 | 2014 in % to 2000 |
|---|---|---|---|---|---|---|
| The total of formed and disposed waste in 28 countries | 521/- | 503/185 | 485/154 | 477/143 | 475/132 | 91/- |
| Austria | 580/196 | 562/18 | 579/25 | 578/23 | 565/23 | 97/12 |
| Belgium | 471/91 | 456/8 | 447/5 | 437/4 | 435/4 | 92/4.4 |
| Bulgaria | 612/400 | 554/411 | 460/318 | 432/298 | 442/307 | 72/77 |
| Great Britain | 577/468 | 509/234 | 477/177 | 482/164 | 482/134 | 82/29 |

| Country | 2005 | 2010 | 2012 | 2013 | 2014 | 2014 in % to 2000 |
|---|---|---|---|---|---|---|
| Hungary | 446/366 | 403/284 | 402/263 | 378/244 | 385/221 | 86/60 |
| Germany | 642/167 | 602/3 | 619/1 | 609/8 | 618/9 | 96/5.4 |
| Denmark | 610/66 | 673/23 | 750/16 | 752/13 | 759/10 | 124/15 |
| Spain | 658/337 | 510/318 | 468/284 | 454/253 | 435/240 | 66/71 |
| Italy | 509/385 | 547/253 | 504/468 | 491/181 | 488/154 | 96/40 |
| Netherlands | 598/57 | 571/9 | 549/8 | 526/8 | 527/8 | 88/14 |
| Poland | 320/313 | 316/195 | 317/198 | 297/157 | 272/143 | 85/46 |
| Romania | 355/295 | 313/238 | 252/171 | 254/175 | -/- | 72/59 |
| Finland | 502/305 | 470/212 | 506/166 | 493/144 | 482/124 | 86/28 |
| France | 514/219 | 533/166 | 523/139 | 517/134 | 511/132 | 99/60 |
| Sweden | 428/97 | 439/4 | 450/3 | 451/3 | 438/3 | 102/3.1 |
| Switzerland | 656/54 | 708/- | 694/- | 702/- | 730/- | 111/- |
| Russia, take out Municipal Solid Waste | 207 | 330 | 371 | 347 | 389 | 188 |

As for Russia, there was an increase in the volume of solid waste disposal. If we compare solid waste disposal management in the EU, the USA and Russia, it becomes obvious that Russia needs serious work to be done regarding a number of indicators. In 2014–2015 the government of the country undertook legislative, organizational, administrative, fiscal, economic and other measures intended to provide a specific reserve for the further development and increase of efficiency of production and consumption waste, including the scope of the MSW. One of the most urgent question is the creation of a modern infrastructure for the processing (recycling) of various waste to produce secondary (recurrent) and/or by-products [2].

## 3. Modern methods of processing municipal solid waste

Although there is a variety of methods processing municipal solid waste, there is some generally accepted classification of technologies. Among the basic methods are: 1) deposition of landfilled solid waste; 2) sorting; 3) mechanical biorefining; 4) direct combustion; 5) anaerobic fermentation and composting; 6) pyrolysis and gasification.

It should be noted that there are no universally accepted standards or regulations to strictly specify or apply the above-mentioned methods. Waste processing issues are addressed in different ways in different countries, regions, cities and towns. Much depends on the local conditions, financial possibilities and other factors. Therefore, each case is to be treated individually depending on the goals and objectives to be achieved. The authors of the paper have not determined any particular solution to the problem of solid waste processing in Ekaterinburg.

In this paper, the authors discuss pyrolysis as a highly effective way of waste recycling and disposal. It helps to make the process of waste processing useful in terms of its transformation into heat or electrical energy. Besides it helps to significantly lower pollutant emissions, which are a way higher than in the normal combustion in the incinerator. It is necessary to partially or fully prepare waste for the process of pyrolysis.

The pyrolysis process includes the following stages:

- drying up to a certain moisture degree;
- pyrolysis process itself;
- gasification.

The following components are formed in the pyrolysis process: synthesis gas, resins, water vapor, solid carbon residue, which can also be called coke. Pyrolysis is a process of decomposition of the substance into simpler components, which is implemented under high temperatures stress with the lack of air. During gasification process the solid residue transforms into gas under the influence of atmospheric oxygen or steam [6]. The inorganic residue received after the gasification should be

IOP Conf. Series: Earth and Environmental Science **72** (2017) 012015 doi:10.1088/1755-1315/72/1/012015

disposed in a special landfill. However, the overall proportion of such residues does not exceed 10 % of the original mass of the overall load.

The process can be divided into three types in accordance with the pyrolysis temperature of [6, 7]

- low-temperature pyrolysis presupposes the temperature of 450–500 °C. The solid waste is said to release the least of gas, the maximum number of different oils, tars or resins and solid residue;
- medium temperature pyrolysis has the temperature up to 800 °C. This type of pyrolysis is known for the increase in the gas generation and the volume of resins, oils and reduced residue;
- high temperature pyrolysis implies temperatures above 800 °C. During the high-temperature pyrolysis gases are generated to the maximum degree; formation of resins, oils and solid is minimal. High temperature pyrolysis is limited by 1050–1100 °C, since this temperature range is connected with the process of melting residues – slag, which subsequently complicates the work of slag removal system.

The pyrolysis process includes four successive stages:

- waste sorting and selection of metals and the subjects above the allowable size. This procedure is carried out with the help of electromagnets and induction separation;
- gasification process, which is aimed to generate synthesis gas, slag;
- cleaning of synthesis gas in a scrubber with an alkaline solution of sulfur, fluorine, chlorine, cyanide;
- burning of purified synthesis gas in the waste heat boiler in order to produce electricity and heat.

This method of processing of municipal solid waste is environmentally friendly and promising to date. The resulting syngas can be effectively used. High temperatures help to decompose harmful substances. The sorting process helps to select metal, which may also be reused.

High-temperature pyrolysis is a cost-effective, environmentally sound and technically feasible way to dispose municipal solid waste without any substantial preparation, sorting, drying, since there are no stringent requirements to the stock (which means that even unsorted garbage can be disposed).

The process of high temperature pyrolysis efficiency reaches up to 95 %. It means that it is possible to process little combustible wastes with up to 90 % ash content and a high humidity of 60 % to [6, 7].

Advantages of high-temperature pyrolysis are as follows:

- High temperature pyrolysis allows to process high ash and the wet components;
- Exhaust gases are generated with the low solids content, as they have to undergo a kind of a filter through the raw material layer. This helps to reduce costs for the purchase of gas treatment plants, such as cyclone filters;
- The process is connected with the generation of valuable synthesis gas that can be used to generate heat and electricity;
- High temperature pyrolysis forms slag, which can be used in the construction industry;
- There is a garbage sorting process, which helps to select non-ferrous and ferrous metals, available-for-sale.

Thus, the usage of this technology can completely change the current approach in the field of waste management and significantly reduce the amounts of waste location and begin to effectively process and use municipal solid waste.

## 4. Feasibility study of the project for the processing of municipal solid waste with the help of high-temperature pyrolysis in Ekaterinburg

Ekaterinburg is a large city with the population over 1.5 million people. A human being produces about three kilograms of waste daily or about a ton of waste yearly. Every year the city processes and recycles more than half of human waste. The municipal economic entity Ekaterinburg was founded in 2015 1150.02 ths. Tons of waste (0.64 % of the total waste generation by region as a whole). The municipal solid waste amounted to about 446.2 thousand tons. Waste recycling and disposal amounted to 578.77 thousand tons, which is 50.3 % of the total waste generated in the municipal entity, and 0.69 % of the volume of waste and disposal waste in the whole region.

In 2015 Ekaterinburg Municipal Unitary Enterprise Specialized Depot delivered about 109.1 thousand tons of municipal solid waste (in 2014 – 116.2 thousand tons) to Shirokorechensky Waste Sorting Complex of Solid Waste Landfill. Waste sorting averaged 6.6 % (in 2014 – 8.7 %; in 2013 – 11.7 %) [8].

It is obvious that the city of Ekaterinburg faces quite a challenge in the sphere of municipal solid waste management. The reform in the sphere of municipal solid waste management requires the improvement of mechanisms and methods of municipal solid waste management within the territory of Ekaterinburg municipal entity [9].

The undertaken research helped to work out a pilot project, which has been proposed for the processing of municipal solid waste into a combustible gas with the help of high-temperature pyrolysis.

To implement the project, it is necessary to build a hangar, equip it with magnet conveyor belts, crushers and fine strainer, the pyrolysis reactor, heat exchangers, condensers and gas, installation of scrubbing and a pipeline to transport gas. The total investment for construction of municipal solid waste processing plant amounts to 249755 thousand rubles.

Prospective investor is a regional operator, acting under the public-private partnership and aiming at creating infrastructure for the management of municipal solid waste.

According to the envisaged changes in the Federal legislation (Art. 1 of the Federal Law No. 89-FZ) [10], regional operators are the ones to carry out activities for the collection, transportation, processing, recycling and disposal of municipal solid waste within the territories of the Russian Federation.

Regional operators are to carry out their activities in accordance with the regional and territorial programs in the sphere of waste management [3, 4], which makes regional operators to be interested in the municipal solid waste processing and recycling.

The plant will allow to process 103920 tons of waste per year. Experts believe that one ton of waste processed with the help of high-temperature pyrolysis will generate the average of 450–500 $m^3$ of synthesis gas. To calculate the production capacity of the plant it is taken the average of 475 $m^3$/ton of waste. Production capacity of the plant will amount to 49312 thousand $m^3$ per year.

The generated gas has the following composition (Table 3):

**Table 3.** Heat of combustion product gas

| Component | Net Calorific Value, kJ / m3 | Gas share, % | Heat of combustion, kJ |
|---|---|---|---|
| $H_2$ | 12640 | 40 | 5056 |
| CO | 10820 | 17 | 1839.4 |
| $CH_4$ | 35850 | 11 | 3943.5 |
| $CO_2$ | - | 32 | - |
| Total | - | 100 | 10838.9 |

The calculations show that the cost of 1000 $m^3$ of gas will be 1073.9 rubles, on the whole issue – 53009.7 thousand rubles.

At the moment, the cost of 1000 $m^3$ of natural gas for wholesale buyer is about 3646 rubles.

Table 3 shows that the heat of combustion of the generated gas is 3.3 times lower than the heat of natural gas combustion. Thus, it becomes possible to calculate the market price of the product gas by the specific heat ratio. The price should not exceed 1102 rubles. The project proposes to sell gas at its cost.

Novosverdlovskaya Central Heating and Power Station is considered to become one of the consumers of the generated gas. This station has a capacity of 557 MV. The efficiency of the gas turbine generator is 65–70 %, the volume of the gas consumption is about 800 million $m^3$

per year. These calculations are made for a constant maximum load. Though, in reality, the station consumes about 500–600 million m$^3$ per year. Therefore, it is going to be not a problem for the station to consume gas generated in the amount of approximately 49 million m$^3$.

Part of the generated gas will be used for the needs of the plant itself. This technology involves the heating of the raw material. This requires energy, heat, which can be produced by burning the generated synthesis gas. It is expected to use about 15 % of the generated gas. Thus, about 41.957 thousand m$^3$ of gas can be sold.

Also, mechanically separated sweepings reach about 6 %, the amount of separated metal is about 4 %. Thus, the reactor is fed with about 90 % of the waste incoming to the plant. This means that the plant must process about 115466 tons of waste per year. The income sources of the project are shown in Table 4.

**Table 4.** Income of the Plant for Processing Municipal Solid Waste

| The Source of Income | The amount | Price per Unit, rub. | Amount, thous. rub. |
|---|---|---|---|
| Gas distribution, thousand m$^3$ | 41957 | 1073.9 | 45057.6 |
| Sale of Ferrous Scrap, tons | 4618 | 6000 | 27708.0 |
| Payments for the disposal of municipal solid waste, tons | 115466 | 150 | 17319.9 |
| Total | – | – | 90085.5 |

Further on, the researchers carried out feasibility study of the investment project. The project implementation period is 16 years.

Key indicators of economic efficiency of the investment project were calculated at a discount rate of 12 %, which is shown in Table 5.

**Table 5**. Key Indicators of Economic Efficiency of the Investment Project

| № | Name of the Indicator | Unit | Rated Value |
|---|---|---|---|
| 1 | Net income | thous. rub | 415380 |
| 2 | Net present value | thous. rub | 47668 |
| 3 | Discounted payback period | years | 10.9 |
| 4 | Index of profitability investment | – | 1.19 |
| 5 | Maximum cash outflow | thous. rub | 220000 |
| 6 | Internal return rate | % | 15.95 |
| 7 | Break-even point | thous. m$^3$ of gas | 18621 |
| 8 | Financial safety margin | % | 55.6 |

The analysis of the indicators of economic efficiency of the project proves that the project of the construction of the plan for the processing of municipal solid waste with the help of high-temperature pyrolysis is attractive and appropriate for implementation. The main environmental benefit of the project is the recycling of accumulated and newly generated waste. It helps to prevent the accumulated environmental damage and prevent damage to the environment. Though payback period of this investment is large enough, we can confidently talk about the need to invest in the future in terms of achievements of the current and deferred environmental impact for the future generations.

## 5. Conclusion

As a result, the implementation of the research leads to the following conclusions, which are crucial to justify the choice of technology for the processing of municipal solid waste.

1. At present, there are several technologies of municipal solid waste utilization excluding the process of direct incineration. In each case, the technology is determined in accordance with local conditions and requirements. As for the municipal entity of Ekaterinburg the researchers have not determined any universal analogue, fully consistent with the existing conditions terms (municipal solid waste composition, sorting and processing line, capacity, tariffs, etc.).

2. The high-temperature pyrolysis technology is technically feasible and can be considered as industrial for the municipal solid waste in the city of Ekaterinburg.

3. The technical efficiency of the pyrolysis technology in the first place depends on the morphological composition of municipal solid waste, which is determined by several organizational and social factors: the level of living standards, waste collection and transportation system as well as other factors to be examined.

4. The syngas generated in the pyrolysis process may have a calorie, which is about three times lower than that of calorific gas.

5. The synthesis gas pyrolysis can be considered as a substitute fuel for several energy and processing plants running on gas.

6. The project will reduce the load of municipal solid waste landfills. It will help to recycle about 10 % of the municipal solid waste generated in the city of Ekaterinburg and to improve the environment. Moreover, the project corresponds with the Integrated Strategy for Municipal Solid Waste Management within the territory of Sverdlovsk region up to 2030.

**References**

[1] Federal Law No. 458 of 29.12.2014 On Amending the Federal Law "On Production and Consumption Waste, the Russian Federation Certain Legislative Acts and Russian federation Annulment of the Certain Legislative Acts (Legislative Acts provisions) Available from: http://www.consultant.ru/document [Accessed 18th April 2017]

[2] State report On the State and Environmental Protection of the Russian Federation in 2015 Available from: http://www.mnr.gov.ru/regulatory/detail.php?ID=286341 [Accessed 18th April 2017]

[3] Berezyuk M, Rumyantseva A and Lapina A 2016 Municipal solid waste management in a new legislation: comprehensive approach *E3S Web of Conferences* **6** article number 1006

[4] Berezyuk M and Rumyantseva A 2016 The new system of municipal solid waste management: the innovative approach *The innovative development of the economy: research and practical and theoretical journal* **5 (35)** p 19–29

[5] Official website of Eurostat [in Russian] Available from: http://data.trendeconomy.ru/eurostat/env_wastrt [Accessed 18th April 2017]

[6] Rehraha D, Bansodeb R, Hassanb O and Ahmednaa M 2016 Physico-chemical characterization of biochars from solid municipal waste for use in soil amendment *Journal of analytical and applied pyrolysis* **118** p 42–53

[7] Bugayan S A 2015 Disposal of solid waste: foreign and domestic experience *Science and education: agriculture and economics; entrepreneurship; law and governance* (Rostov-on-Don) **7 (62)** p 27–31.

[8] State report On the Condition and Environment Protection in the Sverdlovsk Region in 2015 Available from: http://mprso.midural.ru/article/show/id/1084 [Accessed 18th April 2017]

[9] The Municipal Solid Waste Treatment Integrated Strategy at the Territory of the Sverdlovsk region till 2030 Available from: http://docs.pravo.ru/document/view/51687348/79831482 [Accessed 18th April 2017]

[10] Federal Law No. 89 of 24.06.1998 On the Production and Consumption Waste Available from: http://base.garant.ru/12112084/ [Accessed 18th April 2017]

# Fiscal instruments for regulating the sustainable development of urban transport systems in Russia

**I Mayburov[1, 2] and Y Leontyeva[3]**

[1] Department of Financial and Tax Management, Ural Federal University named after the first President of Russia B. N. Yeltsin, Ekaterinburg, 19 Myra str., 620002, Russia
[2] Finance and Credit Department, Far Eastern Federal University, Vladivostok, 8 Suhanova str., 690091, Russia
[3] Department of Financial and Tax Management, Ural Federal University named after the first President of Russia B. N. Yeltsin, Ekaterinburg, 19 Myra str., 620002, Russia

E-mail: mayburov.home@gmail.com

**Abstract**. The article explains the role of public transport pricing. It proves the need for a systemic approach to building a modern public transit system. The authors argue that the main objective of the approach should be to reduce the use of private vehicles in the urban environment and increasing public transport use. It is proven that for the consumer of transport services the price per trip is an important factor when deciding whether to travel by car or by public transport. The authors analyze the available literature assessing the effects of widespread car ownership on users of the city transit system. Conflict situations that occur due to the unabated desire of city residents to travel by car are analyzed. A research method is proposed. It is shown that public transport fares have been growing in Russia at an accelerated pace when compared to the overall increase in prices of all goods and services, including motor vehicles, petrol and oils. The fare growth has resulted in a 3.6-fold drop in demand for public transport services over the 15 years being analyzed. Over the same period, the number of privately owned cars grew 120 percent. A conclusion is drawn that regular fare hikes have encouraged urban population to gradually opt against travelling by public transport. That resulted in higher demand for car travel and, eventually, in an accelerated growth in car usage. One can conclude that a persistent institutional trap has taken shape in Russian metropolises. Essentially, it means that higher public transport fares have led to lower demand for public transit services. As ridership goes down, public transport operators have to again increase prices, thus driving the demand for their services down. It is proven that escaping the trap will require restoring the ratio of prices to make sure that the price charged for a public transport trip is far lower than the cost of travelling by car. The aim of this study is to assess the influence of the factor of public transport fares on demand for private car usage and public transport ridership.
**Key words.** Public transport, private transport, transport pricing.

## 1. Introduction

All too often, we underestimate the role of the transport system in the harmonized development of the urban area we live in. We usually view the city transport system as a local system for transporting passengers and cargo within a specific area. Meanwhile, the real function of the city transport system is much broader; the system defines the look and viability of the modern city.

As Vuchic rightly points out, transport is the "lifeblood" of cities, connecting all other subsystems and functions (economic, recreational, social etc.) Consequently, the efficiency of the urban transport system determines the effective performance of other systems in a modern megalopolis.

As a rule, city authorities confuse cause and effect when trying to address emerging transportation problems. Congestion is a typical example of that. City authorities believe that it is the main problem of the transport system and try to deal with it in a traditional way by increasing road capacity. However, the more roads there are and the high their capacity, the more cars take to the roads, creating more traffic jams. Eventually, such solutions only have a short-term effect, and congestion will only get worse over time.

That being said, traffic jams are not the cause, but the effect of an incorrect urban transport policy. Systemic measures are needed to eliminate traffic jams. The measures should be primarily aimed at giving people an incentive to use public transport and discouraging them from travelling by car. Applying these measures holistically will increase the effectiveness of the city transport system in the future.

A systemic approach to building a modern urban transport system consists in setting the right priorities for the development of different modes of transportation. Urban transport modes include bicycle, car, bus, tram, or metro. In a majority of Russian cities the bicycle can be used only a few months a year due to weather conditions. As a result, the strongest competition for passengers occurs between public and private transport.

There is a burning need in all Russia's cities to work out effective fiscal and administrative measures for regulating the development of different modes of urban transport. When doing this, one should carefully assess the implications of changing the structure of transport and the short-term and long-term effects of changing the structure in terms of environmental impacts. The main goal of such regulation should be reducing the use of private vehicles in the urban environment and increasing the use of various modes of public transport.

Analysis of pollutant emissions into the atmosphere shows that, in general, during the period of 1995–2015 these emissions have not increased. However, the structure of the emissions changed. The share of vehicle emissions in the total emissions in Russia has increased to 45 % (Table 1).

**Table 1.** Dynamics of pollutant emissions into the atmosphere

| Indicator | 1995 | 2000 | 2005 | 2010 | 2015 | Increase,% |
|---|---|---|---|---|---|---|
| Emissions of pollutants into the atmosphere, million tons | 32.3 | 32.3 | 35.8 | 32.7 | 31.3 | 97 |
| Emissions of air pollutants from stationary sources, million tons | 21.3 | 18.8 | 20.4 | 19.1 | 17.3 | 81 |
| Emissions of air pollutants from motor vehicles, million tons | 11 | 13.5 | 15.4 | 13.6 | 14.0 | 127 |
| Share of vehicle emissions in the total emissions of pollutants into the atmosphere | 34.06 | 41.80 | 43.02 | 41.59 | 44.70 | 131 |
| Emissions of pollutants from motor vehicles per unit of land area, t/km$^2$ | 0.64 | 0.79 | 0.90 | 0.80 | 0.82 | 128 |

The most significant contribution to the increase of emissions has been made by private cars. It is obvious that the traffic of private cars should be reduced. The most effective tool is a fiscal impact on car owners.

The price per trip is one of the most effective instruments for regulating demand for public and private transport. For the consumer of transport services the price of per trip is an important factor when deciding whether to travel by car or by public transport. If the cost of a car trip is lower than the fare paid for travelling to the same destination by public transport, the city dweller will opt for riding in a car.

The aim of this study is to assess the influence of the factor of public transport fares on demand for private car usage and public transport ridership.

## 2. Literature review

A rapid growth in car ownership rates in the middle of the 20th century is considered to be the bifurcation point in the evolution urban transport systems. It was that process that redefined the traditional idea of the perfect structure of urban transport systems because the processes that were actually taking place in cities were unrelated to conventional standards of transport planning. Mass car ownership resulted in a number of conflict situations that started to occur in all big cities.

Such situations stem from city dwellers' unabated desire to use their private cars and the limited capacity of the road network. Another conflict is due to motorists' desire to park their vehicles in a walking distance from their terminal destination (workplace, school, leisure facility) and limited parking space. Mass car ownership also gives rise to a conflict between public and private transport because they start to compete for access to the road network. Mass car ownership reduces the travel speeds of surface modes of public transport and makes them less attractive to passengers. Mass car ownership generates a serious conflict between downtown and suburban areas when well-off inhabitants move to suburbs, cities sprawl outward, and downtown areas see a decline. For a long time, these conflicts were not treated theoretically.

Pioneering studies of urban transport systems amid rapidly growing car ownership were conducted by Wardrop [1] and Hollatz [2]. Wardrop studied an equilibrium distribution of public and private transport flows within a section of a road network. The point of equilibrium was determined by comparing total disutility (total costs) of all travellers. Wardrop proved that when each road user chooses their preferred means of transport, this choice is not socially optimal. It is necessary to implement incentives encouraging the use of public transport. At the same time, it is necessary to adopt measures to discourage the use of private transport. Hollatz and Tamms showed that high concentrations of cars in a city not only lead to lower efficiency of the entire transport system of the city, but generally decreases the quality of life and safety for the entire urban community.

Burrington proved that mass car ownership results in time loss and other effectiveness tradeoffs for all users of the transport system. Johnson proved that growth in car ownership is detrimental to the natural and manmade urban environment [3, 4].

Bruun and Vuchic [5] used the two-dimension measure of "time-area" to illustrate the efficiency of public transport. The measure expressed in square meter-minutes per trip was used to produce a simple and clear comparison of the amount of space required by various travel modes – walking, driving and bus transit.

Vuchic [6] also proposed an original graphic method using the time-area coordinates to demonstrate differences in modal capacities by a sketch of facilities required for transporting 15,000 persons/hour. Vuchic proved that automobiles take up far more space in the city streets than other modes of transport: transporting that number of passengers by automobile will require 17 lanes plus 34.5 ha of parking area.

Vuchic [7] proved that encouragement of public transport use and pedestrian traffic will give a city an edge over suburbs in terms of variability, comfort and overall transit costs. The more car-dependent a city is and the more it ignores public transport alternatives, the more likely its downtown areas are to fall into decline. Burchell [8] substantiated the idea that urban sprawl and the appearance of so-called edge cities have proofed ineffective not only in terms of initial area development costs, road building and other infrastructure, but in the purely functional sense too.

Pucher [9] and Holtzclaw [10] performed a comparative analysis of public expenditures and subsidies associated with urban travel. They claim that car users pay only a portion of the transportation costs. They do not cover social, environmental and other indirect costs. Car users only defray around 60 percent of total urban travel costs. These subsidies to automobile users substantially exceed the government subsidies to public transit. All modes of public transport require public investment in infrastructure and rolling stock and partial subsidies to cover their operation expenditures. Users' expenditure is limited to the amount of the fare paid. In Western European cities and the USA the farebox recovery ratio usually ranges from 20 to 90 percent. In the USA, federal investments in public transport, including planning, research and funding for city infrastructure amounted to a mere 3bn to 5bn dollars a year between 1970 and 1990 and showed practically zero growth.

Gómez-Ibáñez [11] proved that demand for public transport services is very sensitive to fares. Federal and local subsidies, however, did not result in fare decreases, but led to ineffective spending.

Small [12] substantiated a connection between road pricing and higher quality of bus transit. Mayburov and Leontyeva [13] substantiated the need for cardinal changes in Russian cities' approach to long-term transport planning. These include programs encouraging people to switch to public transport and fiscal instruments discouraging car use [14].

## 3. Materials and methods

The hypothesis of this research is that disproportionate growth of public transit fares, as compared with overall increases in consumer prices, leads to a substantial shit in user preferences in the long-term run: passengers decide against using public transport in favor of private automobiles.

We proceeded from the following facts when formulating our hypothesis:

(1) The mobility of urban population is high and will only grow further. Urban mobility is supported with either with public or private transport. At the same time, public transport and private vehicles compete against each other for the same road network that was paid for by the entire local community.

(2) The use of public and private transport in the urban environment are two interrelated and interdependent indicators. The more people travel by private transport, the lower the need for public transport. And vice versa: the more people use public transport, the less they need private vehicles.

(3) There is a global trend toward fiscal regulation of the costs of car usage with the aim of increasing them through the introduction of transport-related charges (tolls, parking fees, fuel taxes etc.) Government fare regulation is poorly developed and is usually linked to the processes of approving fares that will ensure a substantiated profit margin for public transport operators.

(4) We believe that in a modern city public and private transport are equally accessible, so we do not take the accessibility factor into account further on. The time of travel factor and comfort of various modes of transport in the city environment are neglected, too.

When developing a method of studying the impact of the pricing factor of public transport services on the demand for various modes of transportation, we took the following parameters into account.

The analyzed period is from 2000 to 2015. The database for the analysis was provided by the data of the Federal State Statistics Service that reflects national average figures.

To estimate the intensity of public transport use we employ ridership data for electric propelled public transport (tram, trolleybus and subway). This information is available on the website of the Federal State Statistics Service (www.gks.ru).

The intensity of private car use was estimated through the number of privately owned automobiles. Unfortunately, the Federal State Statistics Service does not employ the car use indicator. The application of this indicator is accompanied with certain reservations because the ownership of a vehicle is not the equivalent of using it, although there is a direct correlation between them, which enables us to use this measure in our analysis.

The pricing factor of public transport services was estimated through the base fare growth index for electric propelled modes. This index is calculated by multiplying the chained (annual) fare indices for electric propelled modes of public transport in corresponding years. The value of the base fare growth index reflects the general trend in fares for electric public transport over the period of analysis. The chained (annual) index describes the trend in fares for electric public transport services throughout a year. Chained (annual) fare indices can be found on the website of the Federal State Statistics Service.

The growth in electricity powered public transport fares unadjusted for inflation was estimated by using the base fare growth index that is unadjusted for inflation. The measure was calculated as the ratio of chain (annual) public transport fare indices to the values of the chained (annual) consumer price index in corresponding years. The chain (annual) consumer price index is available on the website of the Federal State Statistics Service

Values of the consumer price growth index indicate inflation rates over the analyzed period. We calculated the measure by multiplying the chained (annual) consumer price index levels over relevant years.

The obtained values were put in the same coordinate system. Dependencies were identified by processing the data with Microsoft Excel.

## 4. Analysis of results

The results of the performed calculations for some years are shown in Table 2.

One can see that over the analyzed period between 2005 and 2015 the base fare growth index increased ten-fold. At the same time, the base consumer price growth index increase five times. Consequently, the prices of public transport services grew twice as fast as overall inflation. This means that the growth in public transport fares in Russia has outpaced the increase in prices of all goods and services, including cars, petrol and motor oil.

A similar situation was observed in the USA where the average bus fare grew from 39 cents to 88 cents between 1980 and 1992 in constant dollars. Over the same period, the price of gasoline went down 2 percent from 122 to 119 cents per gallon (from 31 to 30 cents per litre) [15]. The downward trend in petrol prices and the upward trend in public transport prices were clearly targeted against low-income people. Higher petrol taxes alleviate the anomaly at least partially, especially if the tax revenue is spent on improving alternatives to car travel.

**Table 2.** Values of the public transport pricing factor and demand for various modes of transport in Russia's cities

| No | Indicator | 2000 | 2005 | 2010 | 2015 |
|---|---|---|---|---|---|
| 1 | Base consumer price growth index | 1.2 | 2.28 | 3.72 | 5.64 |
| 2 | Base fare growth index for electricity powered public transport | 1.42 | 3.48 | 7.02 | 10.43 |
| 3 | Base fare growth index for electricity powered public transport unadjusted for inflation | 1.18 | 1.53 | 1.88 | 1.85 |
| 4 | Number of passengers carried on electricity powered public transport in cities, billion persons | 22.9 | 12.34 | 7.58 | 6.44 |
| 5 | Number of privately owned automobiles, million units | 19.1 | 24.1 | 32.6 | 42.3 |

The growing public transport fares have resulted in a 3.6-fold drop in the demand for public transport services over the 15 years of analysis. Over the same period, the number of automobiles owned by individuals has grown 120 percent. It is possible to conclude that public transport services have a high price elasticity of demand. Regular fare hikes encourage people to gradually stop using public transport, replacing it with higher demand for car rides. As a result there has been observed an accelerated growth in case use.

It would be wrong to state with confidence that the growing car ownership rates are exclusively the result of erroneous urban transport policies. The growth was spurred on by the mass ownership of cars that swept Russia's cities 30 to 40 years later than developed countries. However, the accelerated growth in public transport fares has undoubtedly had a considerable negative impact on the structure of demand by public transit and made the situation with growing car ownership worse.

The graph below illustrates the results of the performed calculations (Figure 1).

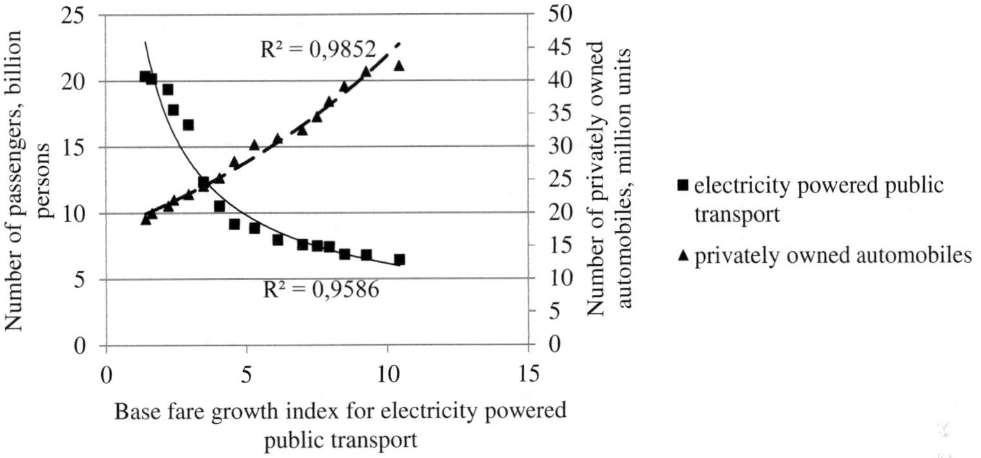


**Figure 1.** Impact of the pricing factors of public transport services on demand for various modes of transport

One can see that there is a power-law dependence between a drop in the demand for public transport and the fare growth index. Meanwhile, there is an exponential dependence between the growing demand for private car travel and the fare growth index.

Since one of the causes of fare hikes is the inflation-driven rise in price levels, we recalculated the considered dependences after eliminating the inflation impact in the Russian economy. The resulting dependences shown on Figure 2 use the base fare growth index unadjusted for inflation.

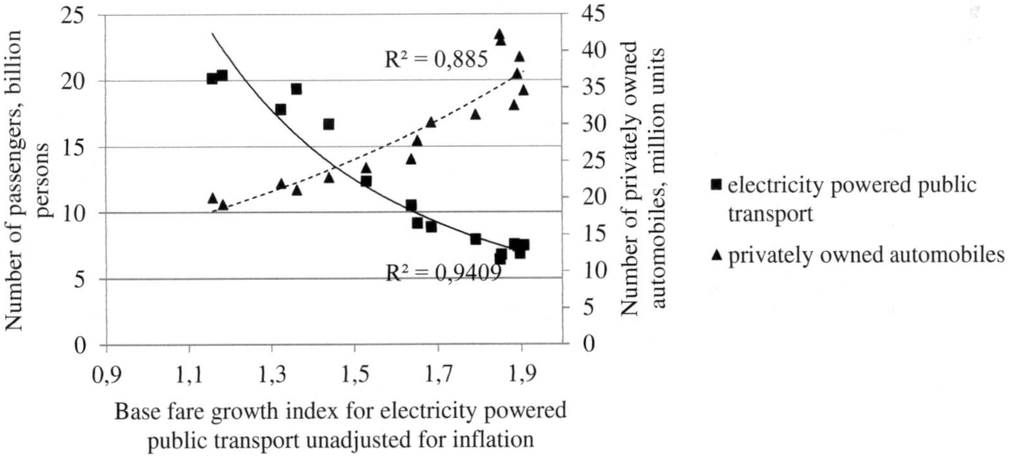


**Figure 2.** Impact of the pricing factors of public transport services unadjusted for inflation on demand for various modes of transport

The graph provides a lot of information. It makes it possible to see that the dependence of the decrease in the demand for rides on public transport on fares is clearly manifested until the index reaches certain high levels (around the 1.8-fold point). Further growth in public transport fares does not lead to such a sharp drop in demand. When the fare growth index reaches high values (around the 1.7-fold point), the previously obvious exponential dependence of car ownership rates fades away.

This effect might be attributed to the fact that objectively there is some demand for public transport services that does not depend on the price of these services. The minimum demand for the services is provided by people, who do not have driving licenses, by young people, tourists and other categories of passengers. For them, the price elasticity of demand is very low. As a rule, these categories of passengers do not change their transport choices.

## 5. Conclusion

As we can see, urban transport policy aimed at ensuring the break-even performance of public transport operators and attracting private business into the sector results in profit making becoming the key goal of the provision of public transport services. Consequently, all transport companies start to increase their prices (or seek an approval for the increase by the regulator) in order to ensure their competitive edge amid falling demand.

As a result, we witness a persistent institutional trap has taken shape in Russian cities over the past 15 years. Essentially, it means that higher public transport fares have led to lower demand for public transit services. As ridership goes down, public transport operators have to again increase prices of their services, thus driving the demand further down. This ineffective urban transport policy brings about a considerable change in the structure of demand for city trips. An increasing number of city residents prefer to travel by car, which leads to a sharp growth in car use in the city environment and associated negative impacts.

It is very difficult to escape the institutional trap. In order to reverse the observed negative trends in the structure of the demand for urban transport it is necessary to restore the right price ratios when the cost of travelling by public transport is much lower than the cost of a car ride. To achieve that, an adequate city transport police needs to be developed that would result in lower public transport fares and make it costlier to use a private car in the city.

Additionally, high public transport fares, by contrast with a fairly low cost of travelling by car, leads to uncontrolled spread of car use and eliminates the dependency of the demand for public transport services on the cost of the services.

## References

[1] Wardrop J G 1952 *Proceedings of the Institution of Civil Engineers* **1** p 325–62

[2] Hollantz and Tamms 1965 *Die Kommunalen Verhehrsprobleme in der Bundesrepublik Deutschland.* (Essen, Germany: Vulkan-Verlag)

[3] Burrington S H 1994 *Road kill: how solo driving runs down the economy. Conservation Law Foundation (CLF)* (Boston, Massachusetts)

[4] Johnson E W 1993 *Avoiding the collision of cities and cars: urban transportation policy for the twenty-first century* (Chicago IL: American Academy of Arts and Sciences and the Aspen Institute)

[5] Bruun E C and Vuchic R V 1995 *The time-area concept: development, meaning, and application. TR Record 1499* (Washington, DC: Transportation Research Board) p 95–104

[6] Vuchic V R 1981 *Urban public transportation: systems and technology* (Englewood Cliffs, NJ: Prentice-Hall)

[7] Vuchic V R Kikuchi S Krstanoski N and Shin Y E 1995 *Negative impacts of busway and bus lane conversions into high-occupancy vehicle facilities. TR Record. 1496* (Washington DC: Transportation Research Board) p 75–86

[8] Burchell R W et al. 1999 *Eastward Ho! development futures: paths to more efficient growth in Southeastern Florida* (Report prepared by the Center for Urban Policy Research, Rutgers University, for the Florida Department of Community Affairs and the U. S. Environmental Protection Agency)

[9] Pucher J 1995 *Budget cutters looking at wrong subsidies. Passenger Transport* (Washington DC: American Public Transit Association)
[10] Holtzclaw J 1995 America's autos and trucks on welfare: a summary of subsidies *Mobilizing the Region* **15**
[11] Gómez-Ibáñez J A 1996 Big City Transit Ridership, Deficits and Politics: Avoiding Reality in Boston *Journal of the American Planning Association* **62** p 30–50
[12] Small K A 2003 *Road Pricing and Public Transport* (University of California at Irvine) p 1–32
[13] Mayburov I and Leontyeva Y 2016 Forming the social optimum of passenger flows in urban agglomerations in Russia *International Conference on Sustainable Cities (ICSC 2016)* E3S Web of Conferences **6**
[14] Golubeva A and Magaril E 2013 Improved economic stimulation mechanism to reduce vehicle CO2 emissions *Urban Transport XIX* C A Brebbia WITPress p 485–95
[15] *Urban Transportation Monitor (UTM). Numerous statistical data and information items* 1995 (Burke, VA: Lawley Publications)

# The chance of transport behavior paradigm in the largest cities of Russian Federation

**Denis Vlasov**

National Research Moscow State University of Civil Engineering, 129337, 26 Yaroslavskoe shosse, Moscow, Russia

E-mail: vlasych@mail.ru

**Abstract**. This article discusses the issue of transit scenarios (or paradigm) changes in the largest Russian cities. Carrying out research and evaluation studies the authors surveyed the passengers of the new rapid off-street transit system, the Moscow Central Circle (MCC). The survey was prepared using the MCC reports data on passengers landing measured in the morning peak hours. Analysis of the survey results obtained by the authors, reveals shifts in pedestrian traffic and permits to estimate its size and main directions. It was noted a significant change in common behaviour regarding the increased length of the walking distance to the stations of rapid off-street transport. In view of the fact that the issue of the walking distance standard governs a huge range of further decisions, the research results can be used in sustainable transit travel planning and urban territorial development as well as in preparation of the regulatory documentation.

## 1. Introduction

There is no doubt that development of transport infrastructure during the reconstruction of the urban environment is aimed at the improvement of the environmental situation. Today, urban transport is a major source of urban air pollution in many cities of the world. However, we cannot ignore the obvious fact that it is impossible to create a favorable conditions for human life, sustainable socioeconomic and spatial development of the city without the developed transport infrastructure.

The capital of the Russian Federation was no exception among the cities. If in the middle of 90-ties the share of air pollution by motor transport in the total volume of emissions amounted to 83.2 %, in 2000 year this share already amounted to 93.7 percent, and, unfortunately, over the past 15 years, the situation has not changed significantly [1]. Fuel combustion contributes up to 500 types of pollutants. Major share of them are oxides of carbon, nitrogen, sulphur, hydrocarbons, soot, etc.

Aiming to reduce (or completely eliminate) the harmful effects of urban transport infrastructure pollutions on the environment, transport policy should consider the following activities:

1. restrictions on the movement of freight transport;
2. maximum increase of the road network capacity;
3. strict requirements for the environmental characteristics of transport;
4. developing networks of rapid off-street rail public transport;
5. developing various types of urban ground-electric transport;
6. enhancing competitiveness of urban land passenger transport;
7. promotion of small cars drivers to make individual trips and development of non-motorized vehicles.

Moscow Central Circle Rail Project implemented nowadays in Moscow, meets all the necessary requirements and methods to reduce harmful effect of air pollutants on the urban environment. The main purpose of this research is to determine the changes that Moscow Central Circle Rail has brought to the development of non-motorized trips and, particularly, to study the dynamics of the walking movements distances caused by the development of rapid off-street modes of transport.

The world practice shows that Transit Oriented Development approach (based among others on prioritizing urban public transport) and creation of safe and comfortable urban environment in the areas adjacent to the off-street transport stops have led to significant increase of non-individual vehicle travel. Thus, according to the US High Speed Rail Association data, the introduction of the TOD principles have reduced the dependence on individual transport for approximately 57 % [2]. The changes in population mobility behavior are indicated by acceptable length of walking access trip to the off-street transport station.

Moscow Central Circle (MCC) – is an urban rail line built at the beginning of the XX century and used mainly for freight service. Over the past 10 years, the road has been rebuilt and electrified, which made it possible to organize on the rout the rapid passenger traffic line. The length of the ring is 54 km, there are located 31 stops. When designing the ring, it was foreseen the possibility to transfer to the radial lines of the subway and to the railway.

It was decided to conduct a survey designed to seek passenger views on the introduction and effectiveness of the new passenger system and particularly, among others tasks, to define which types of transport are used or preferred to arrive to the MCC station and to determine the walking distance for the pedestrian access to the station.

## 2. Methods

Survey methodology was chosen as the main technics to gather information approaching passengers at the MCC stations with further statistical processing of the results obtained.

The reference values were those required by the following regulatory rules:
- The federal Code of Rules "Town planning" – 500 m (with possible increase, under certain conditions up to 800 m);
- Regional standards of Moscow in the field of transport – 500–700 m.

The survey is a sampling observation conducted in the morning «peak» hours on typical days (the days with the maximum size of passenger traffic – Tuesday, Wednesday and Thursday). To create sample collections, we'll use self-random selection and repetition-free sampling, since this allows us to minimize the sampling error [2, 3] ensuring its representativeness, for which it is necessary to ensure:
- sampling from homogeneous population subgroup;
- study of the characteristics of the general population;
- equal probability for each unit of the general population to be included in the sampling;
- sufficient number of units for the sample size.

Estimation of the sample size required to determine the representative number of TTH.

Calculations are carried out on the basis of a formula for repetition-free sampling.

$$n = \frac{N \times \sigma_x^2 \times t^2}{(\Delta_x^2 \times N) + (t^2 \times \sigma_x^2)}$$

where:
n – necessary sample size, people;
N – sample size or general population (determined on the basis of full-scale surveys);
σ –standard deviation, variance;
t – the confidence coefficient. In calculations, the value of the confidence coefficient is assumed to be 3, which corresponds to a probability value of 0.997;
Δx – the sampling error limit .

The conducted calculations showed that the estimated value of the interviewed passengers was 350 people, with a total volume of the general population of about 35 thousand people. (The maximum amount of landing for all MCR stations per «peak» morning hour). The required number of passengers for the survey at each station was determined in proportion to the size of the passengers' landing.

The main task of statistical analysis is to identify homogeneous subgroups of passengers. We assume that the subgroup is homogeneous if the distribution of values in it is subject to the Gauss-Laplace law for ordinary distribution.

$$f(x) = \varphi' = \frac{1}{\sigma\sqrt{2\pi}} e^{\frac{(X-\overline{X})^2}{2\sigma^2}}$$

where:

φ′ – the relative density of the distribution (the ordinate of the normal distribution curve);

π = 3.14, e = 2.72 – mathematical constants;

$x$ – the average value of the characteristic in the distribution;

σ – square deviation.

## 3. Implementation

On February 2017, the structure of arrival to MCR stations is as follows:
- on the underground – 22.6 %;
- by rail – 10.5 %
- on land passenger transport – 22.2 %;
- by taxi – 1.6 %;
- on the individual car – 1.7 %;
- on foot – 41.3 %.

The absence of cyclists in the structure is explained by the fact that at the time of the survey the MCR functioned only in the autumn-winter period, when cycling in Moscow is not possible.

Thus, today, walking is the most simple, free and easy way to get to the stop points of the MCR. Which, in turn, makes it particularly interesting to study the possible length of a pedestrian access routes using the Central Ring as an example.

The total number of passengers arriving to the MCC on foot (from a representative sample) was 150 people. For statistical processing, all passengers were divided into subgroups – intervals, relative to the length of access routes. The interval value was taken equal to 0.25 km. The results of the distribution are shown in the Table 1.

**Table 1.** Evaluating pedestrian routs length distribution

| Interval of the access length, km | Number of the passengers, people | Rate, % |
|---|---|---|
| 0–0.25 | 7 | 4.67 |
| 0.251–0.5 | 36 | 24.00 |
| 0.51–0.75 | 50 | 33.33 |
| 0.751–1.0 | 31 | 20.67 |
| 1.0–1.25 | 9 | 6.00 |
| 1.251–1.5 | 6 | 4.00 |
| 1.51–1.75 | 6 | 4.00 |
| 1.751–2.0 | 2 | 1.33 |
| 2.01–2.25 | 1 | 0.67 |

| Interval of the access length, km | Number of the passengers, people | Rate, % |
|---|---|---|
| 2.251–2.5 | 1 | 0.67 |
| 2.51–2.75 | 1 | 0.67 |
| Total | 150 | 100.00 |

The distribution obtained from these values is shown in Figure 1.

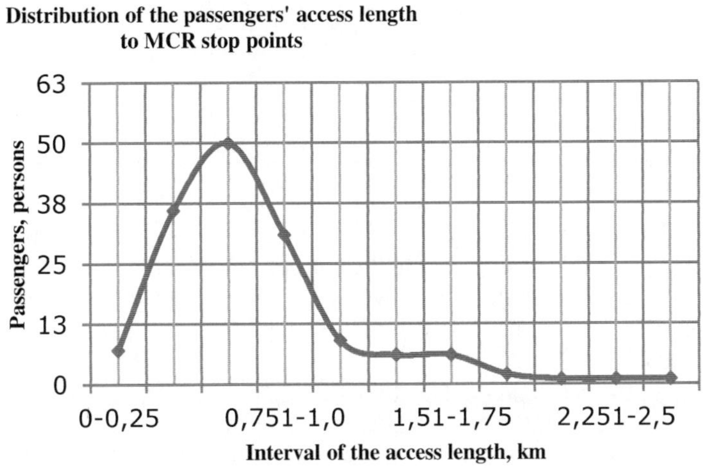


**Figure 1.** Distribution of the passengers' access length to MCC stop points

The performed calculations showed that the curve obtained, actually, is an ordinary distribution curve, which indicates the homogeneity of the resulting population. At the same time, it is obvious that analyzing it by using the law of variation of the individual characteristic values (Pearson's law), the interests of pedestrians following long (in comparison with the average) distances will not be taken into account sufficiently. In this connection, in future we'll use a cumulative curve.

## 4. Experiments

When processing survey data, we have established the routes from the departure points to the stop points of the MCC. This allows us to determine not only their air access distance but also actual distances of the pedestrian road network. The grouped survey results and calculations for the construction of cumulative curves are given in the following Table 2.

**Table 2.** Survey results and calculations

| № | Interval value, km | Air distances | | | Actual walking distances | | |
|---|---|---|---|---|---|---|---|
| | | Number of the passengers, pass. | Rate, % | Cumulative frequency, % | Number of the passengers, pass. | Rate, % | Cumulative frequency, % |
| 1. | 0–0.25 | 7 | 4.67 | 4.67 | 6 | 4.00 | 4.00 |
| 2. | 0.251–0.5 | 36 | 24.00 | 28.67 | 24 | 16.00 | 20.00 |
| 3. | 0.51–0.75 | 50 | 33.33 | 62.00 | 28 | 18.67 | 38.67 |
| 4. | 0.751–1.0 | 31 | 20.67 | 82.67 | 33 | 22.00 | 60.67 |
| 5. | 1.0–1.25 | 9 | 6.00 | 88.67 | 15 | 10.00 | 70.67 |
| 6. | 1.251–1.5 | 6 | 4.00 | 92.67 | 24 | 16.00 | 86.67 |

| № | Interval value, km | Air distances | | | Actual walking distances | | |
|---|---|---|---|---|---|---|---|
| | | Number of the passengers, pass. | Rate, % | Cumulative frequency, % | Number of the passengers, pass. | Rate, % | Cumulative frequency, % |
| 7. | 1.51–1.75 | 6 | 4.00 | 96.67 | 4 | 2.67 | 89.33 |
| 8. | 1.751–2.0 | 2 | 1.33 | 98.00 | 5 | 3.33 | 92.67 |
| 9. | 2.01–2.25 | 1 | 0.67 | 98.67 | 5 | 3.33 | 96.00 |
| 10. | 2.251–2.5 | 1 | 0.67 | 99.33 | 2 | 1.33 | 97.33 |
| 11. | 2.51–2.75 | 1 | 0.67 | 100.00 | 1 | 0.67 | 98.00 |
| 12. | 2.751–3.0 | – | – | – | 2 | 1.33 | 99.33 |
| 13. | 3.01–3.25 | – | – | – | 0 | 0.00 | 99.33 |
| 14. | 3.251–3.5 | – | – | – | 1 | 0.67 | 100.00 |
| | Total | 150 | 100 | | 150 | 100 | |

The obtained cumulative curves are shown in the figure. To determine the maximum length of a comfortable access route, by analogy with the current standards of JV "Urban Development" on the maximum travel time in settlements, we use the requirement of 90 % of the security of the results. The calculations performed show the following:

- the air distance is equal to 1.2 km;
- the actual distances of the pedestrian road network is equal to 1.7 km.

A comparison between the results obtained from the experiments shows that the actual distances are longer than the air line for more than 40 %, which indicates the potential for increasing the standard distances, when implementing the principles of TOD and creation of comfortable and permeable urban environment [4–8].

## 5. Results

The main factors for increasing of the standard access distance by more than 2 times are the following:

- the requirements of JV "Urban Development" are based on outdated researches of the 60s – 70s of the last century;
- intensive implementation of infrastructure, improvement and beautification projects in Moscow causes changes in the paradigm of transport behavior, in particular, in increasing the time and distance of the walking access to stopping points of out-of-street transport;
- six months of operation evidently are not enough for the completed formation of the route network of land passenger transport and the main consumers of MCR services are residents of the adjacent construction;
- relatively high fare in public transport.

The results obtained can be used as for the territorial development, dynamics planning and preparation of regulatory documentation [9], so in scientific research [10, 11]

## References

[1] 2003 *Moscow at the turn of XX–XXI century: collective monograph under the General editorship of Professor A Kuzmin* (Moscow: Moscow textbooks and kartolitografiya) 214 p
[2] Batrakova L 2009 *Theory of statistics* (Moscow: Knorus) 528 p
[3] Samoylov D 1998 *Sociology in urban and regional planning* (Moscow: MSCU) 97 p
[4] Sherbina E, Danilina N and Vlasov D 2015 *Int. J. of Appl. Engin. Res.* **10** p 43131–138
[5] Vlasov D and Danilina N 2013 *Public transport transit hubs system and "Park-and-Ride" facilities* (Lap Lambert Publ. Germany) 81 p
[6] Vlasov D 2009 *Transport transit hubs of big cities (on the Moscow example)* (Moscow: ACV) 96 p

[7] Bakhirev I 2016 *Gradostroitel`stvo* **2 (42)** p 12–19
[8] Mihailov A and Kopylova T 2015 *Vestnik IrGTU* **7 (102)**, p 168–74
[9] Vlasov D 2016 *International Conference on Sustainable Cities* (*ICSC 2016*) **6** Available from: http://dx.doi.org/10.1051/e3sconf/20160601012 [Accessed 6th May 2017]
[10] Vlasov D, Menshutin A and Danilina N 2016 *International Conference on Big Data and its Applications* (*ICBDA 2016*) **8** (Electronic Edition) Available from: http://dx.doi.org/10.1051/itmconf/20160801002 [Accessed 6th May 2017]
[11] Danilina N and Vlasov D 2016 *International Scientific Conference "Integration, Partnership and Innovation in Construction Science and Education* (MATEC Web Conf.) **86** Available from: http://dx.doi.org/10.1051/matecconf/20168605017 [Accessed 6th May 2017]

# Simulation model for assessing the efficiency of a combined power installation based on a geothermal heat pump and a vacuum solar collector

**Ya I Vaysman[1], AA Surkov[2], Yu I Surkova[3] and AV Kychkin[4]**

[1] Doctor of Medical Sciences, professor, supervisor, Institute for the Environmental Protection, Perm National Research Polytechnic University
[2] PhD in Technical Sciences, docent, Institute for the Environmental Protection, Perm National Research Polytechnic University
[3] Senior lecturer, Institute for the Environmental Protection, Perm National Research Polytechnic University
[4] PhD in Technical Sciences, docent, Institute for the Automation Microprocessors, Perm National Research Polytechnic University

e-mail: alex.a.surkov@gmail.com

**Abstract.** The article is devoted to the use of renewable energy sources and the assessment of the feasibility of their use in the climatic conditions of the Western Urals. A simulation model that calculates the efficiency of a combined power installations (CPI) was (RES) developed. The CPI consists of the geothermal heat pump (GHP) and the vacuum solar collector (VCS) and is based on the research model. This model allows solving a wide range of problems in the field of energy and resource efficiency, and can be applied to other objects using RES. Based on the research recommendations for optimizing the management and the application of CPI were given. The optimization system will give a positive effect in the energy and resource consumption of low-rise residential buildings projects.

## 1. Introduction

Despite the wide practice of using renewable energy sources (RES), there is practically no information on the effectiveness of the use of geothermal heat pumps (GHP) and vacuum solar collectors (VCS) in the regions of Russia with a cold climate, including in the conditions of the Western Urals. To solve the problem of objective efficiency assessment of heat supply systems through a combination of heat pump equipment and VCS is considered to be possible due to the development of information technologies, energy monitoring and energy modeling tools. The complex use of RES and the combination of various complementary, i.e. mutually complementary sources on the basis of an integrated simulation assessment of their potential leads to a significant increase in the efficiency of not only individual power installation, but also power systems of different scale. At the same time, the definition of effective modifications for involving RES in the energy balance is a complex problem that has not yet been resolved. At the moment, there are no reliable models for the effectiveness assessing of combined power installations (CPI) based on RES, taking into account the random processes of energy consumption caused by environmental factors. To create such a model is proved to be actual [1, 2].

## 2. Research on the development of a simulation model on the basis of laboratory research module with HTN and VCS

A project for the creation of power efficient stand-alone research module with a building dispatching system was developed and implemented at Perm National Research Polytechnic University jointly with the Ministry of Industry of the Perm Region [3]. The main source of heat supply of the module is GTN Buderus Logatherm WPS 22, consuming up to 9 kW of electric energy and issuing up to 20 kW of thermal energy. To increase the hydraulic stability of the system and to make the operation of the pump more stable, the battery-tank PS 300 is installed. This heat pump works in conjunction with the VSC, forming CPI for the thermal management of the building, Figure 1.

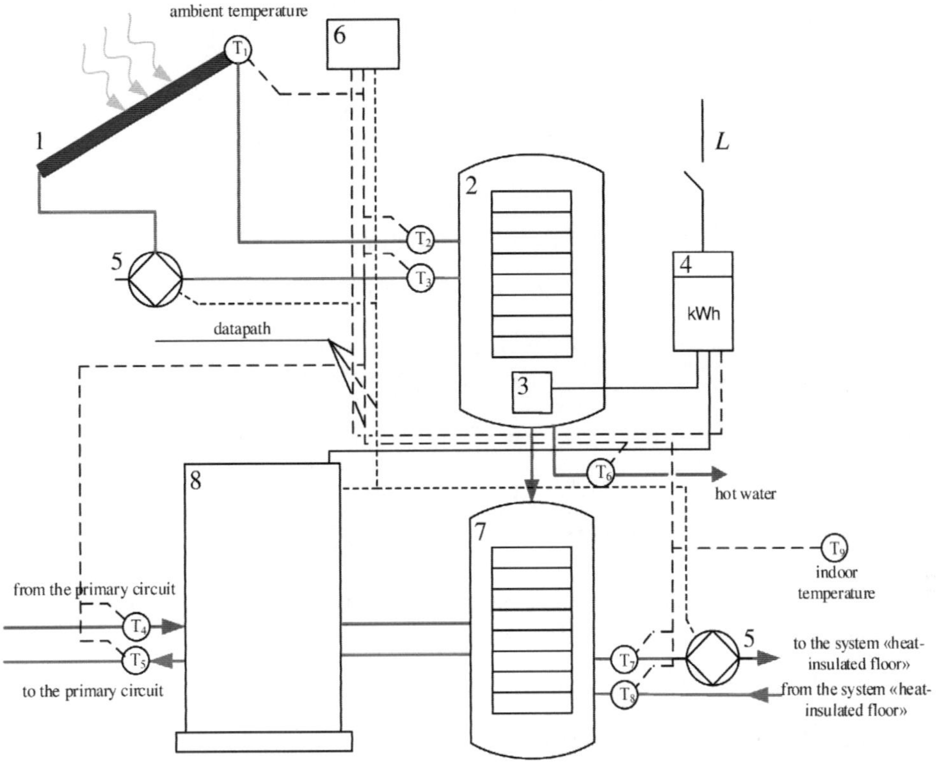


**Figure 1**. CPI scheme: 1 – VSC; 2 – boiler; 3 – thermoelectric heater; 4 – electric energy meter; 5 – pump; 6 – controller; 7 – heat exchanger tank; 8 – heat pump

Control over the operation of the CPI is carried out with the help of multifunction controllers and sensors of the dispatch and automation system that allows continuous monitoring and control of various parameters for assessing the effectiveness of this installation:

- climatic parameters of the environment (temperature, velocity of air, insolation level – $T_1$);
- flow and temperature of the heat-transfer agent along the circuits of the wells and the «heat-insulated floor» ($T_2 – T_8$);
- heat and power resources consumption;
- microclimate control in each placement ($T_9$).

## 3. Mathematical model for the evaluation the efficiency of CPI

The model for evaluating the efficiency of power installations based on integral criteria is most appropriate. Such criteria integrate energy balances estimations, degree of efficiency, unit cost of the produced energy, costs of electric energy per unit of thermal, capital expenditures. Despite the versatility of integrated criteria, their applicability for evaluating the effectiveness of CPI on the basis of GTN and VSC in the climatic conditions of the Western Urals is low. This is mainly due to the climatic features of the region during the heating season, sharp changes in the ambient temperature in the autumn and spring, which are of a random nature. In addition, in the framework of the integral criteria under consideration, the excessive thermal power is accumulated, which is not entirely correct when providing a constantly maintained indoor temperature [4, 5]. In this regard, the following calculated indicators are proposed for the study of the efficiency of the CPI with the GTN and VSC within the heating season $I = 1..N$ based on the simulation model:

- energy provision with thermal power, generated by a heat pump $Q_i^{\mathrm{gl}}$, exceeding the heat load on the premises $Q_i^{\mathrm{p}}$:

$$E_1 = \frac{1}{N}\sum_{i=1}^{N} F_i ; F_i = \begin{cases} 1 & Q_{\mathrm{i}}^{\mathrm{gl}} \geq Q_{\mathrm{i}}^{\mathrm{p}} \\ 0 & Q_{\mathrm{i}}^{\mathrm{gl}} < Q_{\mathrm{i}}^{\mathrm{p}} \end{cases}, \tag{1}$$

- energy provision with thermal power, generated by VSC $Q_i^{\mathrm{g2}}$, exceeding the heat load on the premises:

$$E_2 = \frac{1}{N}\sum_{i=1}^{N} F_i ; F_i = \begin{cases} 1 & Q_{\mathrm{i}}^{\mathrm{g2}} \geq Q_{\mathrm{i}}^{\mathrm{p}} \\ 0 & Q_{\mathrm{i}}^{\mathrm{g2}} < Q_{\mathrm{i}}^{\mathrm{p}} \end{cases}, \tag{2}$$

- energy provision with thermal power, generated by CPI, exceeding the heat load on the premises:

$$E_3 = \frac{1}{N}\sum_{i=1}^{N} F_i ; F_i = \begin{cases} 1 & Q_{\mathrm{i}}^{\mathrm{gl}} + Q_{\mathrm{i}}^{\mathrm{g2}} \geq Q_{\mathrm{i}}^{\mathrm{p}} \\ 0 & Q_{\mathrm{i}}^{\mathrm{gl}} + Q_{\mathrm{i}}^{\mathrm{g2}} < Q_{\mathrm{i}}^{\mathrm{p}} \end{cases}, \tag{3}$$

- absolute indicator characterizing the increase in the energy provision indicator using CPI compared to the energy provision achieved by using a heat pump:

$$\mathrm{E4} = \mathrm{E3} - \mathrm{E1}. \tag{4}$$

To determine $Q_i^{\mathrm{p}}$ it is necessary to use the following:

$$Q_i^{\mathrm{p}} = \mathrm{V} \cdot \Delta \mathrm{T_i} \cdot \frac{\mathrm{K}}{860}, \tag{5}$$

where V – volume of the heated placement of the research module, 600 m$^3$; $\Delta\mathrm{T_i}$ – difference between indoor temperature and ambient temperature (°C) in the *i*-th time; K – coefficient of thermal loss of the building, taken as 1 (range 1.0–1.9 corresponds to the average thermal insulation: standard construction, double brickwork, small number of windows, standard roof); the value 860 is used to convert to kWh.

The thermal power of the geothermal heat pump is chosen from the relation: $Q_i^{\mathrm{gl}} \geq Q_i^{\mathrm{p\_av}}$, where $Q_i^{\mathrm{p\_av}}$ – average value of the thermal losses of the in the *i*-th time.

Then the expression for calculating the thermal power of the pump is the following: $Q_i^{\mathrm{gl}} = Q_i^{\mathrm{gr}} + Q_i^{\mathrm{el}}$, where $Q_i^{\mathrm{gr}} = L \cdot K_i^{\mathrm{gr}}$ – thermal ground energy in the *i*-th time; L – wells length required to collect the required heat output; $K_i^{\mathrm{gr}}$ – value of the specific heat output of the ground in the investigated area is within 30..45 W/m; $Q_i^{el}$ – electric power of the heat pump in the *i*-th time, amounts to ¼ of $Q_i^{gl}$ .

Taking into account sufficient conditions $Q_i^{gl} = Q_i^{p\_av}$, we obtain an expression for calculating the total length of a wells` group:

$$L = \frac{3}{4}\frac{Q^{p\_av}}{K_{gr\_\min}} . \quad (6)$$

The expression for calculating the thermal power produced by the VSC:

$$Q_i^{g2} = \eta_0 S^{g2} E_i , \quad (7)$$

where $E_i$ – insolation in the *i*-th time, W/m²; $S^{g2}$ – VSC square, m²; $\eta_0$ – coefficient of the installation efficiency in normal conditions, for VSC with heat pipes it is experimentally calculated and is 64,9 % [6].

The insolation, as the distribution function of the intensity of the sun's rays can be written as follows: Ei = Ai·CI · sinα, where CI, (W/m²) – solar constant; Ai – coefficient reflecting the percentage of the solar radiation passage as a function of cloudiness which varies from 1 for a clear day to 0.6 for an overcast day; α – angle of sunlight incidence (from 0° to 90°).

The larger the angle of sunlight incidence, the more energy is transferred to the earth's surface. In the north, with each degree of latitude, the degree of midday sun height over the horizon decreases. Knowing the geographical coordinates of Perm: 56.285° E and 58.017° N (in decimal degrees), we obtain the values of the midday sun height: in the spring – 32°; In the summer – 55.5°; In the autumn – 32°; In winter – 8.5°. The collector is fixed on the southern wall of the building at an angle of 45°. This tilt angle is optimal for medium latitudes, first, to absorb maximum solar radiation, and secondly to reduce the amount of snow accumulating on the reservoir itself [7].

## 4. Experimental study of the CPI operation of the research module using simulation methods

According to SP 131.13330.2012 "Construction climatology", the climate of Perm is moderate continental, there are 250 sunny days per year in the territory of the region, and only 29 sunny days are completely clear (in Perm 145 sunny days), the average January temperature is -12,8 °C. The average temperature in July is +18.6 °C.

For each of the months, the duration of the light day is determined, during which the earth's surface in Perm gets direct solar radiation required for the solar collector. In the case of cloudy and overcast days, diffuse (weakened) solar radiation comes to the surface. The values of diffuse radiation were obtained by multiplying the value of direct solar radiation by the coefficient of a cloudy day per hour when there was a weak cloud cover, and for an overcast day – by multiplying by the cloudy day coefficient for each of the hours when there was high cloudiness. In the calculations, the correction coefficient $k_1$ of a cloudy day and $k_2$ a overcast day for each of the months are adopted. For taking into account the change in the angle of sunlight incidence, the coefficients possess large values in the autumn/spring period and smaller in the winter. The solar constant $C_I$ varies with time, as the distance between the Earth and the Sun changes throughout the year (annual variation of 6.9 % – from 1.412 kW/m² in early January to 1.321 kW/m² in early July), solar activity is also changing. Also, the duration of a light day *T (*hour) is taken into account [8].

The statistical estimation of $N_0$ – the number of solar days, $N_1$ – the number of fair days with some clouds, $N_2$ – the number of overcast days and total solar energy by months for Perm was carried out for the heating period 2016–2017, Table 1.

**Table 1.** Climatic parameters of the region for the heating season 2016–2017

| Month | $C_I$, $W/m^2$ | $T_{av}$, °C | $N_0$ | $N_1$ | $N_2$ | *T, hour* |
|---|---|---|---|---|---|---|
| October | 1367 | 1.20 | 4 | 4 | 23 | 10.3 |
| November | 1382 | -3.70 | 2 | 5 | 23 | 7.8 |
| December | 1397 | -10.70 | 1 | 2 | 28 | 6.5 |
| January | 1412 | -14.55 | 3 | 5 | 23 | 7.4 |
| February | 1397 | -11.05 | 3 | 4 | 21 | 9,5 |
| plan for March (forecast) | 1382 | -6.30 | 1 | 3 | 27 | 12 |
| plan for April (forecast) | 1367 | 3.25 | 4 | 7 | 19 | 14.5 |

Taking into account the statistics of the ambient temperature, the thermal losses of the building *K* and the added random variation of the parameters depending on the type of construction and the placement insulation required to maintain the operating temperatures, the amount of necessary heat energy for the period from October, 2016 to April, 2017, Figure 2. The figure shows the volumes of heat losses of the building at temperatures held indoors: 1 – 21 °C; 2 – 18 °C; 2 – 15 °C; 2 – 12 °C; 2 – 9 °C.

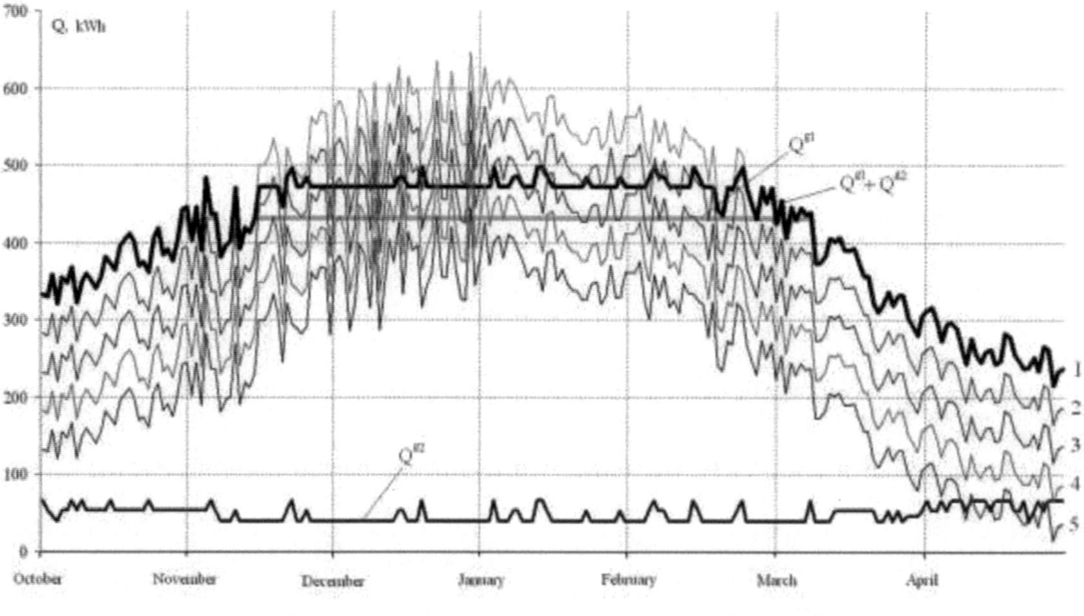


**Figure 2.** Visualization of the simulation results of the heat and power balance of the research module with full heating of 100 % of the building area

The power of the heat pump is determined based on the average thermal loss of the building $Q^{p\_av}$ = 18 kW. To create a comfortable microclimate in the building within heating period of 212 days, it is necessary to following volume of heat energy:

$$\sum_{i=1}^{229} Q_i^p = 212 \cdot Q^{p\_av} \cdot 24\ (\text{ч}) = 91584\ \text{kWh}.$$

Taking into account the heat pump's consumption of electric power, which is approximately ¼ of the heat output, Buderus model of 20 kW was chosen. The consumed electric power is 5 kW, therefore the required value of heat power of 15 kW can be provided by a geothermal source – a set of wells with the total length according to the expression (6) of at least L = 500 m. Figure 2 shows the graphs of the heat energy generated by the heat pump $Q^{g1}$.

From October to mid November and from March to the end of April, the heat energy $Q^{g1}$ completely covers the building's losses while maintaining a comfortable temperature of 21 °C. However, within 3.5 months of this energy is not enough, for example: during December, the heat pump can provide only 9 °C indoors.

The power of the VSC $Q^{g2}$ is determined according to the expression (7). With an installation area of 10 m$^2$, we obtain an average power $Q^{g2}$ = 1.86 kW. Consequently, this collector produces about 44.77 kWh of heat per day, see Figure 2. According to the meter, the volume of the consumable coolant (*Vc*, l), which is heated by the collector during the day, was determined. To increase the temperature of 1 l of coolant by 1 degree approximately 1.16 W of power is required. Then, we find that the heat energy produced by VSC will be sufficient to heat 1 m$^3$ of the heat-transfer agent of the system «heat-insulated floor» up to 38 °C [9].

Figure 2 clearly shows the periods when the heat power of the CPI exceeds the thermal load of the building, and when not.

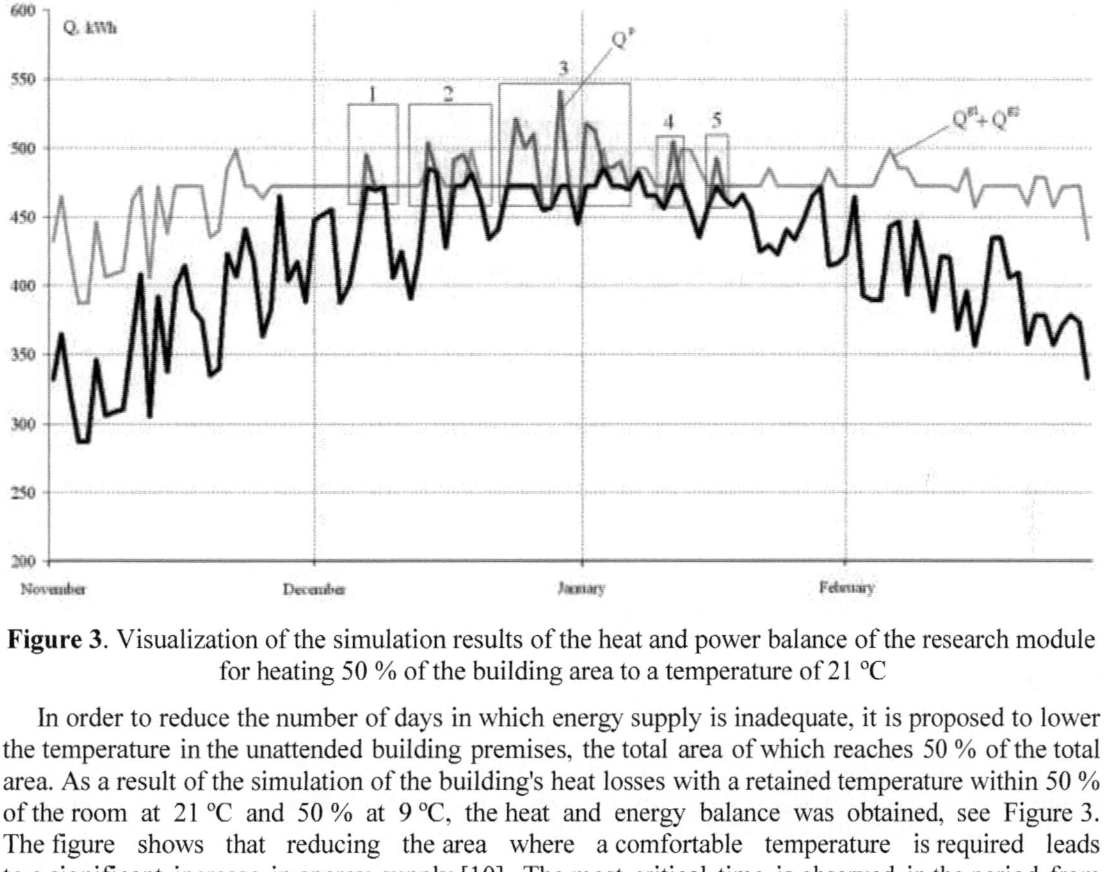


**Figure 3**. Visualization of the simulation results of the heat and power balance of the research module for heating 50 % of the building area to a temperature of 21 °C

In order to reduce the number of days in which energy supply is inadequate, it is proposed to lower the temperature in the unattended building premises, the total area of which reaches 50 % of the total area. As a result of the simulation of the building's heat losses with a retained temperature within 50 % of the room at 21 °C and 50 % at 9 °C, the heat and energy balance was obtained, see Figure 3. The figure shows that reducing the area where a comfortable temperature is required leads to a significant increase in energy supply [10]. The most critical time is observed in the period from the second half of December to the middle of January. Intervals 1 to 5 are observed where the building's losses exceed the heat power of the CPI. In this case, the additional heating of the heat-transfer agent must be carried out by means of an electric heater.

Taking into account the expressions (1)–(4), an estimation of the CPI efficiency was made, see Figure 4. The diagram shows the variation of indicators $E_1$–$E_4$ depending on the required temperature in the building (model situations 1–7) and the results of simulation of the generated heat capacities and building losses for the heating period 2016–2017, which consists of 212 days. Model situation 1 corresponds to maintaining the indoor temperature equal to 9 °C; Situation 2 – temperature equal to 12 °C; Situation 3 – temperature equal to 15 °C; Situation 4 – temperature equal to 18 °C; Situation

5 – temperature equal to 21 °C; Situation 6 – a 50 % area temperature is 21 °C and another 50 % area temperature is °C; Situation 7 is a 50 % area temperature is 18 °C and another 50 % area temperature of 9 °C.

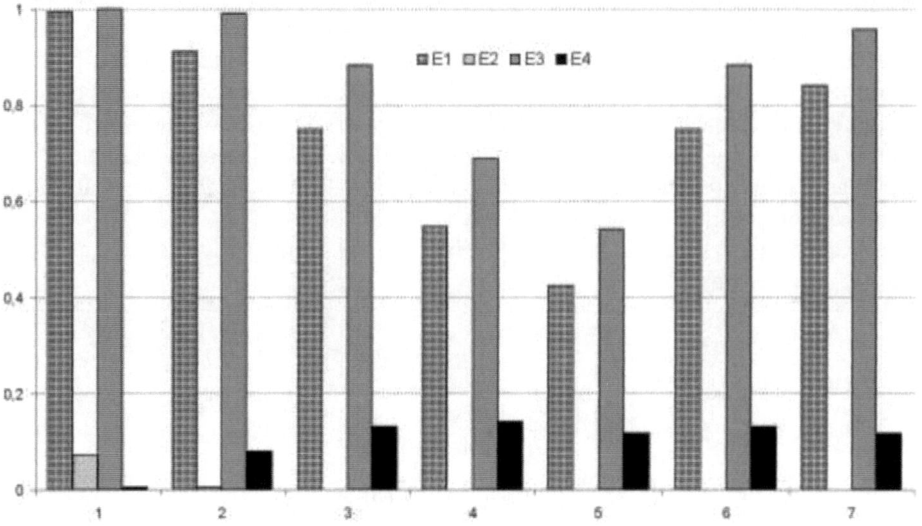


**Figure 4**. Visualization of calculation results of CPI` efficiency indicators

Despite the low efficiency of independent use of VSC – $E_2$, it provides a good efficiency gain in the structure of the CPI – $E_4$, reaching 16 %.

## 5. Conclusion

The developed simulation model allows solving a wide range of problems in the field of energy and resource efficiency, and also can be applied to other objects using RES.

Based on the simulation results, the following recommendations can be made:

- to reduce the amount of heat losses in the building under study, it is necessary to reduce the glazing area of the object and to apply the more efficient insulation;
- the dispatch and automation systems for the existing building allows efficient distribution of heat energy in residential and business premises, while reducing the need for heat power, and therefore in the electrical energy required for the heat pump;
- the model permits to optimize the control of the CPI.

The calculations show that the application of GHP and the VSC in the climatic conditions of the Western Urals is technically and economically feasible. GHP allows reducing the energy consumption of a residential building and can be used as an alternative to traditional sources of heat supply, including gas boilers, in areas where there is no centralized heat and gas supply system. In conditions of the middle and central Russia, solar water heating systems can be effectively used by various consumers for household purposes for 6–8 months a year (March/April – September/October).

## Acknowledgement

This work was supported by PNRPU grant, agreement No. 2016/PI-2 "Development of a methodology for monitoring and utilization of the heat flows as low-grade resource of the enterprise".

## References

[1] Claridge D E 1998 A Perspective on Methods for Analysis of Measured Energy Data from Commercial Buildings *Journal of Solar Energy Engineering* **120** (3)

[2] Jacobsen F R 1985 Energy signature and energy monitoring in building energy management systems *Proceeding of CLIMA 2000 World Congress vol. 3: Energy Management*

[3] Surkov A and Vaisman J 2014 The research module for an assessment of effectiveness of application of energy-saving technologies in housing and communal services *Bulletin of the Perm national research polytechnical university Applied ecology Urbanism* **1** (13) p 64–72

[4] Lyakhomskii A V, Perfil'eva E N and Kychkin A V, Genrikh N A 2015 A software-hardware system of remote monitoring and analysis of the energy data *Russian Electrical Engineering* **86 (6)** p 314–19

[5] Seem J E 2007 Using intelligent data analysis to detect abnormal energy consumption in buildings *Energy and Buildings* **39 (1)** p 52–58

[6] Prerez-Lombard L, Ortiz J and Pout C 2008 A review on buildings energy consumption information *Energy and Buildings* **40 (3)** p 394–98

[7] Faizrakhmanov R A, Frank T, Kychkin A V and Fedorov A B 2011 Sustainable energy consumption control using the My-JEVis energy management data system *Russian Electrical Engineering* **82 (11)** p 607–11

[8] Kychkin A V 2016 Synthesizing a System for Remote Energy Monitoring in Manufacturing *Metallurgist* **59 (9–10)** p 752–60

[9] Kychkin A V and Mikriukov G P 2016 Applied data analysis in energy monitoring system *Regional Energy Problems* **2 (31)** p 84–92

[10] Il'in R A 2010 Algorithm of assessment of efficiency at designing and usage of heat power installations of different types *Vestnik of Astrakhan state technical university Series: marine engineering and technologies* **2** p 79–82

IOP Conf. Series: Earth and Environmental Science **72** (2017) 012019 doi:10.1088/1755-1315/72/1/012019

# Optimization of Power Distribution Networks in Megacities

**V Z Manusov[1], P V Matrenin[1], J S Ahyoev[1] and L Sh Atabaeva[2]**

[1] Department of Power Supply System, Novosibirsk State Technical University, Novosibirsk, Russia, 630073
[2] Department of Foreign Language, Novosibirsk State Technical University, Novosibirsk, Russia, 630073

E-mail: javod_66@mail.ru

**Abstract**. The study deals with the problem of city electrical networks optimization in big towns and megacities to increase electrical energy quality and decrease real and active power losses in the networks as well as in domestic consumers. The optimization is carried out according to the location selection and separate reactive power source in 10 kW networks of Swarm Intelligence algorithms, in particular, of Particle Swarm one. The problem solution based on Particle Swarm algorithm is determined by variables being discrete quantities and, in addition, there are several local minimums (troughs) to be available for a global minimum to be found. It is proved that the city power supply system optimization is carried out by the additional reactive power source to be installed at consumers location reducing reactive power flow, thereby, ensuring increase of power supply system quality and decrease of power losses in city networks.

## 1. Introduction

Optimization problems of city electric networks include the definition of steady-state electrical network mode where technical constraints on electrical energy quality have been sustained. Three basic problems of optimization have been highlighted in the distribution networks where each problem requires its own optimization method: selection of optimal allocation and source power of reactive power, optimization of transformation coefficients and selection of optimal points of circuits' interruption. The optimization takes into consideration the voltage deviation constraints in all nodes including load nodes with no controlling means available per reactive power of sources generated and per currents controlled within the lines [1–2].

Reactive power optimization is defined as the best location of installation and selection of compensating units power value (CU). CU comprises of a capacitor bank, electric switchgear, protection and controlling means.

With partial or complete reactive power compensation consumed by current-using equipments the current and the voltage drop in power lines is decreased and at the same time real power losses are also decreased in the network.

## 2. Statement of Problem

The function analysis of power supply systems in big cities and megacities shows that their essential part consists of distribution networks of 10–35 kW, and in many developing countries the networks have 20 kW voltages [3]. Significant influence on sustainable development of city electrical networks has the effectiveness increase and the quality of city power supply system functioning. Thus, the urban dweller's quality life index is determined by three factors: ecological environment, education and health care services and technical support comfort of vital communication functions including quality and accessibility of electrical and thermal energy. The vital importance of sustainable urban development is power consumption increasing of not only real but active power, which is necessary for domestic electrical appliances to be operational. The operating principal is based on magnets fields' applying. Reactive power consumption in urban area is conditioned by three factors:

- Rapid growth of urban population and urban settlements' area.
- Increasing number of electrical appliances (with electrical engines) used by population, in particular, washing machines, fridges, air-conditioners, dishwashers, vacuum-cleaners, fans, pumps and others.
- Significant increase of city Power Transmission Lines (PTL) length corresponds to the urban expansion area what increases real and reactive power losses while transporting electrical power at a distance.

This problem, in its turn, is resolved by two necessary solutions, in particular, selection of optimal nodes of compensating units allocation with minimal of the reduced costs and selection of optimal values of reactive power in operating condition including minimal losses. In this case, the CU's installed capacity takes the form of a constraint.

The attempt of taking two decisions is undertaken in this study with their further combination made. The first stage considers the calculation of steady-state mode with parameters set in the network and loads. The second one is used by the Swarm Intelligence algorithm where at each stage the access towards the objective function value is implemented being defined by the first problem. Genetic algorithms do not guarantee much better solution to be found but they usually allow getting "fast enough" of "reasonable" problem solution. It is the composition of these two problems solution which introduces the basic research novelty of the study presented [4–5].

## 3. Mathematical Model of Problem

Currently, in order to solve problems related to sustainable development of big cities and megacities, the more attention is paid to the energy savings and to production processes efficiency increasing. Thus, the technical and economic performance of city power grid enterprises has been significantly improved being resulted in the tariff reduction of electrical power and other energy carriers. It, in its turn, reduces harmful emissions produced by power stations and, thereby, the quality of ecological environment is improved. Power losses decrease contributing to the tariff reduction improves economic viability of urban dwellers and reduces the poverty alleviation countrywide.

To decrease electric power losses the following methods are available:

- Design and application of energy savings technologies;
- Allocation optimization and power sources selection of reactive energy;
- Smart grid design using new principles, innovations and process control.

The selection of mathematical model to solve the optimization problem of power supply for the cities to be sustainably developed is defined by the fact that the problem has discrete variables and can contain several local minimums out of which the selection of the global one is crucial. Widely applied gradient methods are not suitable for this task as they require function derivation. This requirement is not observed in this task, moreover, for big cities with big quantity of adjunctions the system of equations has high dimension and under sustainable city development the "curse of dimensionality" is being increased in quadratic degree. Due to this, more effective method of discrete optimization was chosen in this study with the possibility of defining of global minimum – the Swarm Intelligence method.

The aim of optimization is the minimization for real power losses and for CU costs. Thus, power factor tgφ for grids with 10 kW is limited by 0.4 variables. As the result, the optimization problem is formulated:

$$W(Q) = Z_{\Delta P}(Q) + Z_{CU}(Q) \rightarrow \min$$

where $Z_{\Delta P}$ is financial losses of real power , $Z_{CU}$ is financial costs provided to CU installation and defined as:

$$Z_{\Delta P} = C_p \cdot \tau \cdot \Delta P,$$

with the following constraints: $0 \leq \text{tg}\varphi \leq 0,4$
where $Q$ is a CU power vector;
$Q_i$ is CU power in n-node (if $Q_{maxi} = 0$, CU is not put down in n-node);
$n$ – is a number of nodes which can CUs can be installed in;
$\tau$ – time period/interval of system operation considered per hour/annum corresponds to 8760 hours;
$\Delta P_\Sigma$ – is total active power losses in the network;

In order to solve the problem the fragment of megacity's city distribution network has been selected consisting of 36 – nodes and 48 – arms (Figure 1).

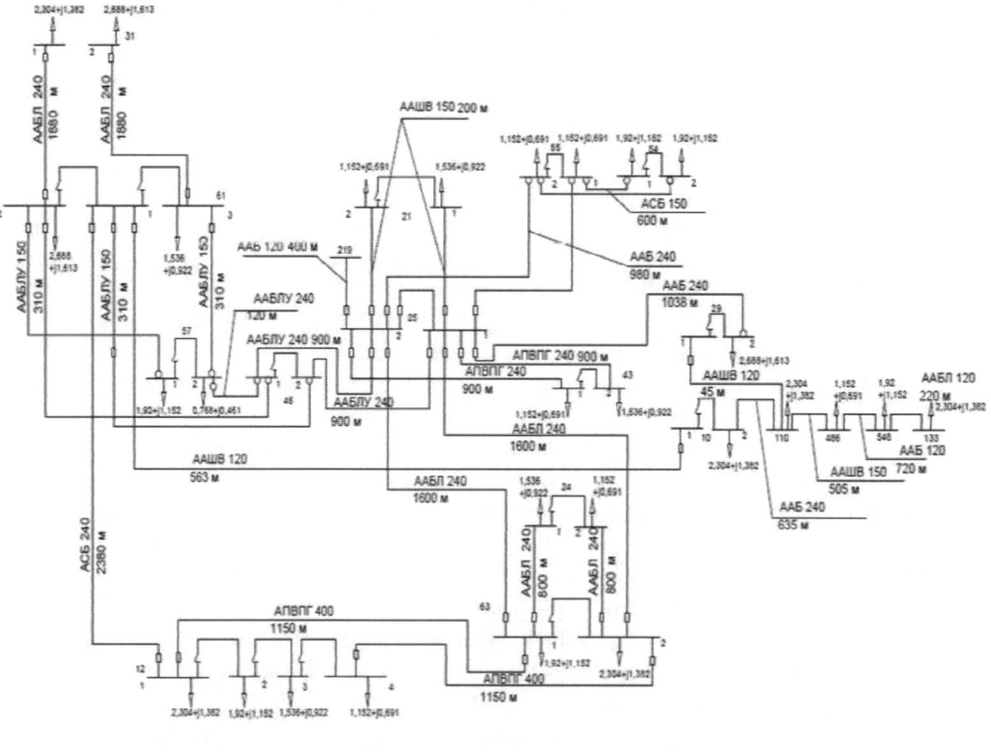


**Figure 1**. Fragment of Electric Distribution Network

## 4. Particle Swarm Algorithm

The following regulations have been introduced:

- Particles existing in the world where time is discrete;
- Particles estimate their position by means of fitness-function;
- Each particle knows its position in the area where it has found the most amount of food (its best position);
- Each particle knows its position in the area where the most amount of food among all positions is and where all particles have been available (the best common position);
- Particles strive to take the best positions where they have been already themselves and occupy the best common position;

- Particles, at random, change their speed in such way that the described tendency defines only average motion of particles;
- Particles have inertia and that is why their speed is dependable each time on the previous time speed;
- Particles cannot abandon the limited searching area.

The main idea of the method is in particles movement in the space of solutions. Let the problem of finding the minimum (maximum) of a function $f(X)$ is solved, where X is a vector of different variables that can take values from some area of D. Then each particle at any time is specified by value of X parameters from D area (points data in solution space) and value of optimized function $f(X)$ (being attractive to this point). In this case, the particle "remembers" the best point being in the solution space and strives to come back but it should obey the law of inertia tending towards some stochastic variation of movement direction. Nevertheless, these rules are not enough to be transformed to the system as the links among elements have not been set. As a link, so called, general memory is used due to which each particle is aware of the best point coordinates being among all ones where any particle has been earlier. Finally, it is striving to be engaged in the best position among all the particles available being influenced by the particle movement among other particles, inertia, and random deviations [6–7].

The algorithm is completed at reaching of certain allocated number of iterations either with reaching of the satisfactory solution or upon completion of the operation after certain period of time. Consequently, algorithm can be written as follows:

1. To randomly distribute particles in the solution area and set zero initial speeds.
2. Optimized function values with each particle renewal if local and global best solutions are required.
3. To calculate new values of speeds per each particle.
4. To calculate new coordinates of particles.
5. If the termination condition is satisfied, the algorithm should be completed: otherwise, step 2 needs to be taken.

The result of algorithm operation is the best global solution.

According to the formula, the Particle Swarm algorithm is $PSO = \{S, M, A, P, I, O\}$.

1. Set of agents (particles) $S = \{s_1, s_2, ..., s_{|S|}\}$, $|S|$ is a number of particles . On $j$-iteration i particle is specified by the state of $s_{ij} = \{X_{ij}, V_{ij}, X^{best}_{ij}\}$, where $X_{ij}=\{x^1_{ij}, x^2_{ij}, ..., x^l_{ij}\}$ is a vector of variable parameters (particle's position), $V_{ij}=\{v^1_{ij}, v^2_{ij}, ..., v^l_{ij}\}$ – is a vector of particle speeds, $X^{best}_{ij}=\{b^1_{ij}, b^2_{ij}, ..., b^l_{ij}\}$ is better in value of fitness-function of the particle position among all the positions which was taken by the algorithm operation from the 1st to $j$ iterations, $l$ is a number of variable parameters.

2. The vector $M = X^{best}j$ is the best value of variable parameters vectors which has been obtained among all the particles from the 1st to $j$-iteration of algorithm. This vector provides the indirect experience exchange among the particles.

3. A algorithm describes the Particle Swarm functioning. There are different modifications of this algorithm. Further, the basic algorithm description is provided.

3.1. Generation of initial positions and speeds are (j = 1):

$$X_{i1} = rand(G(X)), i = 1, ..., |S|,$$

where r and $(G(X))$ – are a vector of equally distributed random values meeting the constraints onto the searching area;

$$V_{i1} = rand(-V_{\max}, V_{\max}), i = 1, ..., |S|,$$

where r and ($-V_{max}$, $V_{max}$) – are a vector of equally distributed random values in the range of ($-V_{max}$, $V_{max}$).

$$X_{i1}^{best} = X_{ij}, i = 1, \ldots, |S|.$$

The best position is selected on a random basis (while calculating fitness-functions the real best position will be defined):

$$X_1^{best} = X_{11}.$$

3.2. Fitness-function calculation and definition of the best position are as follow:

$$X_{ij}^{best} = X_{ij}, \phi(X_{ij}^{best}) < \phi(X_{ij}), i = 1, \ldots, |S|,$$

$$X_j^{best} = X_{ij}, \phi(X_j^{best}) < \phi(X_{ij}), i = 1, \ldots, |S|.$$

The calculation of $\varphi(X) = f(X)$ takes place in the external environment through data exchange with feedback of $(I_{oc}, O_{oc})$.

3.3 Particles' movement:

$$V_{ij+1} = V_{ij}\omega + \alpha_1\left(X_{ij}^{best} - X_{ij}\right)rnd_1 + +\alpha_2\left(M - X_{ij}\right)rnd_2, \; i = 1, \ldots, |S|,$$

$$V_{ij+1} = \begin{cases} V_{ij+1}, -V_{\max} \le V_{ij+1} \le V_{\max} \\ -V_{\max}, V_{ij+1} \le -V_{\max} \\ V_{\max}, V_{ij+1} \ge V_{\max} \end{cases}, i = 1, \ldots, |S|,$$

$$X_{ij+1} = \begin{cases} X_{ij} + V_{ij+1}, G(X_{ij} + V_{ij+1}) = 1 \\ X_{ij}, G(X_{ij} + V_{ij+1}) = 0 \end{cases}, i = 1, \ldots, |S|,$$

where $rnd_1$ and $rnd_2$ are random numbers equally distributed in the interval [0,1), $G(X)$ is used as the predicate showing whether the $X$ is belonged to the $D$ area of possible values.

3.4. If on j iteration the termination condition is fulfilled, the value $X_{final}^{best} = X_j^{best}$ is supplied to the output $O_1$. Otherwise, the transition towards the iteration 2 is taken place.

4. The vector $P = \{\alpha 1, \alpha 2, \omega\}$ are coefficients of $A$ algorithm which are used in the formula and correspondingly influences the particles' movement in searching area. Correspondingly, $\alpha_1$ and $\alpha_2$ coefficients define the influence the agent's individual and group experience. The coefficient ω specifies the inertial properties of particles.

5. The $I$ and $O$ identifiers, which have described earlier the input and output of swarm, are not dependent on the Swarm Intelligence algorithm implementation.

## 5. Electrical Network Optimization

The optimization is carried out by means of algorithm based on the Swarm Intelligence, the optimal distribution of reactive power sources is obtained and the reactive power value of the selected nodes is obtained as well [8]. The parameters are set for the optimization to be required: the cost of electricity for 1 kW/hour is $\beta = 2.6$ rub, the cost for 1 kVAr of the installed capacity is $Q = 550$ rub. The obtained values of the reactive power are tabulated in 1. Total losses of real power are $\Delta P = 0.888$ MW. The required value of the compensated reactive power is $Q = 7.886$ MVAr.

**Table 1.** Reactive distribution and value

| Node | Q MVAr |
|---|---|
| 133 | 1.244 |
| 241 | 0.830 |
| 242 | 0.622 |
| 571 | 1.037 |
| 572 | 0.264 |
| 541 | 0 |
| 542 | 0 |
| 291 | 0 |
| 121 | 1.194 |
| 311 | 1.244 |
| 312 | 1.452 |

## 6. Operational Task for Electrical Network Optimization

The distinctive feature of project problem being considered in the sub-section 3.3 presents the constraints being imposed on CU, in particular, on the value of the installed capacity [9]. The constraints set have been presented in the Table 2. If the value is $Q_{cu\ p}<Q_{max.est}$, the optimization program in the assigned node sets up the value being equal to the reactive power $Q_{cu}=Q_{cu\ p}$, otherwise with $Q_{cu}>Q_{max.est}$ towards the assigned node, the maximum reactive power is set up equaling to $Q_{cu}=Q_{max.est}$.

**Table 2**. Maximum installed reactive powers

| Node | $Q_{cu\ p}$, MVAr | $Q_{max.est}$, MVAr |
|---|---|---|
| 133 | 1.244 | 1.250 |
| 241 | 0.830 | 0.850 |
| 242 | 0.622 | 0.600 |
| 571 | 1,037 | 1.000 |
| 572 | 0.264 | 0.400 |
| 541 | 0 | 0 |
| 542 | 0 | 0.700 |
| 291 | 0 | 0 |
| 121 | 1.194 | 1.200 |
| 311 | 1.244 | 1.250 |
| 312 | 1.452 | 1.450 |

Real $tg\varphi_{nat}$ of load and $tg\varphi_{opt}$ after optimization are calculated

$$tg\phi_{nat} = \frac{Q_n}{P_n},$$

After optimization:

$$tg\phi_{opt} = \frac{Q_n - Q_{cu}}{P_n},$$

For example, for the node 133, where $P_n$ = 2.314 MW, $Q_n$ = 1.382 MVAr, $Q_{cu}$ = 1.244 MVAr,

$$tg\phi_{nat} = \frac{1.382}{2.304} = 0.59,$$

$$tg\phi_{opt} = \frac{1.382-1.244}{2.304} = 0.059$$

The other calculations have been presented in Table 3. Total losses of active power are $\Delta P$ = 0.883 MW. The required value of the compensated reactive power is $Q$ = 8.667 MVAr.

Thus, in the conditions of real maintenance and presented constraints per compensated devices, the real power total losses increase by 0.001 MW, i.e., by 1 kW. In this case, the power of mounted batteries is increased with rounding of numbers of their power towards down by 106 кVAr. Thus, the optimization results confirm the correctness of the problem solution in the project formulation.

**Table 3**. Optimization results by nodes

| Node | $P^n$, MW | $Q_n$, MVAr | $Q_{cu}$, MVAr | $tg\varphi_{nat}$, r.u | $tg\varphi_{opt}$, r.u |
|---|---|---|---|---|---|
| 133 | 2.304 | 1.382 | 1.244 | 0.59 | 0.05 |
| 241 | 1.536 | 0.922 | 0.830 | 0.6 | 0.04 |
| 242 | 1.152 | 0.691 | 0.600 | 0.59 | 0.08 |
| 571 | 1.92 | 1.152 | 1.000 | 0.6 | 0.08 |
| 572 | 0.768 | 0.461 | 0.400 | 0.6 | 0.08 |
| 541 | 1.92 | 1.152 | 0 | 0.6 | 0.6 |
| 542 | 1.92 | 1.152 | 0.700 | 0.6 | 0.23 |
| 291 | 0 | 0 | 0 | 0 | 0 |
| 121 | 2.304 | 1.382 | 1.200 | 0.59 | 0.08 |
| 311 | 2.304 | 1.382 | 1.250 | 0.59 | 0.08 |
| 312 | 2.688 | 1.613 | 1.450 | 0.6 | 0.06 |

## 7. Conclusion

1. The sustainable development of cities and megacities, by all means, requires a new approach in the task of increasing technical and economic performance of city power grid enterprises and power supply system at large. The quality of city power supply system functioning must correspond to the improvement of technical comfort of urban habitat. Due to it, the city development is accompanied by significant changes of domestic electrical appliances' composition and requirement towards the quality of electrical power being necessary for their operation.

2. The study shows that one of the abovementioned conditions for technical development becomes an innovation task for selection of optimal placement reactive power compensating units and further optimization exploitation of values selection providing the reactive power into the network which controls the city electrical network modes. To reduce power losses in power networks and to meet the quality criteria of electrical power on voltage deviation the concrete values of reactive power source values are being optimally selected within daily time interval.

3. The optimization was carried out by Swarm Intelligence algorithm as the problem had discrete functions and it was necessary to find global minimum of target function.

It is proved that as the result of optimization the best variant on minimum of losses requires CU' installation with power of Q = 8.773 MVAr. Total real power losses are reduced from 1.063 to 0.882 MW making 7.7 %. The payback period of this exercise makes 1.6 years. Thus, the reactive power compensation in city power networks is quite efficient, cost effective and significantly improves technical and economic performance of electrical grid enterprises and promotes the sustainable development of cities.

## References

[1] Glazunov A A, Leschinskaya T B and Shvedov G V 2005 *Multicriterial optimization of deep input parameters in city power supply system based on the development uncertainty of power load* (Moscow: Agrokonsalt) 111 p

[2] Matrenin P V, Grif M G and Sekaev V G 2016 *Stochastic optimization Methods* (Novosibirsk) p. 63

[3] 2012 *Reference book on electrical networks design* ed by Faibisovich D (Moscow: ENAS) p. 376

[4] Manusov V Z, Kirgizov A and Ahyoev J 2016 Construction and optimization of a power complex with a distributed generation on the basis of renewables and methods of artificial intelligence (on the example of the Republic of Tajikistan) *E3S Web of Conferences* **6** article number 03006 Available from: https://doi.org/10.1051/e3sconf/20160603006 [Accessed 18th April 2017]

[5] Pavlyuchenko D and Shevtsov D 2013 Reactive power compensation using fuzzy logic controlled capacitors banks *The 8 international forum on strategic technologies* (IFOST 2013) (Mongolia, Ulaanbaatar) **2** p 563–66

[6] Manusov V Z, Matrenin P V and Tretiakova E S 2016 Implementation of Population Algorithms to Minimize Power Losses and Cable Cross-Section in Power Supply System *International Journal of Electrical and Computer Engineering* (IJECE) **6 (6)** p 2955–61 Available from: http://dx.doi.org/10.11591/ijece.v6i6.11361 [Accessed 18th April 2017]

[7] Kennedy J and Eberhart R C 1995 Particle Swarm Optimization *Proc. of IEEE International Conference on Neural Network* (Piscataway, NJ) p 1942–48

[8] Manusov V Z and Sodnomdorzh D 2006 *Problem-solving Methods modes of electric power systems under uncertainty* (Ulaanbaatar: ADMON Publ) 200 p

[9] Guzhov N, Ol'khovskii V and Pavliuchenko D 2006 *Power supply systems* (Novosibirsk: NSTU Publ) 154 p

IOP Conf. Series: Earth and Environmental Science **72** (2017) 012020 doi:10.1088/1755-1315/72/1/012020

# Radon safety in terms of energy efficiency classification of buildings

**A Vasilyev[1], I Yarmoshenko[2] and M Zhukovsky[3]**

[1] Researcher, IIE UB RAS, Ekaterinburg, Russia.
[2] Vice-director, IIE UB RAS, Ekaterinburg, Russia.
[3] Director, IIE UB RAS, Ekaterinburg, Russia.

E-mail: vav@ecko.uran.ru

**Abstract**. According to the results of survey in Ekaterinburg, Russia, indoor radon concentrations above city average level have been found in each of the studied buildings with high energy efficiency class. Measures to increase energy efficiency were confirmed to decrease the air exchange rate and accumulation of high radon concentrations indoors. Despite of recommendations to use mechanical ventilation with heat recovery as the main scenario for reducing elevated radon concentrations in energy-efficient buildings, the use of such systems did not show an obvious advantage. In real situation, mechanical ventilation system is not used properly both in the automatic and manual mode, which does not give an obvious advantage over natural ventilation in the climate of the Middle Urals in Ekaterinburg. Significant number of buildings with a high class of energy efficiency and built using modern space-planning decisions contributes to an increase in the average radon concentration. Such situation contradicts to "as low as reasonable achievable" principle of the radiation protection.

## 1. Introduction

Despite of elaborated energy efficiency policy in construction industry, a number of issues of the radiation safety of construction are still unresolved due to the introduction of energy-efficient solutions in this sector [1–3]. The problem of the increased accumulation of pollutants in buildings, in particular the problem of indoor radon, is of current interest. Radon is a naturally occurring radioactive gas found in soil and building materials. According to the epidemiological data [4, 5], there is a direct relationship between the lung cancer and indoor radon. Radon is able to migrate and accumulate in buildings. Features of design and operation of various buildings can result in high radon concentrations.

In Russia, the requirements for energy-efficient construction have been established by the government since 1996. In the following years, new standards have been introduced to regulate the resistance to heat loses and air infiltration aimed to decreasing energy consumption. The establishment of new standards led to the application of building solutions that create conditions for increased accumulation of pollutants in new buildings. These solutions include the heat-shielding of building envelope, replacing obsolete doors and windows in dwellings, special requirements for plan decisions for rooms that are part of common household use. This work is focused on the problem of radon exposure in modern energy-efficient buildings in Ekaterinburg, where the energy consumption potential being extremely high. The research focuses on the influence of space-planning decisions of modern multi-storey buildings of various energy efficiency classes on radon concentration in dwellings.

## 2. Materials and methods

The radon survey data [6] supplemented by measurements of radon concentration in energy-efficient buildings in Ekaterinburg during the last 5 years was used in the analyses. The total number of buildings in the combined sample was 452. Measurements of the radon concentration were carried out by the track detectors (LR 115). The measurement period was 2–3 months.

To determine the average annual radon concentration, the temperature normalization was carried out when it was necessary [6]. Assigned energy efficiency class, the living and total area, and the number of rooms were found out for each dwelling. According to the data of the companies managing housing and communal services, the total number of buildings with assigned energy efficiency class was 65. The energy efficiency class of a building in Russia is determined by the deviation of the value of the specific heat consumption for heating, ventilation and hot water supply of the building from the standardized level (Table 1).

**Table 1.** Energy-efficiency buildings classification in Russia

| Class | | Deviation of the value of the specific heat consumption from the standardized level, % | Recommendations |
|---|---|---|---|
| Name | Symbol | | |
| In the case of design and operation of new, reconstructed, modernized buildings | | | |
| Very high | A+ | below -60 | Economic incentives |
| | A | from -45 to -59,9 | |
| High | B++ | from -35 to -44,9 | Economic incentives depending on the year of construction |
| | B+ | from -25 to -34,9 | |
| | B | from -10 to -24,9 | |
| Normal | C | from +5 to -9,9 | – |
| In the case of operation of existing buildings | | | |
| Reduced | D | from +5,1 to +50 | Modernization of the building after 2020 is advisable |
| Low | E | more than +50 | Immediate insulation of the building |

More detailed study was conducted in one dwelling with the highest energy efficiency class "A". This building is distinguished by absence of the water heating system and presence of the mechanical ventilation system with heat recovery. Heating in the house is carried out by means of warm electric floors. In this building, measurements of the radon concentration and the temperature difference between indoor and outdoor atmosphere were carried out in an apartment on the first floor for 1.5 months in step of 1 hour.

The AlphaGUARD radon monitor was used as a measuring instrument. The measurements were carried out in the cold season (January to March 2017). The temperature difference between the indoor and outdoor atmosphere varied from 25 to 45° C. It allowed us to determine the dominant radon entry mechanism and radon entry rate by methods developed earlier [7].

In addition, the methods mentioned allow to determine the air exchange rate in the stationary mode of room use. The air exchange rate in the active mode of room use was adopted equal to one exchange per hour, according to the characteristics of the ventilation system.

## 3. Results and discussion

Based on the data obtained during the analysis, a diagram showing the distribution of radon concentration values in the dwellings of Ekaterinburg, depending on the energy efficiency class was obtained (Figure 1). According to the achieved results, there is a significant increase in radon concentration in the rooms situated in buildings with high energy efficiency class (B, B+, B++).

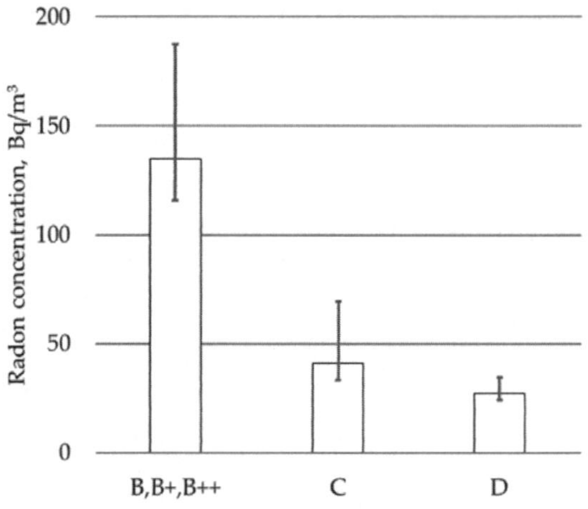


**Figure 1.** Distribution of radon concentration values in the dwellings in Ekaterinburg, depending on the energy efficiency class

Increase of radon concentration in modern buildings is associated with new construction technologies. The use of various construction technologies and various types of building materials (reinforced concrete panels, brick, monolithic concrete etc.) can lead to significant differences in radon concentration levels. As was shown earlier in the work [7], the lowest radon concentrations were achieved in panel block buildings built between 1970 and 1989. The houses of the specified years of construction are distinguished by the highest value of the ratio of living to total area of the building.

At the same time, the areas of common use and office areas in new buildings are usually characterized by low air exchange rate, especially in the cold period of the year. Thus, the leakiness of the entrance doors to apartments, their high air permeability, as well as the ways of utility lines input from public areas create conditions for reducing the contribution of the fresh air to the general air exchange of dwellings. The indirect proof of this hypothesis is the dependence on Figure 2, displaying the effects of space-planning decisions on the achieved values of the radon concentration in the surveyed buildings.

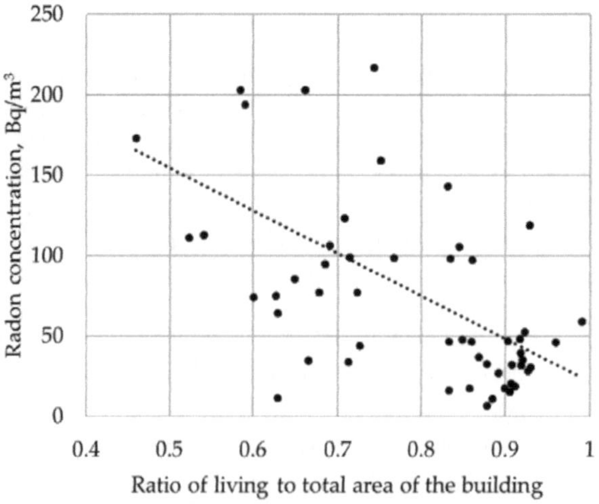


**Figure 2.** Ratio of living to total area of the building vs radon concentration

According to [7], the part of modern panel block apartments built during the period from 2001 to 2013 does not exceed 30% of the total number of buildings constructed during the period. Thus, the significant number of buildings with a high class of energy efficiency and built using modern space-planning decisions, contributes to an increase in the average radon concentration, which contradicts the concept of systematic decrease of radiation exposure.

Based on the results of long-term measurements of radon concentrations in a dwelling with energy efficiency class "A", the average value of radon concentration 130 Bq/m$^3$ was obtained. Also, the radon entry rate was analyzed. The radon entry rate was about 35 Bq/(m$^3$·h) in average and almost unvaried with ΔT in the range from 25 to 45° C. Such estimation fully corresponds to the values obtained earlier for other energy-efficient buildings in Ekaterinburg.

The air exchange rate between the indoor and outdoor atmosphere has a significant effect on the radon concentration. The main driving forces of the air exchange under absence of the mechanical ventilation are the temperature difference between indoor and outdoor atmosphere and wind pressure. As was shown in the works [7, 8], the value of the air exchange rate in buildings with natural ventilation system is sufficiently low (from 0.2 to 0.3 h$^{-1}$ in the stationary mode of room use). For a dwelling with the energy efficiency class "A", a similar result was obtained (Figure 3).

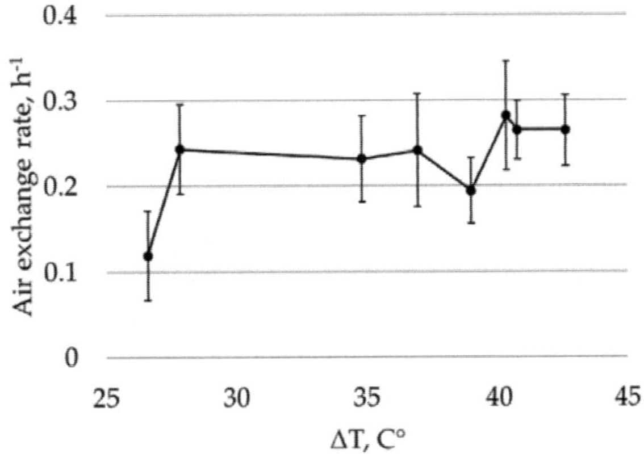


**Figure 3.** Dependence of air exchange rate on the temperature difference between indoor and outdoor atmosphere

According to [2, 3] the use of mechanical ventilation with heat recovery is considered as the main scenario for reducing elevated radon concentrations in buildings. However, in the climate of the Middle Urals in Ekaterinburg, the use of such systems did not show an obvious advantage in terms of reducing concentrations of pollutants.

## 4. Conclusion

According to the results obtained, there is a significant increase in radon concentration in buildings with a high energy efficiency class. Based on the results of the research, the air exchange rate in energy-efficient buildings remains insufficient due to low air permeability of building envelope, entrance doors and windows. The main type of ventilation in new residential buildings in Ekaterinburg is natural ventilation. Residents are forced to control the air exchange in rooms by opening windows or ventilation flow valves (if provided). In turn, outreach work with the public regarding decreasing of energy consumption is conducted in the terms of reducing costs for housing and communal services. In particular, with regard to ventilation systems, the public is advised not to leave the ventilation routes permanently open and mechanical ventilation to be in operation constantly.

Often, the required air exchange is not ensured by the occupants for comfort reasons (during the cold season) or for reasons of reducing costs (even in the case of mechanical ventilation with heat recovery systems). To increase the air exchange rate and reduce energy consumption without decline of living conditions, it may be recommended to use the heat recuperation in ventilation systems (i.e., a system that allows to save thermal energy significantly, by heating incoming fresh air by return air without mixing them). However, the presence of such ventilation systems is not sufficient, since the mechanical ventilation system is often not used properly both in the automatic and manual mode, which does not give an obvious advantage over natural ventilation.

**References**

[1] Pérez-Lombard L, Ortiz J and Pout C 2008 A review on buildings energy consumption information *Energy and Buildings* **40** p 394–8

[2] Milner J, Shrubsole C, Das P, Jones B, Ridley I, Chalabi Z, Hamilton I, Armstrong B, Davies M and Wilkinson P 2014 Home energy efficiency and radon related risk of lung cancer: modelling study *BMJ* **348** p 1–12

[3] Santos Hugo R R and Leal Vítor M S 2012 Energy vs. ventilation rate in buildings: A comprehensive scenario-based assessment in the European context *Energy and Buildings* **54** p 111–21

[4] Darby S, Hill D, Auvinen A, Barros-Dios J M, Baysson H et al 2005 Radon in homes and lung cancer risk: a collaborative analysis of individual data from 13 European case-control studies *BMJ* **330** p 223–7

[5] Krewski D, Lubin J H, Zielinski J M, Alavanja M et al 2005 Residential radon and risk of lung cancer: a combined analysis of seven North American case-control studies *Epidemiology* **16** p 137–45

[6] Yarmoshenko I, Onishchenko A and Zhukovsky M 2013 Establishing a regional reference indoor radon level on the basis of radon survey data *J Radiol Prot* **33** p 329–38

[7] Yarmoshenko I, Vasilyev A, Onishchenko A, Kiselev S and Zhukovsky M 2014 Indoor radon problem in energy efficient multi-storey buildings *Radiat Prot Dosim* **160** p 53–5

[8] Collignan B, Lorkowski C and Améon R 2012 Development of a methodology to characterize radon entry in dwellings *Building and Environ* **57** p 176–83

# Energy from Biomass for Sustainable Cities

**D Panepinto[1], M C Zanetti[1], L Gitelman[2], M Kozhevnikov[2], E Magaril[3] and R Magaril[4]**

[1] Department of Environment, Land and Infrastructure Engineering, Politecnico di Torino, Corso Duca degli Abruzzi, 24, 10129, Torino, Italy
[2] Department of Energy and Industrial Management Systems, Ural Federal University, Mira str., 19, Ekaterinburg, 620002, Russian Federation
[3] Department of Environmental Economics, Ural Federal University, Mira str., 19, Ekaterinburg, 620002, Russian Federation
[4] Department of Oil and Gas Processing, Tyumen Industrial University, Volodarskogo str., 38, Tyumen, 625000, Russian Federation.

E-mail: np.fre@mail.ru

**Abstract**. One of the major challenges of sustainable urban development is ensuring a sustainable energy supply while minimizing negative environmental impacts. The European Union Directive 2009/28/EC has set a goal of obtaining 20 percent of all energy from renewable sources by 2020. In this context, it is possible to consider the use of residues from forest maintenance, residues from livestock, the use of energy crops, the recovery of food waste, and residuals from agro-industrial activities. At the same time, it is necessary to consider the consequent environmental impact. In this paper an approach in order to evaluate the environmental compatibility has presented. The possibilities of national priorities for commissioning of power plants on biofuel and other facilities of distributed generation are discussed.

## 1. Introduction

Over the past four decades considerable changes have occurred in the world economy as regards material extraction, trade and the welfare of the world's population. Figure 1 shows trends in GDP, population, material extraction, exports and RME exports. According to [1], global consumption of mineral resources, metals and other materials has tripled and continues to grow at an accelerating pace. On the one hand, it matches the population growth and improving living standards overall. On the other hand, the growing use of material resources is accompanied with an increasing environmental footprint, faster changes in climate and air pollution. The situation in urban areas causes the biggest concern as the clustering of transport, energy and production facilities results in pollutants entering the breathing zone, which has a negative effect on people's health.

The implementation of the Sustainable Development Goals set by the Resolution, adopted by the UN General Assembly on 25 September 2015 [2], requires that nations independently develop their own specific sustainable development strategies, projects and programs. At same time, provisions should be made for alternative options of creating new technologies that combine economic growth and the minimum environmental impact.

Priorities in national solutions addressing the issue of reducing toxic emissions from stationary and mobile sources vary in regions that have a shortage of hydrocarbons, and in countries that boast sufficient fossil fuel reserves. In the former case it is necessary to intensify research on technologies that increase the accessibility of alternative energy sources. It has to be noted that in general the production and use of biofuels have been constantly growing in Europe and Eurasia, increasing 4.3 times over a decade to 13.726 mln. tonnes of oil equivalent in 2015, as reported by [3]. Countries with hefty conventional energy reserves should focus on making technological processes and the economies in general more energy efficient and on improving the quality of traditional fuels. This is not in disagreement with the need to develop other industries, diversify production and to further embrace alternative energy sources. Figure 2 illustrates the contribution of some countries to global extraction of conventional energy resources and production of biofuels in 2015.

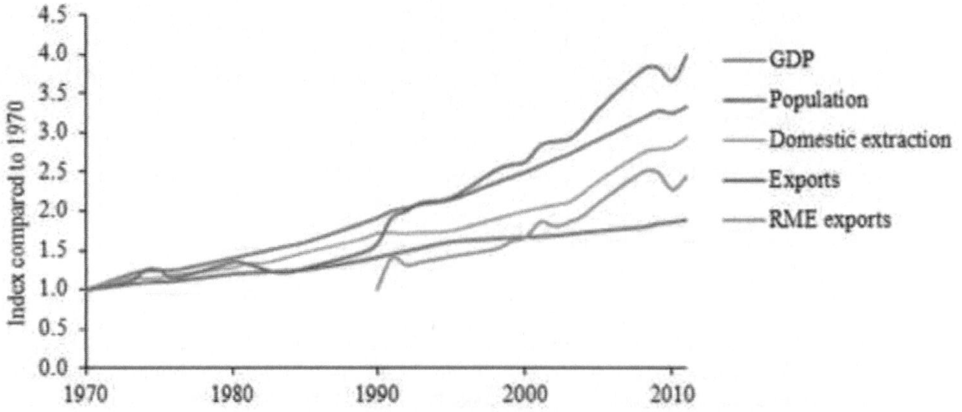


**Figure 1.** Trends in GDP, population, domestic extraction (DE), exports and raw material equivalents (RME) of exports, 1970–2010 (based on [1])

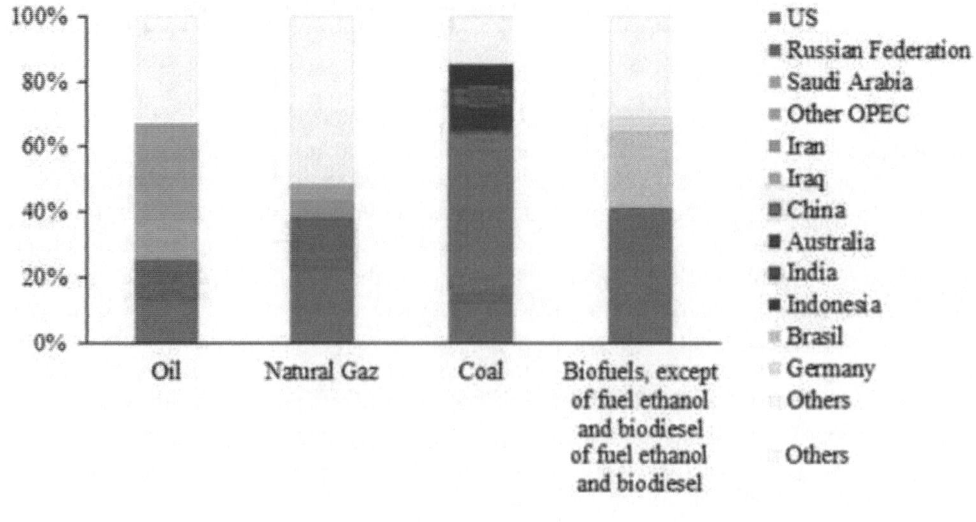


**Figure 2.** The main contributors to the production of traditional energy resources and the production of biofuels in 2015 (based on [3])

The objective of European Union Directive 2009/28/EC is to obtain 20 % of all energy from renewable sources – for example, wind power, solar, biomass, etc. – by 2020. Of these energy sources, the bioenergy obtained from renewable energy plants (based on biogas produced by the anaerobic digestion of manure, and by burning energy crops such as vegetable oil, wood, and solid biomass) are being strongly encouraged under the European legislation, but their effects on air quality raise serious concerns.

In fact, on the one hand there are important positive aspects such as limitation of greenhouse gas (GHG) emission, use of secondary materials, improvement of local activities, right destination of organic sludges, and limited cost. On the other hand, the impacts on land use need for chemical and imported resources, and, above all, formation of secondary pollutants (dust, nitrogen oxides, and VOC) should be carefully considered in terms of their lower prosperity, higher costs, and specific local conditions. Other forms, such as hydro or solar energy have not been considered at all, even if they could contribute to the total renewable energy budget. Therefore, in the present work, the authors evaluate not only the potential energy production from biomass plants (and its use as substitution of the fuel for the domestic house boiler) but also their environmental compatibility. In some cases, the two goals of renewable energy production and improved air quality conflict with each other, and it is necessary to consider both the avoided emissions of GHGs and the effect of biomass combustion on local air quality in the vicinity of the plant itself. In literature, there is a lot of material on the subject ([4–6]) concerning also both biofuel [7] and biodiesel [8]. In this study a general methodology in order to define the environmental compatibility and in order to define future energy compatibility planning has been shown.

## 2. Energy from biomass: state of art

Bioenergy production plants must be evaluated on the basis of their capacity to satisfy both local and regional needs for thermal and electrical energy. Such evaluation requires the consideration of a given plant's contribution to reducing GHG emissions and climate change, and of its effect on local environmental quality (air pollutant concentrations, use of natural resources, re-emission of residuals [9]).

In order to study the ability of different forms of biomass to satisfy these general requirements, a number of scenarios has been constructed, to cover large spans of both time and land area [10, 11]. Two key inputs to such analyses are the amount of woody biomass available for large-scale energy production and its price. One short-term study, [12], describes public and private initiatives for investigating regional energy scenarios, based on agricultural biomass in two countries in the south west of Germany. That study focused on the development of more sustainable approaches to energy crop cultivation and on the use of cogeneration systems. The most practical and economical approach to bioenergy production is to find sustainable methods of satisfying the energy production requirements. A study with this aim, [13], discussed the different aspects of sustainability: the utilization of natural resources (land, water, and fertilizers), the effect on climate change, the potential introduction of local air pollutants, and the necessity of disposal, or if possible, reuse, of residual wastes from biomass production. To analyze the effects on land use and the transformation of soils, the carbon balance and $CO_2$ emissions data must be considered. A detailed analysis with reference to bioethanol production from corn, [14], shows that, considering the complete cycle of biomass production and transformation, it is possible to define the global impact of a particular strategy.

Using a more general approach, [15] discussed all the aspects of energy utilization and its major environmental impacts from the standpoint of sustainable development, including anticipated patterns of future energy use and subsequent environmental issues. Renewable energy technologies and efficient energy utilization were identified as the most effective potential solutions to current environmental impacts. Those practical examples of implementation are useful due to their conclusions and recommendations.

## 3. Environmental compatibility: methodological approach

In order to determine the potential energy from local biomass and the resulting environmental impacts, it is necessary to use the following tools:

- energy balance;
- environmental balance.

Firstly, by using the tool of *energy balance*, we can know if the use of local biomass could satisfy the local energy requirements and if it is possible to export it. Using the currently available biomass it is possible to calculate the available power (Figure 3).

Secondly, in order to evaluate the environmental benefits deriving from biomass use, it is necessary to estimate both the atmospheric emissions of the existing boilers, and the expected emissions after the installation of biomass plants. The *environmental balance* can be computed according to the following formula: **local / global emissions (added / eliminated) = biomass plant emissions – substituted emissions.**

The environmental balance is divided between two scales, one on the local scale and the other one on the global scale. More specifically, at the local level we considered the emissions avoided for the thermal energy produced by the plant and used for district heating. On the global scale, we considered two different components: the value of avoided emissions for the thermal energy and the value of avoided emissions for the electrical energy produced by the plant. For the environmental balances it is necessary to take into account emission factors for the various analyzed pollutants, for the different types of biomass used (liquid or solid), and for the various steps in the process. An emission factor is defined as the weight of pollutant issued by a source referred to the entity of energy production (MJ, kWh).

For these calculations, the generally considered parameters are:

- on the local scale: dust, nitrogen oxide, and sulphur oxide,
- on the global scale: dust, nitrogen oxide, sulphur oxide and carbon dioxide.

However, in an environmental balance, the assumption that the amount of $CO_2$ absorbed during plant growth was equivalent to the greenhouse gas emissions released during energy conversion is an excessive hypothesis. In fact, it should be noted that, for a detailed analysis of the environmental effects in terms of $CO_2$, we have to consider the emissions of other gases such as $CH_4$ emissions and $N_2O$ produced during pre and post storage of biomass to produce biogas and the indirect emissions deriving from agricultural processes, such as harvesting and transport. In this way, we take into account the environmental effects at the level of combustion of biomass and the whole process: from harvesting to the collecting of biomasses to transportation and use of biomasses. The assessment is only referred to $CO_2$.

Generally for the emission factors used to estimate the emission of $CO_2$, it is possible to refer [16]. Regarding the transport of biomasses, it is possible to consider the emission factors of transport derived by European model COPERT 3, which is about 413.5 g $CO_2$/km with a capacity of 9 tons of transportable biomass (Figure 4).

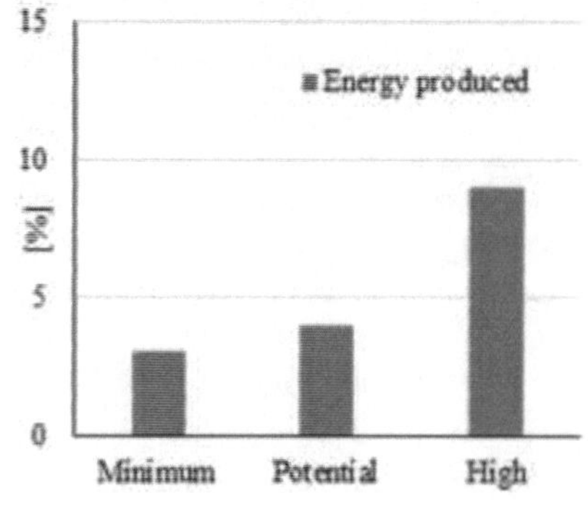


**Figure 3.** Example of results of the energy balance

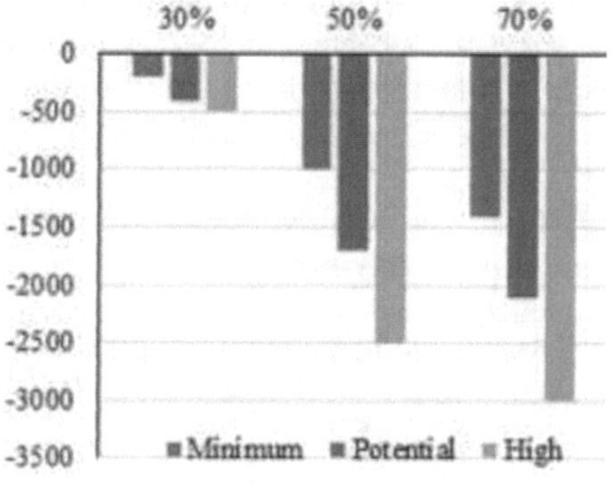


**Figure 4.** Example of results of the environmental balance

## 4. Taking national priorities into account

Biofuel power plants and other types of renewable energy sources constitute distributed (small-scale) generation installations that are particularly flexible and adaptable to changing demand and location. Such installations are in the closest proximity to the load center, which ensures a high level of reliability and cost effectiveness of the grid. In many cases they require less per unit investment; it takes months rather than years to build them and their payback period does not exceed three years. Full automation makes it possible to keep them running in optimum modes and, as a result, to ensure high energy and environmental effectiveness. The development of distributed generation in regions leads to a decrease in demand for backup capacity in energy systems.

It is necessary to emphasize that, when incorporated in regional power grids, such facilities, cannot be viewed as an alternative to the grid; they must not be set in opposition to one another as their functions complement each other. At the same time, the optimal share of the regional grid capacity in electric power supply will differ across the country depending on such factors and limitations as operating mode, availability of resources, climate and environment (e.g., peak loads, demand for electricity where there is no heat load at cogeneration plants, possibility of developing generation based on renewable energy sources). The key function of regional grids in the structure of the power supply system is to compensate for the excessive economic inertia of the national grid that is built around large thermal power plants, NPPs and hydropower plants. As a result, the installed capacity put into operation in the regional and national grids is aligned as much as possible with the pace of demand growth, while the cost, reliability and environmental characteristics improve across the national grid.

Standing at 1%, the share of RES in the energy system of Russia is insignificant yet (figure 5). When assessing the effectiveness of small-scale generation development, one has to bear it in mind that Russia has an excess capacity of over 20GW (around 8% of total capacity) [17]. There is, therefore, no need for new power plants, be it fossil fuel ones or renewable installations, in the country. This does not mean, though, that some regions of Russia do not have to deal with a pressing issue of optimizing the structure of electricity generation capacity in order, for example to meet peak demand or improve the environmental situation.

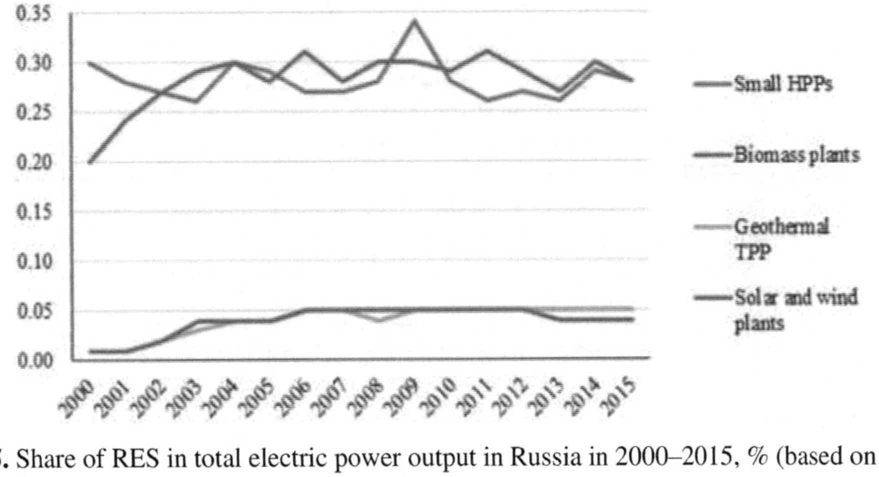


**Figure 5.** Share of RES in total electric power output in Russia in 2000–2015, % (based on [18])

In terms of greenhouse gas emissions and Russia's obligations under the Paris Agreement, Russia is in a drastically different situation than other countries: over 80 % of its energy is produced from zero-carbon and low-carbon fuels (NPPs, gas-fueled TPPs, hydropower plants). In the countries that have taken the lead in embracing renewable energy sources, the situation is different: the share of coal in China's energy balance is over 60 %; in the USA and Germany around 40 %.

One has to remember that the areas of decentralized energy supply (15 % of total generation) cover over two thirds of the territory (remote areas of the Far North, Siberia, the Far East). The development of the electric power industry there naturally implies the predominant use of local energy sources and RES.

It should be understood that despite the decentralized nature of the energy systems, it is in these regions that active industrial development is bound to start soon. Work on specific projects of integrating renewable energy with large-capacity installations should start today. The task of designing local grids that include renewable energy installations is focused on hybrid energy systems (HES) that consist of an array of primary energy sources, a system of installations transforming primary energy into electrical energy, mechanical energy and heat; various types of consumers, energy transmission and distribution systems. HESs should be created as a coherent combination of consumers cum power generators who form the technological foundation of smart micro-grids [19].

The authors' research on the prospects of renewable energy in Russia shows that the list of relevant institutional, organizational and economic decisions includes:

- forming a clear energy policy of a region, effective mechanisms of its implementation and establishing professional bodies of administration and regulation;
- project design and forecast support of the regional energy sector development;
- creating a mechanism of attracting investment, providing funding for motivational offers (incentives) and making sure that appropriate technological solutions are available on the energy equipment market;
- implementing a set of technical measures to adapt the electric power distribution system of the region to the connection of additional generators;
- developing an accessible equipment market;
- deployment of information systems of operating mode control, equipment monitoring and cost accounting.

## 5. Conclusion

In the context of energy consumption growing worldwide, one of the priorities of the development of civilization is to find ways of minimizing the impact of power generating facilities and transport on the environment by means of adopting the most effective technologies.

When building a strategy of small-scale generation development on the basis of RES, it is necessary to take into account specific national priorities that stem from climate conditions, the vastness of territory and industrial architecture. Every region of a country that is as diverse in this sense as Russia has its own optimum share of renewable energy sources. At the same time, using renewable energy resources might be particularly relevant for big cities, especially those ambitiously embracing smart technologies, as an element of the green economy and environment.

In terms of investment and operational arrangements, it seems highly promising to establish specialized local grids with hybrid energy systems that operate in collaboration with "pro-active consumers".

## Acknowledgement

The work was supported by Act 211 of the Government of the Russian Federation, contract No. 02.A03.21.0006.

## References

[1] Schandl H et al 2016 *Global Material Flows and Resource Productivity: Assessment Report for the UNEP International Resource Panel* (Paris: UNEP) p 6–32

[2] Transforming our world: the 2030 Agenda for Sustainable Development A/RES/70/1, Resolution adopted by the UN General Assembly Sept. 2015 Available from: http://www.un.org/ga/search/view_doc.asp?symbol=A/RES/70/1&Lang=E [Accessed 26th April 2017]

[3] BP Statistical Review of World Energy June 2016 Available from: http://www.bp.com/content/dam/bp/pdf/energy-economics/statistical-review-2016/bp-statistical-review-of-world-energy-2016-full-report.pdf [Accessed 26th April 2017]

[4] Prabhakar SVRK and Elder M 2009 Biofuels and resource use efficiency in developing Asia: back to basics *Appl Energy* **86** p 30–36

[5] Rada E C 2017 Environmental pollution from waste and biomass energy generation *International Journal of Energy Production & Management* **2** p 1–7

[6] Ragazzi M, Ionescu G and Cioranu S I 2017 Assessment of environmental impact from renewable and non-renewable energy sources *International Journal of Energy Production & Management* **2** p 8–16

[7] Sadeghinezhad E, Kazi S N, Badarudin A, Togun H, Zubir M N M, Oon C S and Gharehkhani S 2014 Sustainability and environmental impact of ethanol as a biofuel *Rev Chem Eng* **30** p 51–72

[8] Sadeghinezhad E, Kazi S N, Badarudin A, Oon C S, Zubir M N M and Mehrali M 2013 A comprehensive review of bio-diesel as alternative fuel for compression ignition engines *Renew Sust Energ Rev* **28** p 410–424

[9] Iglinski B, Piechota G and Buczkowski R 2015 Development of biomass in polish energy sector: an overview *Clean Technol Environ* **17** p 317

[10] Lauri P, Havlık P, Kindermann G, Forsell N, Bottcher H and Obersteiner M 2014 Woody biomass energy potential in 2050 *Energ Policy* **66** p 19–31

[11] Konrad C, Strittmatter J, Grunert A, Brule M, Roth M, Herter M, Gottlicher G, Biehl R and Bott A 2013 Regional energy concepts – based on alternative biomass cultivation for rural areas and its efficient energy usage *International Journal of Sustainable Development and Planning* **8** p 59–74

[12] 2013 *EU bioenergy potential from a resource efficiency perspective* (Luxembourg: Publ Office Eur Union)

[13] Tyner W E, Taheripour F, Zhuang Q, Birur D and Baldos U 2010 *Land use changes and consequent $CO_2$ emissions due to US corn ethanol production: a comprehensive analysis Final Report* (Department of Agricultural Economics Purdue University)

[14] Dincer I 1999 Environmental impacts of energy *Energ Policy* **27** p 845–854

[15] Sadeghinezhad E, Kazi S N, Sadeghinejad F, Badarudin A, Mehrali M, Sadri R and Safaei M R 2014 A comprehensive literature review of bio-fuel performance in internal combustion engine and relevant costs involvement *Renew Sust Energ Rev* **30** p 29–44

[16] Panepinto D, Genon G, Brizio E and Russolillo D 2013 Production of green energy from co-digestion: perspectives for the Province of Cuneo, energetic balance and environmental sustainability *Clean Technol Environ* **15** p 1055–62

[17] Solomin V A 2017 Transition to renewables as a political task Available from: http://www.ng.ru/ng_energiya/2017-01-10/11_6897_sources.html [Accessed 26th April 2017]

[18] Renewable Energy in Russia: Pros and Cons Available from: http://www.ime-mo.ru/ru/conf/2012/22112012/BEZ_22112012.pdf [Accessed 26th April 2017]

[19] Rogalev N D, Tiagunov M G and Shestopalova T A 2015 How to make a power plant more attractive by means of renewable energy sources *Energetik* **1** p 31–33

IOP Conf. Series: Earth and Environmental Science **72** (2017) 012022 doi:10.1088/1755-1315/72/1/012022

# Evaluation of the operation efficiency of solar panels in winter

**A D Burakova, L N Burakova, I A Anisimov and O D Burakova**

Industrial University of Tyumen, 38 Volodarskogo str., Tyumen, 625000, Russia

E-mail: burakova.1992@mail.ru

**Abstract**. The article deals with the issue of increasing the comfort and safety of life in cities by applying an alternative energy source for power supply of transport infrastructure facilities. Due to the peculiarity of the Russian Federation territory location, most cities are characterized by a long winter period, which makes it necessary to consider the features of using solar panels under these conditions. It has been established that the efficiency of solar panels depends on their type and location, the presence of snow cover on their surface, and the ambient air temperature. It has been revealed that flexible solar panels have some advantages that determine their ability to be used for power supply of transport infrastructure facilities. In the paper, the optimum angle of inclination of rigid solar panels in the winter period of the year is determined.

## 1. Introduction

The growth of the population and the increase in the requirements for the comfort of living in cities lead to the need to create an environmental friendly and energy-efficient urban environment [1–7]. This concept is a combination of natural, architectural and planning, socio-cultural and other factors that determine the quality of life of a person in the area under consideration [8]. The most important component of improving the state of the urban environment is spatial organization of the city's territory, which ensures not only the efficiency and convenience of movement, but also the level of safety.

Currently, one of the ways to create an environmental friendly, energy-efficient and safe urban environment is the use of modern technologies, called the "Smart City" [9]. Its distinctive feature consists in using information to provide a stable link between a person and knowledge, which is the driving force for the transition to a new way of consuming energy.

One of the criteria for the quality of life in a "smart city" is safety, in particular, road safety. In February 2014, amendments were adopted to standards in which special attention was paid to ensuring the safety of pedestrians: GOST R 52289-2004, GOST R 52290-2004, GOST R 52605-2006, GOST R 51256-2011, GOST R 52765-2007, GOST R 52766-2007.These regulator documents overview typical schemes of traffic organization on unregulated and regulated pedestrian crossings, and also provide a number of basic requirements that should improve road safety in the areas under consideration. They are: the use of a pedestrian crossing marking on a yellow background, use of "Pedestrian crossing" traffic signs on retroreflective yellow-green shields, installation of duplicate road bumps in front of each pedestrian crossing in populated areas, except for main roads and streets, duplication of warning road signs by road markings on each lane. However, the main measure to improve the level of road safety is equipping pedestrian crossings near educational institutions with a traffic light T7 in a two-way version and a light indication of traffic signs.

These transport infrastructure facilities should be provided with a constant supply of electricity. According to SNiP 2.07.01-89 section 7 "Engineering equipment", in particular, electric wires must be located within the transverse profiles of streets and roads, under sidewalks or dividing strips. Therefore, equipping a pedestrian crossing with a T7 traffic light and a light indication is difficult. The reasons are: placing engineering networks under sidewalks and carriageways; work on the placement of engineering networks (puncture, accounting for other adjacent heat and electricity networks); search for the nearest substation or switchgear.

An alternative option for power supply of transport infrastructure facilities are photoelectric converters. At present, the share of electricity generated by alternative sources is 9.3 %. The use of solar energy is 0.33 % [10]. The sun is an "environmentally friendly" renewable source of energy. The advantages of power supply using photoelectric converters consist in the simplicity and low cost of the installation. For example, the cost of equipment and works on supplying power from the 220V electric grid is 482000 rubles, while the power supply of transport infrastructure facilities with the help of photoelectric converters is 174000 rubles. Despite the economic benefits and ease of installation of this equipment, there are also disadvantages. It are complexity of production and disposal of panels, which is due to the presence of toxic substances in their composition, as well as the dependence of the amount of generated electricity on the spatial location of the solar panel, time of day, as well as natural and climatic conditions. The use of solar energy in winter is challenging due to a number of reasons: a short daylight, low ambient temperature, precipitation in the form of snow, which prevents sunlight from reaching the panel surface.

The purpose of this research is to study the regularities of the current intensity produced by solar panels, depending on the time of day, the spatial arrangement and type of equipment in winter conditions.

## 2. Methodology

Research aimed at studying the ways of using alternative energy sources has been considered by Erik Jarlsby, Michele De Carli, Rada E.C., Gudmestad Ove T., Kjell Traa, Ragazzi M., Ionescu G., Cioranu S. I., Boeri A., Gianfrate V., Longo D. [11–18]. The work of Vitali D., Garbuglia F., D'Alessandro V., Ricci R. presents a method of power supply for led standalone streetlight. This installation is completely autonomous [19]. Its work is supported by solar panels and a vertical wind turbine, which increases the overall efficiency of this equipment. However, the disadvantage of the installation is the rigid fixing of the solar panel, so it is impossible to change the angle of its inclination and increase the efficiency of its operation.

One of the ways to improve the efficiency of photovoltaic converters is to find the optimal angle of inclination of the solar panel depending on the time of day, the period of the year and the geographical location of the territory. These studies were carried out for individual territories, particularly Chisinau and Northern Alberta [20]. The work of Aculinin A., Smicov considers the dependence of the change in total solar radiation on the time of the year and on the angle of inclination of the solar panel, and provides energy calculations of the optimal orientation of the solar panel [21]. As a result of the analysis, the authors establish the dependence of the optimum angle of inclination of the solar panel on the month of the year in Chisinau. However, they indicate that the orientation of the solar panel can be changed twice a year – in the warm and cold period of the year, as monitoring its position during each month is difficult.

As a result of the analysis of the research work from the Institute of Technology of Northern Alberta, it was revealed that the angle of inclination of the solar panel in winter affects not only the amount of electricity produced, but also the amount of accumulated precipitation on its surface that hinders the ingress of solar radiation. In the paper we propose selecting the optimal angle of inclination of the solar panel in winter taking into account its minimal cleaning from snow. The dependence of the increase in electricity generation during snow cleaning on the angle of inclination is shown in Figure 1.

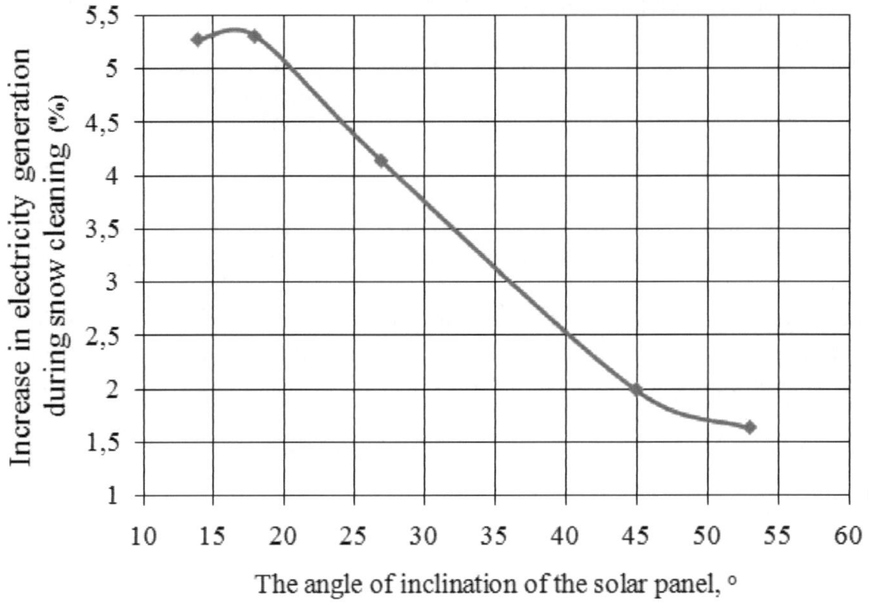


**Figure 1.** Influence of the angle of inclination of the solar panel on the increase in power generation during snow cleaning

Increasing the angle of inclination relative to the horizontal surface reduces the need to clean the solar panel from snow. This is due to an increase in the ability to naturally clean the snow with an increase in the angle of inclination of the solar panels. The authors state that when choosing the angle of inclination relative to the horizontal surface of 90° during 99.5 % of the winter period there is no snow cover on the surface of solar panels, and with a decrease in the angle of inclination from 53° to 14° the difference in the generation of electricity between the cleaned and uncleaned from the snow modules increases.

Ambient temperature is also one of the important factors affecting the efficiency of the solar panel [22, 23]. As a result of the analysis of previous studies it was established that a decrease in air temperature leads to an increase in the power produced by the solar panel. This is due to the fact that the temperature coefficient of the idling voltage exceeds the temperature coefficient of the short-circuit current and causes a significant drop in voltage with increasing temperature, which cannot be compensated by an increase in the current intensity. Power is the product of current intensity to voltage, so increasing the temperature causes a decrease in the efficiency of the solar panel. Thus, low ambient temperatures are a favorable factor that increases the efficiency of solar panels in winter conditions.

The results obtained in the course of the analysis will be the basis for identifying the dependences of changes in the value of the current intensity produced by the solar panel in winter on the time of day, the spatial arrangement and type of equipment.

### *2.1. Equipment and devices used in studies*

In the course of experimental studies, two types of solar panels were used. The first type is the solar panel HH-MONO-80W – a silicon rigid single-crystal panel with a power of 80W and a current intensity of 3.67A. It is characterized by low efficiency, the values of which under ideal conditions can reach 15.7 %. The second type is EP-80W – a silicon flexible single-crystal panel with a power of 80W and a current intensity of 4.04A. Its efficiency values under ideal conditions reach 22 %. The network also consisted of a battery with a voltage of 12V and a capacity of 100A•h and a charge controller with a voltage of 12/24V.

### *2.2. Methodology of research*

The purpose of the experiments was to study the regularities of the current intensity produced by solar panels of various types from the time of day in winter, as well as to reveal the effect of the angle of inclination of the solar panel surface relative to the horizontal surface in winter in Tyumen.

On the tripods, rigid and flexible solar panels were fixed, to which multimeters for measuring the amperage were connected. The assembled installations were placed on a uniformly illuminated platform. The experiment was carried out in several stages.

The essence of the first stage was to determine the dependence of the change in the current produced by the solar panel during the winter daylight. Experimental facilities with a rigid and flexible solar panel were located on a uniformly illuminated site. The rigid solar panel was arranged horizontally, the flexible solar panel was maximally curved to an arc describing a quarter of the sphere. The measurements were taken during the daylight hours in winter – from 8:00 to 18:00. The obtained results made it possible to establish the efficiency of the application of a flexible solar panel and reveal the effect of the Sun's progress during the daylight on the generated current intensity.

The second stage of the experimental studies consisted in revealing the influence of the angle of inclination of the solar panel at the invariable position of the Sun. To do this, the solar panel, located on a uniformly illuminated site, changed the angle of inclination, but the constancy of the location of the Sun was achieved. The measurements were taken several times during the daylight hours. The results obtained during this stage of the experimental studies allowed to determine the influence of the angle of inclination on the current intensity generated with the invariable position of the solar panel, and also to reveal the most optimal position of the solar panel corresponding to the maximum current intensity under the observed conditions.

The third stage of the experiment made it possible to reveal the effect of a change in the location of the solar panel. In this case, initially during the daylight hours, the angle of its inclination was changed so that the Sun's rays hit the surface perpendicularly. Then, according to the compass readings, the panel was directed to the south during the daylight. The obtained results made it possible to determine the influence of the change in the angle of inclination of the solar panel or the constancy of its location on the generated current intensity.

## 3. Results and discussion

The territory of the Russian Federation has some peculiarities: the predominance of its extent along geographical latitudes; location in high and moderate latitudes and under the influence of the Arctic. This explains the presence of a cold winter season almost throughout the country for the exception of the Black Sea coast of the Caucasus. It is characterized by average temperatures that vary depending on the location of the area from 0 to -4° C in the southern regions and down to -45° C in Siberia, the snow cover remaining in different regions from 2 to 7 months, and the short duration of daylight hours [24]. The presented features of the RF territory location make it necessary to consider the principal differences in the use of solar panels for power supply of the municipal transport infrastructure in winter.

### *3.1. Influence of the solar panel type on the current intensity produced during the daylight in winter*

Initially, the types of solar panels (rigid and flexible) were studied to identify their advantages and disadvantages. This established that, together with the positive technical characteristics, the flexible solar panel has such advantages as waterproof construction, lack of glass parts, high wear resistance, and also light weight and thickness, which causes no additional wind resistance. Thus, for power supply of transport infrastructure facilities it is more efficient to use flexible solar panels. To confirm the hypothesis, experimental research was carried out to study the dependence of the change in the current intensity produced during the daylight on the type of solar panel. The results are presented in the form of graphical dependences in Figure 2.

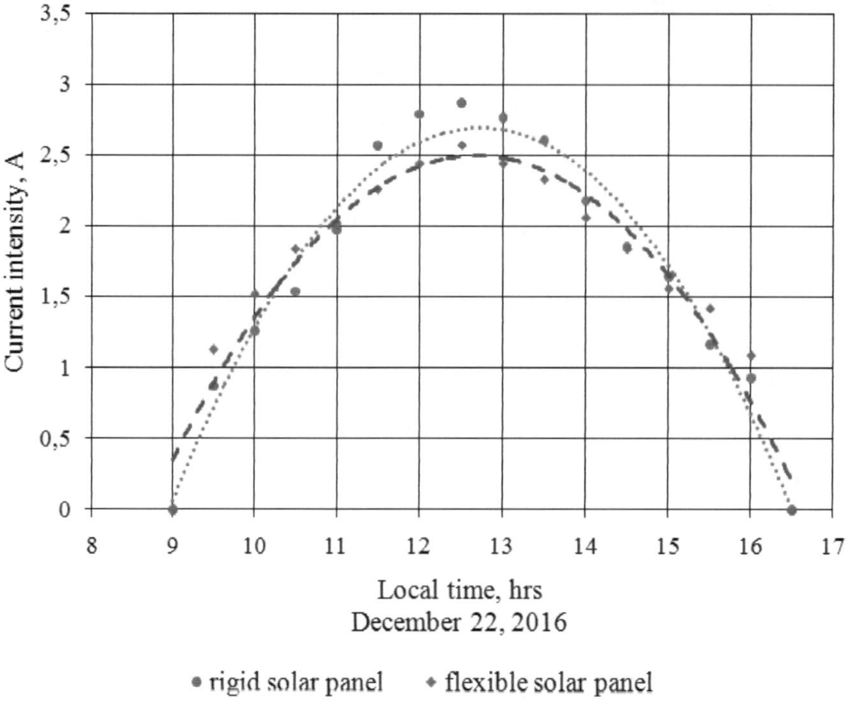


**Figure 2.** Influence of the solar panel type on the change in the current intensity during the daylight if located horizontally

Changes in the current intensity during the daylight are similar to the course of the sun, which is due to the constancy of the location of the solar panel. The maximum values are observed when the Sun reaches its maximum height and the perpendicular location of the solar panel surface relative to its rays. However, giving a flexible solar panel outlines of a quarter of the sphere made it possible to achieve a smoother change in the current intensity depending on the time of day. When calculating the area of the figure restricted by the OX axis and the curve, it is established that the difference between the current intensity produced during the daylight by a rigid and flexible solar panel is 0.45 units. This allows us to conclude that for the power supply of transport infrastructure facilities it is possible to use a flexible solar panel, since its efficiency is not inferior to a rigid solar module and has a sufficient number of advantages in its use.

### *3.2. Influence of the angle of inclination of the solar panel at the constant location of the Sun. The choice of the optimal angle of inclination during the daylight*

Based on the results of the first stage of experimental studies, it was suggested that there is the optimum angle of inclination of the solar panel during the daylight. In the study, the installation with a rigid solar panel was located on a uniformly illuminated site and during the daylight every 30 minutes the optimal angle of inclination of the solar panel was found experimentally. The dependence of the current intensity generated by the solar panel on its angle of inclination at different times of the daylight is presented in Figure 3.

The obtained results made it possible to establish a change in the optimal angle of inclination of the solar panel during the daylight. For example, when the Sun is at the zenith (12:30 local time), the angle of inclination should be from 0 to 10°, while in the morning and evening hours this is 80 to 90° with respect to the horizontal surface. Thus, using the measurements made at different times of the daylight, the dependence of the change in the optimal angle of inclination of the solar panel was constructed, which is shown in Figure 4.

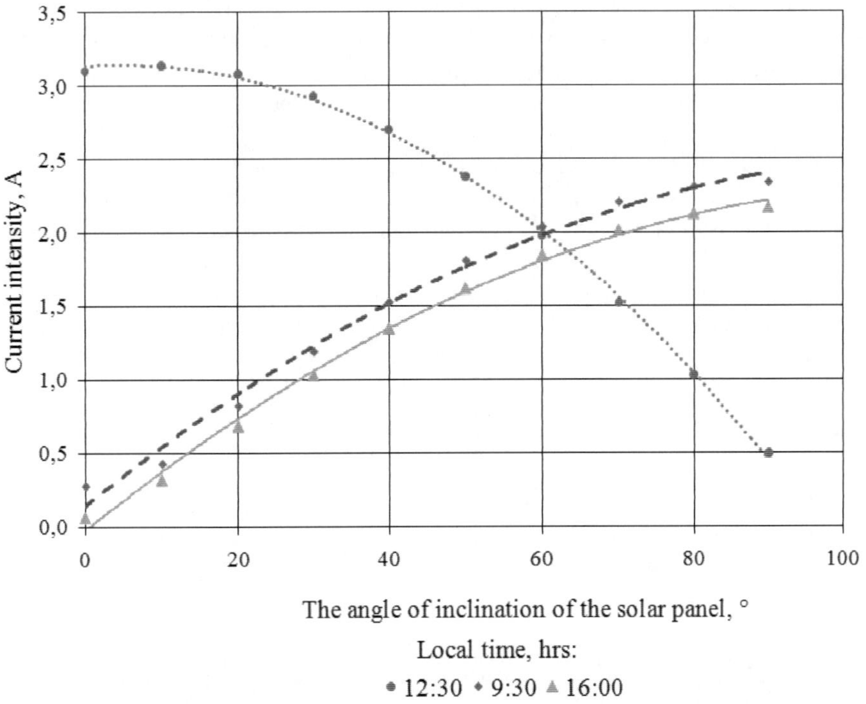


**Figure 3.** Change in the current intensity generated by the solar panel from the angle of inclination at different times of the daylight

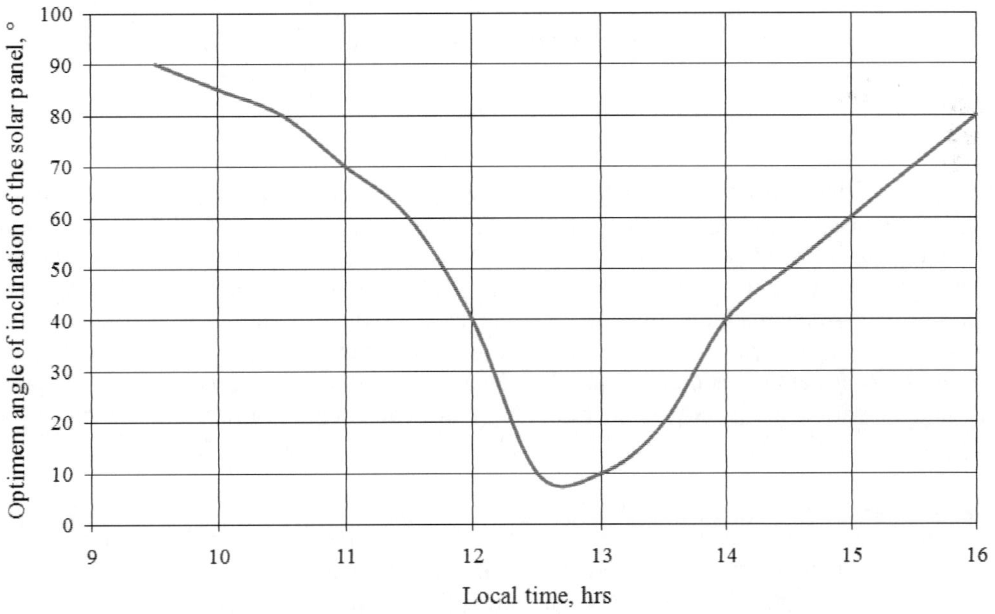


**Figure 4.** Change in the optimal angle of inclination of the solar panel during the winter daylight

The change in the optimal inclination angle during the winter daylight corresponds to the course of the Sun. The minimum value of the angle of inclination of the solar panel relative to the horizontal surface corresponds to noon. However, in winter conditions, as it was established earlier, it is necessary to take into account the presence of snow cover, which does not allow reducing the angle of inclination of solar panels to less than 53°. It is quite difficult to change the location of the solar panel during the day. This necessitates the development of a system for monitoring the location of the solar panel or selection of the optimal angle of inclination.

### *3.3. Effect of the direction of the solar panel on the generated current intensity*

During the third stage of the experiment, the effect of the direction of the solar panel on the change in the generated current intensity is studied. The obtained dependences are presented in Figure 5.

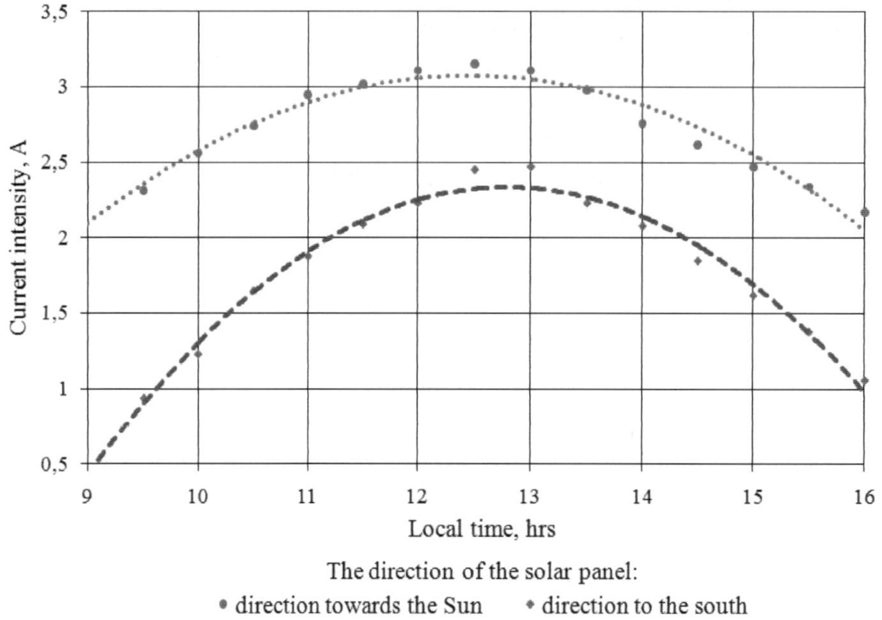


**Figure 5.** Effect of the direction of the solar panel on the generated current intensity.

This stage of the experiment was carried out in winter with the help of an installation with a rigid solar panel located on a uniformly illuminated site. Measurements were carried out every 30 minutes during daylight hours. One of the solar panels was continuously directed to the south during the experiment, the other changed its orientation according to the obtained results of the change in the optimal angle of inclination during the light day. The maximum values of the current intensity and the minimum changes correspond to the curve describing the direction of the solar panel to the Sun. In the case of a constant orientation of the solar panel to the south, the amount of incident solar energy is less, and, consequently, the efficiency of the installation operation is lower and the changes during the daylight are more drastic.

## 4. Conclusions

In the course of experimental studies, the influence of the type and location of the solar panel, the time of day on the current intensity produced has been studied. A flexible solar panel is most suitable for power supply of transport infrastructure facilities. This is due to the features of its design – a small thickness and weight, lack of additional resistance to wind and so on.

As a result of the analysis of the experimental data, the dependence of the change in the optimum angle of inclination of the solar panel during the daylight is constructed. The hypothesis is confirmed about the predominance of the current produced by directing it to the Sun over its orientation to the south. Taking into account the analysis of previous studies, it has been revealed that the angle of inclination of solar panels should not be reduced to less than 53° to maintain the possibility of its natural cleaning.

**References**

[1] Chikishev E, Chainikov D and Anisimov I 2017 Increasing the use of natural gas on motor transport by an efficient location of the NGV RS *IOP Conf. Ser.: Earth Environ. Sci. Rev.* **50** 012004

[2] Chikishev E, Chikisheva A, Anisimov I and Chainikov D 2017 Natural gas use on minibuses, engaged in the carriage of passengers and baggage on the regular routes, as a measure for decrease in harmful environment effects *IOP Conf. Ser.: Earth Environ. Sci. Rev.* **50** 012008

[3] Chainikov D, Chikishev E, Anisimov I and Gavaev A 2016 Influence temperature on the $CO_2$ emitted with exhaust gases of gasoline vehicles *IOP Conf. Ser.: Mater. Sci. Eng. Rev.* **142** 012109

[4] Chikishev E, Ivanov A, Anisimov I and Chainikov D 2016 Prospects of and problems in using natural gas for motor transport in Russia *IOP Conf. Ser.: Mater. Sci. Eng. Rev.* **142** 012110

[5] Ertman J, Ertman S, Anisimov I, Chainikov D and Chikishev Y 2015 Complex characteristic estimation of vehicle adaptability for low temperature operation conditions *Research Journal of Pharmaceutical, Biological and Chemical Sciences* (*Electronic Materials* **6**) (India: RJPBCS) p 1761–70

[6] Anisimov I, Ivanov A, Chikishev E, Chainikov D and Reznik L 2014 Assesment of gas cylinder vehicles adaptabilty for operation at low ambient temperature conditions *WIT Transactions on Ecology and the Enviroment* **190** p 685–95

[7] Ertman S, Ertman J and Zaharov D 2016 Raising of operating a motor vehicle effects on enviroment in winter *Innovative Technologies in Engineering Rev.* **142** 12119

[8] Genon G, Panepinto D and Viggiano F 2016 Sustainability in energy production *International Journal of Energy Production and Management* vol 1 (UK: WIT Press) p 16–32

[9] Anisimov I, Magaril E, Magaril R, Chikishev E, Chainikov D, Gavaev A, ErtmanS, Ertman Yu and Ivanov A 2016 Improving vehicle adaptability to the operating conditions of "smart" cities in the northern regions *E3S Web of Conf. Rev.* **6** 02003

[10] International Energy Agency 2014 *Energy: the answer to – and* the *cause of – some urgent problems* **12** (France: IEA)

[11] Jarlsby E 2016 The global energy challenge: still fuel for progress? *International Journal of Energy Production and Management* **11** (UK: WIT Press) p 33–49

[12] Rada E 2016 Environmental pollution from waste and biomass energy generation *International Journal of Energy Production and Management* **2** (UK: WIT Press) p 1–7

[13] Gudmestad O and Kjell T 2016 Sustainable use and production of energy in the 21$^{st}$ century *International Journal of Energy Production and Management* **1** (UK: WIT Press) p 1–15

[14] Ragazzi M, Ionescu G and Cioranu S 2017 Assessment of environmental impact from renewable and non-renewable energy source *International Journal of Energy Production and Management* **2** (UK: WIT Press) p 8–16

[15] Boeri A, Gianfrate V and Longo D 2016 Green buildings and design for adaptation: strategies for renovation of the built environment *International Journal of Energy Production and Management* **1** (UK: WIT Press) p 172–91

[16] Kolebaje O, Ikusika A and Akinyemi P 2016 Estimating solar radiation in Ikeja and Port Harcourt via correlation with relative humidity and temperature *International Journal of Energy Production and Management* **1** (UK: WIT Press) p 253–62

[17] Giuseppe E, Tisato Ch, Zarrella A and Carli M 2016 Multi-source heat pump coupled with a photovoltaic thermal (ptv) hybrid solar collectors technology: a case study in residential application *International Journal of Energy Production and Management* **1** (UK: WIT Press) p 382–92

[18] Al-Kayiem H 2016 Hybrid Techniques to enhance solar thermal: the way forward *International Journal of Energy Production and Management* **1**(UK: WIT Press) p 50–60
[19] Vitali D, Garbuglia F, D'Alessandro V and Ricci R 2017 The renewable energy in a led standalone streetlight *International Journal of Energy Production and Management* **2** (UK: WIT Press) p 118–28
[20] Matthews T and Sandercock J 2016 *Alternative Energy Program. Solar Photovoltaic Reference Array Report* (Canada: Northern Institute of Technology)
[21] Aculinin A and Smicov V 2016 Evaluation of opportunities of solar exegetics on the basis of accurate ground-based measurements of solar radiation *Problems of the regional energetic* (Moldova: Institute of Power Engineering of the Academy of Sciences of Moldova) p 29–39
[22] Hummel V and Windsperger A 2009 Sustainable energy future for Austria *WIT Transactions on Ecology and the Enviroment* **127**(UK: WIT Press) p 303–11
[23] Kudav G, Panta Y and Yatsco M 2012 Design and testing of wind deflectors for roof-mounted solar panels *WIT Transactions on Egineering Sciences* **74** (UK: WIT Press) p 15–27
[24] Alisov B and Poltaraus B 1974 *Climatology* (Moscow: Publishing House of Moscow University)

IOP Conf. Series: Earth and Environmental Science **72** (2017) 012023 doi:10.1088/1755-1315/72/1/012023

# A sustainable development of a city electrical grid via a non-contractual Demand-Side Management

**Vladislav O Samoylenko**[1] **and Andrew V Pazderin**[2]

[1] Engineer, Ural Federal University, Yekaterinburg, Russia
[2] Head of the department, Ural Federal University, Yekaterinburg, Russia

E-mail: v.o.samoylenko@urfu.ru

**Abstract**. An increasing energy consumption of large cities as well as an extreme high density of city electrical loads leads to the necessity to search for an alternative approaches to city grid development. The ongoing implementation of the energy accounting tariffs with differentiated rates depending upon the market conditions and changing in a short-term perspective, provide the possibility to use it as a financial incentive base of a Demand-Side Management (DSM). Modern hi-technology energy metering and accounting systems with a large number of functions and consumer feedback are supposed to be the good means of DSM. Existing systems of Smart Metering (SM) billing usually provide general information about consumption curve, bills and compared data, but not the advanced statistics about the correspondence of financial and electric parameters. Also, consumer feedback is usually not fully used. So, the efforts to combine the market principle, Smart Metering and a consumer feedback for an active non-contractual load control are essential. The paper presents the rating-based multi-purpose system of mathematical statistics and algorithms of DSM efficiency estimation useful for both the consumers and the energy companies. The estimation is performed by SM Data processing systems. The system is aimed for load peak shaving and load curve smoothing. It is focused primarily on a retail market support. The system contributes to the energy efficiency and a distribution process improvement by the manual management or by the automated Smart Appliances interaction.

## 1. Introduction

An increasing energy consumption of large cities as well as an extreme high density of city electrical loads leads to the necessity to search for an alternative approaches to city grid development. One of the promising approaches is to use the concept of Demand-Side Management to achieve load control for electrical load curve [1–4].

Nowadays the development of market mechanisms and electricity markets of the global energy sector leads to changes in rates system. Advanced metering of electricity and power by accounting tariffs with multiple rates differentiated by times of the day becomes more complicated. There is the perspective for rate changes for medium and small intervals of time. The Smart Metering concept provides for a broad implementation of modern informational and measuring components. A number of government regulations support the course [5–8].

Taking this into account, the common interests of consumers and energy companies are growing. There is a need for a program of market participants' information support and evaluation of electrical parameters, focused on differentiated by time intervals accounting, which is designed to incentive the participants and provide a non-contractual management of power consumption, distributed

generation and load of the grid components [9–12]. The introduction of these informational resources is beneficial to all the market participants:
1) consumers have an opportunity to assess their power consumption and manage it automatically depending upon the rates as well as to reduce their energy costs, to choose their power supplier with the most favorable terms;
2) owners of distributed generation can get the maximum profit from its use;
3) energy companies are able to influence a demand, a proposal for the distributed generation, reducing or changing the load of the grid components. Also they have an opportunity to conduct advanced statistics of consumption as well as automation of financial and contractual processes.

Existing systems of Smart Metering billing usually provide little general information about consumption curve, bills and compared data, but not the advanced statistics about the correspondence of financial and electric parameters. Also, consumer feedback is usually not fully used. So, the efforts to combine the market principle, Smart Metering and a consumer feedback are essential.

## 2. The dual-purpose incentives

Besides of the consumers' efforts to contribute the global idea of energy efficiency and ecology improvement, the main incentive for the consumers is an opportunity to decrease their energy costs

An analysis of the potentials to decrease the costs is aimed at the determination of rates schedule and a consumer's load curve correspondence extent. Also the analysis includes the influence of technical losses.

A case when there are M different rates (or prices of a complex rate) offered to consumer is considered. All the rates (or prices) can be ranged using their values by the principle "cheap – expensive". The most expensive rate $j$ may be assigned to the effectiveness of $\alpha_{max}$ = 0 %, the cheapest – $\alpha_{min}$ = 100 %. Rates with an intermediate price $c_j$ can be assigned to an intermediate value of $\alpha_j$, proportional to the relative price difference between this value and the minimum price to the price difference of these two extreme rates.

The principle of ranking is shown in Figure 1.

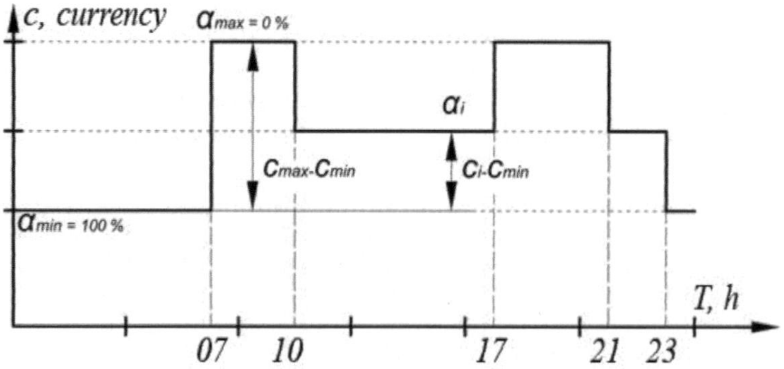


**Figure 1.** The rate efficiency

There are three rates: night (from 23:00 to 07:00), semi-peak (from 10:00 to 17:00 and from 21:00 to 23:00) and peak (from 07:00 to 10:00 and from 17:00 to 21:00). The basic parameter of rate $j$ efficiency a can be defined as normalized relative value obtained by minimum $c_{min}$ and maximal $c_{max}$ prices of night and peak rates, correspondingly:

$$\alpha_j = \frac{1-(c_j - c_{\min})}{(c_{\max} - c_{\min})} \cdot 100\% = \frac{(c_{\max} - c_j)}{(c_{\max} - c_{\min})}, \quad j = 1...M \tag{1}$$

Weight-average rate use efficiency index $e$ is one of the most important parameters. It is based measured energy values $W_j$ (or relative energy values $W_{rj}$ ) obtained by SM:

$$e=\sum_{j=1}^{M}\alpha_j\cdot W_{rj}=\frac{\left(\sum_{j=1}^{M}\alpha_j\cdot W_j\right)}{\left(\sum_{j=1}^{M}W_j\right)} \quad (2)$$

Thus, the coefficient of efficiency e is the weighted average of the rates. Its value represents the opportunity of savings for the consumer. There are two extreme cases:
1) all the electricity was consumed during rate period with $\alpha_{max}$ = 0 %, which means the consumer didn't use the opportunity of savings;
2) all the electricity was consumed during rate period with $\alpha_{min}$ = 100 %, which means the user used all rate for savings.

The actual value of the coefficient of efficiency rates is between 0 % and 100 %.

The weight-average rate use efficiency index e has a triple meaning:
1) utilized economy fact percentage;
2) rates schedule and a consumer's load curve correspondence extent;
3) the elasticity of a demand.

The consumer is able to estimate his savings and choose the most appropriate to him rate. The estimations are suitable for a-priori and a-posteriori dynamic rate selection. The energy company is able to estimate demand-side management efficiency due to the fact that rate use efficiency index is approximately equal to the elasticity of a demand.

There are another components of efficiency rating corresponding the costs for electricity transmission, power market support and other services.

Efficiency for electricity transmission costs is inversely proportional to the share of electricity consumed at the hours of the actual peak: consumption only at the hours of the actual peak leads to $\beta_{max}$ = 0 %, the reduction of consumption during the hours of the actual peak is $\beta_{min}$ = 100 %. The electricity consummated in the hours of the actual peak can has some corresponding intermediate value $\beta_j$

$$\beta_j=(1-\frac{\sum W_j}{W})\cdot 100\%,\quad j=1...N, \quad (3)$$

where $W_j$ – the volumes of electricity consumption during the hours of the actual peak on working days; $W_j$ – volumes of consumption in hours of the actual peak on working days. The rating is aimed at reducing power consumption during the hours of the actual peak.

Rating of the cost effectiveness for power transmission. Powerful consumers can be used to create a regulating effect of the load in hours of planned peak. If the maximum load exceeds the average one during the peak hours over a period of time, the efficiency of using grid power is $\gamma_{max}$ = 0 %, the reduction of consumption in the hours of the planned peak is $\gamma_{min}$ = 100 %. Intermediate power consumption in the hours of the planned peak can be assigned with some appropriate intermediate value $\gamma_k$:

$$\gamma_k=\begin{cases}0\%, & W_k\geq W_{av}\\ (1-\frac{W_k}{W_{av}})\cdot 100\%, & k=1...N,\quad W_k<W_{av},\end{cases} \quad (4)$$

where $W_k$ is the maximum of hourly power consumptions during peak hours on working days; $W_{av}$ is the average hourly power consumption. The rating is aimed at reducing power consumption in the hours of the planned peak.

Rating of the effectiveness of fact-to-plan consumption deviation. The feature of the retail electricity market is that consumers may not be penalized if the electricity consumption deviates at a certain time in proper directions. Therefore, it is proposed to evaluate the effectiveness of fact-to-plan deviation in terms of the deviation of actual costs from potential costs for the case when the planning is not applied at all.

With an increase in electricity costs from the planned level to the level of potential costs with no planning, the efficiency may vary from $\delta_{max} = 0$ %, with the planned costs – $\delta_{min} = 100$ %. Intermediate volumes of consumption in the hours of the planned peak can be assigned some appropriate intermediate value $\delta_l$:

$$\delta_l = \begin{cases} 0\%, \quad C_{l(p)} \geq C_{i(np)} \\ (1 - \dfrac{C_{il(p)} - C_{l(p)}}{C_{i(np)} - C_{l(p)}}) \cdot 100\%, \quad l = 1...N, \quad C_{l(p)} < C_{i(np)}, \end{cases} \tag{5}$$

where $C_{l(p)}$ – costs for planned hourly volumes of consumption; $C_{i(np)}$ – potential costs for the case when the planning is not applied; $C_{il(p)}$ – costs of actual hourly volumes of consumption.

The resulting rating is formed as a weighted average for all indicators instead (2):

$$\varepsilon = \frac{\alpha C_\alpha + \beta C_\beta + \gamma C_\gamma + \delta C_\delta}{C_\alpha + C_\beta + C_\gamma + C_\delta}, \tag{6}$$

where $C_{indexed}$ corresponds to the particular type of costs listed above.

Thus, the resulting rate use efficiency index is a weighted average taking into account the share of costs for various components. Its value characterizes how much the economy has been used by the customer. The real value of the rate use efficiency will be between 0 % and 100 %.

Using the rate use efficiency index $e$, it is possible to determine the economy $E$ and lost savings $E_l$:

$$E = \frac{\varepsilon}{100} \cdot W \cdot (c_{\max} - c_{\min}) \tag{7}$$

$$E_l = \left(1 - \frac{\varepsilon}{100}\right) \cdot W \cdot (c_{\max} - c_{\min}) \tag{8}$$

Another important parameter is the smoothing of load curve. It contributes to equipment lifetime and durability increase as well as peak lines load decrease, so it is very important. From the point of view of Math Statistics the load change can be characterized by standard variation coefficient $v$. But this parameter is not evident even for engineers.

One of the ideas how to estimate the variation using this parameter like an incentive for a consumer is to show him the economy based on the losses decrease. The losses in cables and wires connecting a consumer to electricity mains are considered.

From the theory of electrical power engineering it is well known that the minimum of losses during a period of time corresponds to the case when the load flows are equal at all the parts of the period.

It is useful to estimate relative reduction of the loss for single rate and the relative reduction of losses during of time period of $n$ intervals, correspondingly:

$$r_t = \left(1 - \frac{1}{1 + v^2}\right) \cdot 100\% = \frac{v^2}{1 + v^2} \cdot 100\% \tag{9}$$

$$r = \frac{c\Delta W_{av}}{c\Delta W} = \frac{\sum_{j=1}^{N} W_{avj}^2 \cdot n_j \cdot v_j^2}{\sum_{j=1}^{N} W_{avj}^2 \cdot n_j \cdot \left(1 + v_j^2\right)} \cdot 100\% \tag{10}$$

where $v$ – coefficient of consumption variation; $\Delta W_{av}$ – losses for the case of absolutely uniform consumption; $\Delta W_{av}$ – losses for the case of an arbitrary consumption.

Studies and the calculations show that during the day typical $v$ vary from 0.23 to 0.75 for different rate periods and losses decrease is from 7 % to 30 %, correspondingly. Taking into account that typical losses level is about 10 % for buildings' wiring, savings will be from 0.7 % to 3 %.

Sometimes the problem at hand is not the value of maximum power consumption itself, but the dynamics of load increase at peak times. The concept of easy load curve smoothing by means of Smart Appliances control center and Feedback center interaction is considered (Figure 2).

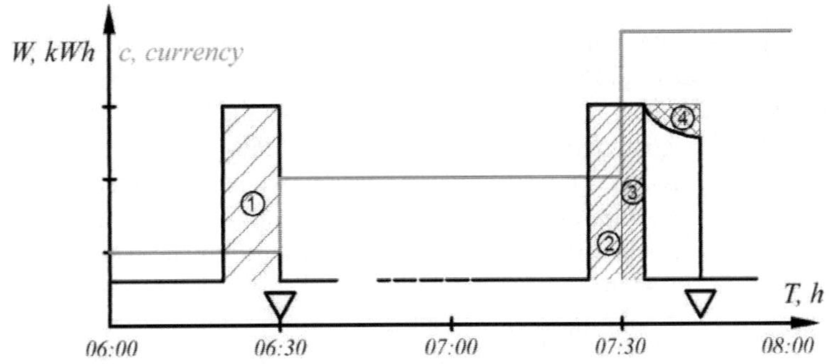


**Figure 2.** The planning principle

The system automatically accounts the load start-up durability which is required for the equipment, and to compare it with the time rate change time. In everyday life for consumers this turns out to be convenient – the opportunity to receive the result of the electrical installations at the time of morning rise or production process run. In the first case (Figure 2) readiness time is equal to the time rate of change. From the point of view of energy company as well as the consumer it is efficient to switch on the device so that the power consumption accounted by cheaper rate (area 1). Operation is planned in such a way that the time of a rate of change is also a readiness time of a process. When the readiness time comes later than time of rate change (the second case), it is profitable that electrical energy is accounted partially by cheaper rate (area 2) and partially by more expensive rate (area 3). The condition of profitable operation:

$$C_4 \leq C_3 - C_2 \tag{11}$$

where $C_4$ are the costs of equipment idle time (e.g., losses), $C_3$-$C_2$ is a profit gained by the rate difference.

## 3. Distributed generation estimation

The particular installations of distributed generation (DG) are under consideration. Such installations are often privately owned and have small power.

First of all, the two effects of the introduction of consumers' generation cause increased attention of energy companies:

1) the effect of masking the load when the sudden shutdown of consumers' own generation may results in feeders overload due to the fact that they was not designed for full capacity;

2) the effect of the consumers' generation far exceeding the demand (e.g. during the night) can cause increased losses and undesirable reverse flows of a great value;

The RMS-imbalance of generation to the load shows the extent of balancing during $M$ periods of time:

$$b = \left( \frac{1}{N} \sum_{i=1}^{N} \left( \frac{W_i - W_{gi}}{W_{gi}} \right)^2 \right)^{1/2} \cdot 100\%, \tag{12}$$

where $W_i$ is load during a single period of time, $W_{gi}$ is generation during a single period of time.

For the case when the imbalance value is far from zero, a permanent connection to the main grid being a damping component is necessary for the potential island of load and generation.

One-time maximum deviation of the generation from load:

$$D = \max\left(\frac{W_{gi} - W_i}{W_i}\right) \cdot 100\%, \quad (13)$$

This parameter shows the extent of instantaneous damping necessity during some particular periods of time and characterize feeders maximum load.

It is known that from the point of view of distribution utilities the benefits from the presence of DG is primarily consist of load coverage and feeders flows reduction. It results at the possibility of postponing capital investments in grid reinforcement. Maximum feeders load corresponds to the period of load curve maximum, especially at the post-failure conditions or at the repair schemes. Demand for DG and own generation of consumers in these conditions increases. That's why the dual-purpose incentives include operation time during the period of the peak rate, the share of electrical energy produced during the period of the peak rate and the income from the generation of electricity:

$$t_{\max} = \left(\frac{T_{\max} - T_{st}}{T_{\max}}\right) \cdot 100\%, \quad (14)$$

where $T_{max}$ is a duration of the peak rate (special or post-failure rate), $T_{st}$ is a idle time of DG installation during the period of the peak rate.

The share of electrical energy produced during the period of the peak rate:

$$d_{\max} = \frac{\sum_{i=1}^{n_{\max}} W_{gi}}{\left(P_g \cdot n_{\max}\right)} \cdot 100\%, \quad (15)$$

where $n_{max}$ is a number of single time periods during the peak rate, $P_g$ is a generation power.

Using the high-price rate $c_{max}$ it is possible to calculate the revenue $R$ as a product of $c_{max}$ and $W_{gi}$ as well as lost revenue due to equipment idle time:

$$R_l = \left(1 - \frac{d_{\max}}{100}\right) \cdot c_{\max} \sum_{i=1}^{n_{\max}} W_{gi}, \quad (16)$$

The mentioned functions of statistics are also available using the Individual Statement.

## 4. The set of parameters

The entire system of the mathematical statistics including the dual-purpose incentives was established (Table 1).

All the information contained in the evaluation is both engineering and motivational (incentive) character. On the one hand, formal recommendations to the consumer are necessary as clarification of opportunities to improve the quality of consumption; on the other hand, a simple formal approach is not comprehensive. The issuance of such recommendations to be implemented with care and some of the issues of engineering psychology and marketing to be used. The recommendations to be linked with the technical, economic, operational and marketing issues facing the energy company at the moment of the recommendations issuance.

The estimation is performed by SM Data processing systems. It can be easily integrated to the Automated Meter Reading and Advanced Metering Infrastructure systems. The cores of these system are shown in Figure 3. Besides energy itself, the measured values may include active and reactive power, currents and voltages, power quality parameters and switched on/off status, if available (for example, using IEC 61850).

**Table 1.** Some DSM important parameters

| Parameter | Sign | Utilization | |
|---|---|---|---|
| | | Energy Companies | Demand Side |
| Rate efficiency | $\alpha, \beta, \gamma, \delta$ [%] | Price signal | Incentives |
| Rates utilization coefficient | $\varepsilon$ [%] | Demand flexibility | Economy potential |
| Economy | $E$ [currency] | - | Utilized |
| Loss of economy | $E_l$ [currency] | - | Missed |
| Average consumption | $W_{av}$ [kW*h] | Balances, statistics | General |
| Consumption variation coefficient | $v$ [%] | Statistics | - |
| Relative losses reduction due to load curve leveling | $r$ [%] | Balances, statistics | Economy potential |
| Percentage of the load covered by DG | $c_v$ [%] | Balances, statistics | Utilized |
| Generation-to-load average unbalance | $b$ [%] | Line av. load | Missed |
| One-time relative generation-to-load unbalance | $D$ [%] | Line max. load | - |
| Absence of idles during the high-price rate | $t_{max}$ [%] | Offer flexibility | General |
| Percentage of generated electrical energy value in comparison with the max. value | $d_{max}$ [%] | Offer flexibility | General |
| Profit gained by DG. during the high-price rate | $R$ [currency] | - | Utilized |
| Loss of profit | $R_l$ [currency] | - | Missed |

## 5. Estimations performed by Smart metering data processing systems

The interaction structure is shown in Figure 3.

The system includes the unit of dynamic rate change and focused primarily on a retail market support. It contributes to the energy efficiency and a distribution process improvement by the motivation (manual management) or by the Smart Appliances interaction (automated management). The feedback of SM can be integrated with Smart Home and Smart Energy. In a manual mode feedback data are outputted to Individual Statement in easy-understandable form, in automatic mode data are used by Smart Appliance control center through WLAN/HAN. Modern systems provide for the opportunity to connect and disconnect loads during grid malfunction.

## 6. A software and hardware complex for an experimental study

A software and hardware complex including SM data processing system was designed and constructed. It was used for Demand-Side Management and its efficiency estimation for different buildings of Ural Federal University campus. A real efficiency of system to be proved by experience of some years practice. Nevertheless, a two of three month application enables to make some notes about non-contractual DSM efficiency.

At the beginning of experiment the following parameters were detected and calculated by the system:

- up to 37 % of customers are not aware about their electricity consumption profile;
- from 25 to 40 % of economy potential is used;

- a bit more than 0.1 % of electricity is consummated in a strict accordance with the planned load curve.

During the experiment the following intermediate results were obtained:

- up to 34.5 % of electricity costs are reduced;
- 42.3 % electricity is consummated by the customers involved in a DSM;
- up to 5.4 % of peak load is shaved, also it is shifted at about 3 h 35 min.

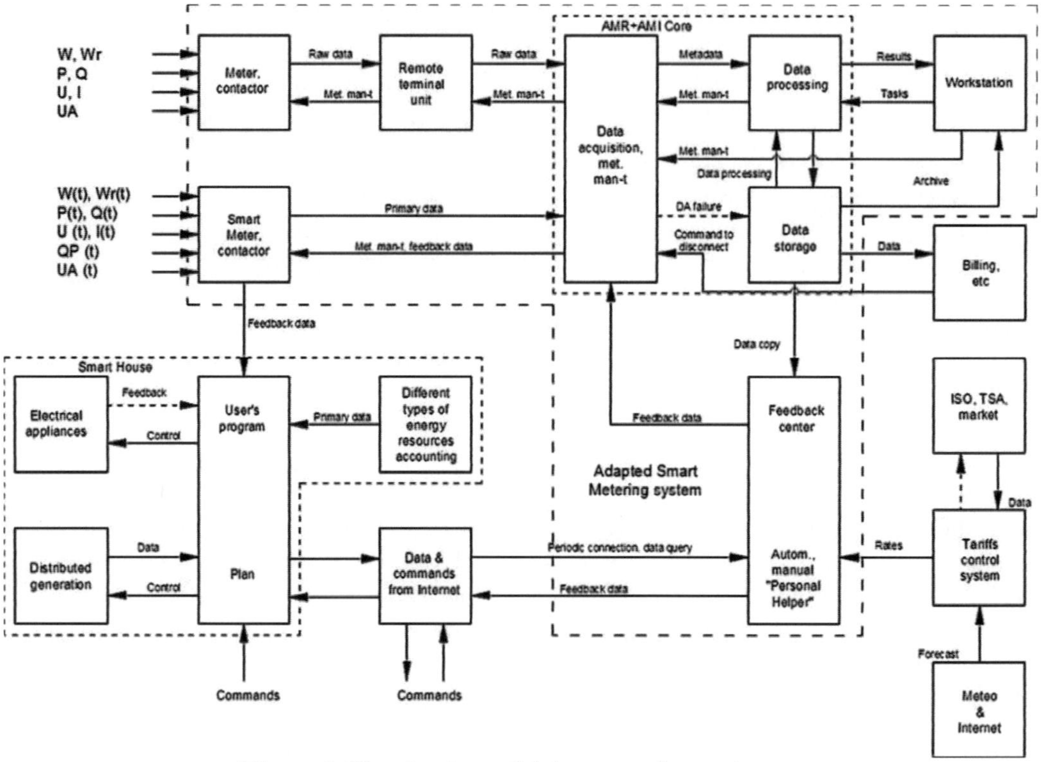


**Figure 3.** The structure of data processing system

## 7. DSM perspectives and further investigations

Although the experiment shows a number of effects listed above, the overall efficiency of DSM for city grids unload is left much to be desired. An analysis reveals a sufficient potential for the problem solution by means of the following measures.

The standard offers also include a widespread DSM units and devices use and an increasing differentiation between electricity prices during the day.

The advanced measures are mostly connected with market pricing model upgrade, but they are supposed to be effective in a case when combined with a well-grounded distribution grid structure complication. It provides an opportunity for dynamic grid reconfiguration. The measures are:

1) the shift of high-price zones of particular city districts or areas. It is aimed primarily at high-voltage grids unload. It requires an advanced statistics for large groups of consumers;

2) the shift of high-price zones of particular feeders connected to particular substations. Supposed to be efficient if a different types of electrical customers in a building connected to two or more feeders. It is aimed primarily at medium-voltage grids unload;

3) the use of a counter regulation by means of decreasing the load;

Also the experiment shows an appreciable load surge during 3–7 minutes of low-price zone beginning, when many consumers are switched on. To avoid this effect, a charge for the value of load ramp up is to be considered.

The listed measures are under investigations now. To incentive the consumers, a virtual prices are going to be used during the experiment study.

## 8. Conclusions

The paper presents the multi-purpose system of mathematical statistics and algorithms of non-contractual DSM and its efficiency estimation useful for the consumers, distributed generation and the energy companies. The estimation is performed by SM Data processing systems. The system of DSM efficiency estimation provides a number of benefits to different parties of electricity market:
1) consumers are motivated by financial incentives and informed by advanced analysis information about rates, economy, and schedule;
2) energy companies are informed about the consumers' response, they have an opportunity to smooth load curve by means of appropriate rate signals, to shave peak load decreasing grid equipment load;
3) the social aspect includes sufficient energy efficiency improvement;
4) distributed generation has an opportunity to gain the maximum profit.

The system shows its efficiency in an experiment for DSM of different Ural Federal University campus buildings. Using the experiment data analysis, the perspective ways of DSM improvement for city grid unload were obtained. They include city grid structure meshing combined with a market pricing model upgrade. The offers are under study to gather a proper statistical data.

## References

[1] Kokin S, Pazderin A and Adarichev E 2009 Ways of decreasing maximum and equalizing curve of big cities' power demand *Proc. of the 1st Int. Conf. on Sustainable Power Generation and Supply* SUPERGEN (Nanjing)

[2] Coletta T, Delabays R, Pagnier L and Jacquod P 2016 Large electric load fluctuations in energy-efficient buildings and how to suppress them with demand side management *Proc. of the IEEE PES Innovative Smart Grid Technologies Conf. Europe* ISGT-Europe (Ljubljana)

[3] Klauw T, Gerards M and Hurink J 2016 Using demand-side management to decrease transformer ageing *Proc. of the IEEE PES Innovative Smart Grid Technologies Conf. Europe* ISGT-Europe (Ljubljana)

[4] Voropai N, Styczynski Z, Komarnicki P, Stepanov V, Suslov K and Stashkevich E 2016 Energy-intensive manufacturing enterprises as active players in demand side management system *Proc. of the IEEE PES Innovative Smart Grid Technologies Conf. Europe* ISGT-Europe (Ljubljana)

[5] Rahman A, Islam R, Sharif K and Aziz T 2016 Developing demand side management program for commercial customers: A case study *Proc. of the 3rd Int. Conf. on Electrical Engineering and Information Communication Technology* ICEEICT (Bangladesh)

[6] Klavsuts I, Rusin G and Klavsuts D 2016 *Proc. of the 11th Int. Forum on Strategic Technology* IFOST (Novosibirsk) p 506–9

[7] Diamantoulakis P, Pappi K, Kong P Y and Karagiannidis G 2016 Game theoretic approach to demand side management in smart grid with user-dependent acceptance prices *Proc. of the IEEE 84th Vehicular Technology Conference* VTC-Fall (Montreal)

[8] Ghazzai H and Kadri A 2017 Joint demand-side management in smart grid for green collaborative mobile operators under dynamic pricing and fairness setup *IEEE Transactions on Green Communications and Networking* **1** p 74–88

[9] Pazderin A, Kokin S, Egorov A and Kochneva E 2009 *Proc. of the 35th Annual Conf. of the IEEE Industrial Electronics Society* IECON (Lisboa) p 1736–41

[10] Pazderin A and Kochneva E 2014 *Proc. of the IEEE Int. Energy Conf.* ENERGYCON (Dubrovnik) p 386–91

[11] Eroshenko S, Khalyasmaa A, Dmitriev S, Pazderin A and Karpenko A 2013 Solution of energy flow problem using state estimation technique *Proc. of 2013 Int. Conf. on Power, Energy and Control* ICPEC (Dindigul)
[12] Zhang W, Cheng G and Dong Z 2015 Demand side management given distributed generation and storage: A comparison for different pricing and regulation scenarios *Proc. of the 10th International Conference on Advances in Power System Control, Operation & Management* APSCOM (Hong Kong)

IOP Conf. Series: Earth and Environmental Science **72** (2017) 012024 doi:10.1088/1755-1315/72/1/012024

# Opportunity and prospect analysis of RES utilization for sustainable development of Ekaterinburg city in Russia

**A Volkov[1] and A Aristova[2]**

[1] Master Student, Department of Environmental Economics, Graduate School of Economics and Management, Ural Federal University named after the first President of Russia B. N. Yeltsin, Mira str. 19, Ekaterinburg, 620002, Russia
[2] Invited Senior Lecturer, Department of Environmental Economics, Graduate School of Economics and Management, Ural Federal University named after the first President of Russia B. N. Yeltsin, Mira str. 19, Ekaterinburg, 620002, Russia

E-mail: volkov.a.v.1994@yandex.ru

**Abstract**. Recently megalopolises have become centres of economy development worldwide. Gradual growth in energy consumption and thereafter – enormous power production and delivery to sustain metropolis' needs entailed, rapid increase in emissions of hazardous substances in quantities, no longer tolerable for secure residence in majority of these cities. Ekaterinburg, is one of them. In order to abridge harmful pollution in Ekaterinburg and further centralize economic importance of the city, this paper proposes to implement the concept of urban sustainable development /ref. / by introducing alternative energy sources, which would progressively displace traditional fossil fuels. A number of actual cases, where the concept was successfully implemented, were studied and analysed to demonstrate how different shares of renewables can become effective substitutes to conventional energy sources in the cities strongly dependent on them: 1. Energy strategy of Pecs (Hungary); 2. International low carbon city (ILCC) project (Shenzhen, China); 3. Electric power system template of Tangshan city (China). Further, regional environmental and economic specifics of Ekaterinburg were studied to understand power consumption needs and energy generation possibilities, which led authors to conclude on the alternative energy sources feasibility, plot specific flow chart for RES implementation in Ekaterinburg's power network and outline recommendations for future works.

## 1. Introduction

Existing statistical estimates of demographic increase and rapid urban development lead to significant changes in structure and volumes of energy consumption. According to R. York [1], 1 % growth in Europe population magnifies energy demands by 2.7 %, thereafter reaching ca. 2.18 Mtoe per year by 2025, which exceeds the levels of 2000 in almost 1.5 times. Along with substantial economic consequences, energy consumption has an impact on quantities of hazardous substances discharged in the atmosphere.

Developing countries have become among the most reflective examples of such a development scenario. Due to intense industrial and urban progression coupled with population increase, China and India are nowadays leaders in carbon dioxide emissions: only in 2010, their share in the global atmosphere pollution with $CO_2$ were 10.4 % and 9.4 % respectively. To compare, Russia, USA and the EU have contributed 5.8 %, 4.1 % and 2.2 % respectively [2].

Furthermore, according to the UNO [3], forecasting the respective decrease in rural population in Russia, China and USA by 38.5 %, 47.3 % and 15.6 %, the emissions numbers are only foreseen to grow.

Progressively cities become main consumers of electrical power, and consequently – ground zero for environmental contamination. With a current share of 75 %, the urban centers are forecasted to reach almost 90 % of the energy use in the nearest future [4]. The numbers allow reasonably assume that replacement of the urban fossil energy with alternative energy sources will substantially improve ecological environment, reducing content of $CO_2$, $SO_2$ and nitrogen oxides ($NO_x$) in the atmosphere.

Thus, it is suggested to investigate how the concept of sustainable development, successfully implemented in a few megapolises, would compensate for increased energy demands and mitigate unwanted ecological disaster and economical outcomes on a national and international levels. This paper outlooks possibility of renewable energy sources implementation in energy networks of Ekaterinburg city in Russia and attempts to build a flowchart of necessary transformations in power infrastructure. It also constructs the basement for future works in need to effectively overcome challenges to achieve sustainable development of a city.

Related research works [5–7] usually focus on finding ways to solve three challenges found relevant to the discussion:

- Complexity of energy transportation utilizing existing network facilities, especially to remote locations [8–10],
- Shortfalls of financial means to build up developed power networks in regions with easy access to vast volumes of renewables [11, 12],
- Control and reduction of hazardous substances emitted into the atmosphere [13, 14].

Taking all of these into account the next section will determine sub-tasks specific to Ekaterinburg conditions.

## 2. Study of Ekaterinburg city specifics

Ekaterinburg is one of the largest and most dynamically growing cities in Russia. Its favorable geographic location determines its national importance as a transportation hub between the Central Russia and Siberia, and one of the most important industrial, transportation, financial, scientific and cultural centers in the country.

As Siemens reported [15] electrical energy feeding the city is produced by power plants working on natural gas. Shortfalls are mainly covered by nuclear energy delivered from neighboring Beloyarsk power station. Detailed distribution of Ekaterinburg's power sources is presented in Figure 1.

Based on available data of the population and total energy demands in the region [16], energy annual consumption per capita was estimated as 10.5 MWh. As reported by the UN [3] the numbers are considerably higher due to the administrative, financial and industrial importance of the city. Based on our estimates and available UN reports, the dependencies of population, energy consumption and IPI is demonstrated in Figure 2.

Ekaterinburg carries extensive network of public transport working on fossil fuels and partly on electricity. Electrical transportation means share equals 42 %. This includes trams (18 %), subway (10 %), trolley busses (8 %), and electrical train (6 %). Municipal lighting network has poor effectiveness where 60 % and 36 % of illumination devices has a light output ratio lower than 50 and 90 lumens per watt accordingly, and only 4 % is above 90 lumens per watt.

Arrest of development takes place due to the overall poor energy efficiency in the city. Besides, unfavorable environmental situation was declared by TROICA project experiment [17] where hazardous pollutant emissions in Ekaterinburg were deeply investigated. It was reported that annual nitrogen discharge N is totaled to enormous 6300 tons, carbon monoxide CO – 133000 tons, methane $CH_4$ – 40000 tons. At low wind conditions in the city, the significant level of air pollution was declared.

Taking all of these into account next section will examine prospects of RES (renewable energy sources) utilization in Ekaterinburg and define key elements of an action plan for urban power network development.

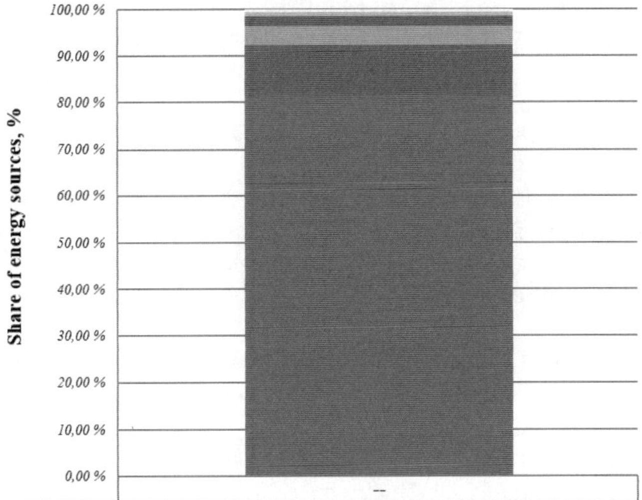


| | |
|---|---|
| Oil | 0,10 % |
| Nuclear Energy | 0,60 % |
| Hydroenergy | 0,80 % |
| Other sources | 2,10 % |
| Oil Products | 4,00 % |
| Coal | 10,90 % |
| Natural Gas | 81,50 % |

**Figure 1.** Primary sources of electricity and heat generation in Ekaterinburg city, %

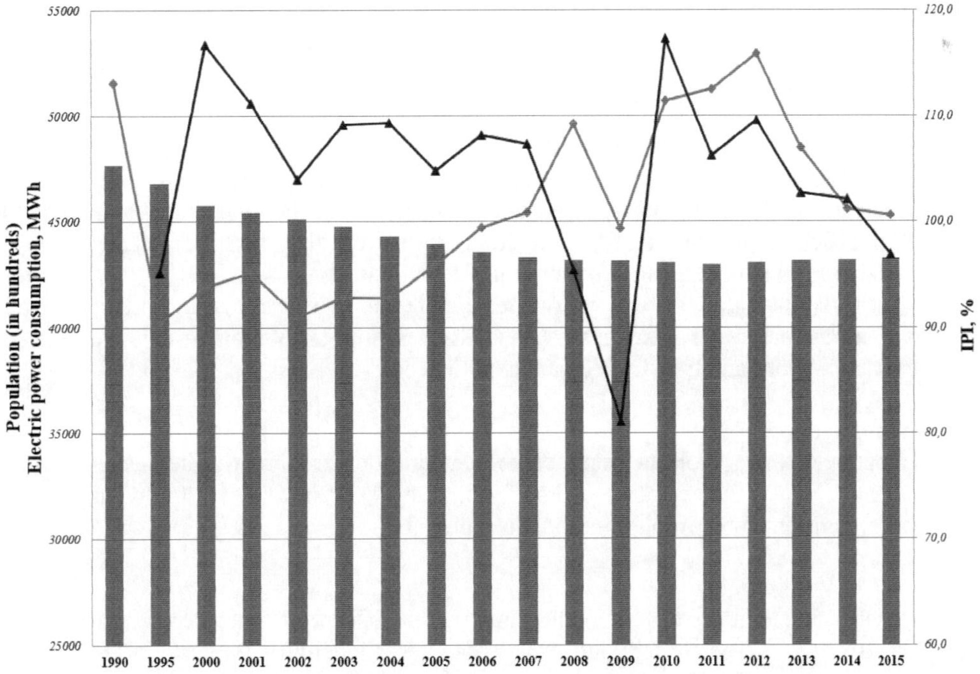


**Figure 2.** Correlation between population, its energy usage and industrial production index in Sverdlovsk region, 1990–2015

Producing ca. 50 % of electricity Novosverdlovskaya thermal power plant is located within the city limits and together with neighboring Sredneuralskaya TTP they form central heating system of Ekaterinburg. Providing 95 % of residential buildings are heated by the network, approximately 3500 km of heating flowlines have spanned in the web all over the city. Due to the high depreciation rate thermal energy losses count on a 25 % level, which governs very low efficiency of the system.

## 3. RES availability in Ekaterinburg city

Unlike conventional energy plants, the energy volumes produced on a renewable power plant are changing with the weather. Understanding of climate conditions in the city is therefore essential to analyze efficiency of the renewables. It is necessary to note, that the research on Ekaterinburg RES potential is still being carried out by the authors. Therefore, the article uses results of existing third party studies, which only touch upon solar and wind energy opportunities relevant for the region.

In the context of how efficient solar energy would be, several researches were carried out [18, 19] and concluded that:

- Sunlight, suitable for solar energy production, is available for 1900–2100 hours per year,
- Maximum solar activity falls on July (over 295 hours), its minimum – on December with only 36 hours,
- Solar energy efficient economic life adds up to 9 months (from April to September),
- Testing of solar concentrators in the city were successfully executed and found efficient with 4560 MJ/m$^2$ of energy produced by solar radiation annually, 645 MJ/m$^2$ as max and 101 MJ/m$^2$ as min of solar radiance in June and December respectively.

It is necessary to notice that clouded sky is a frequent event in the region. However in accordance with the technical acceptability of the solar radiation by Veynberg's criteria, implication of solar energy to the electricity generation network is considered as acceptable if for most of the year flow rate of the solar radiance exceeds 0.42 kWh/m$^2$.

Introducing solar capacitors in the primary central heating system would compensate for heat losses in the heating pipelines and would set conditions for efficient maintenance of the power plant in summer time.

Besides solar energy, highly dependent on whether the sky is clear on a day, hydroelectric power from river Eset, appears to be an attractive alternative indifferent to weather fluctuations thanks to the constant current and deep waters. However due to relatively weak water flow the river can only sustain pumps of the central heating system [20, 21].

Due to the low wind speed, averaged as about 4 m/s at 0–100 m elevations, wind energy generation is deemed efficient only for small, low-power wind mills [22]. For the sake of this study, and despite limited use of efficient wind-powered generators is theoretically feasible, Ekaterinburg is not considered suitable for wind energetics infrastructure development.

To carry out feasibility study of how various renewable energy sources could further be embedded into the city's infrastructure, it is suggested to analyze examples of similar existing international projects to extract methods and transfer relevant experience.

## 4. International practices

This section outlooks three projects implemented in major cities abroad where sources of energy to sustain these cities were replaced from fossil fuels to renewables. The detailed summary of the projects are presented below and outlined in Figure 3.

### *4.1. Pecs*

Based on the national energetics development strategy, the concept of clean Pecs city (Hungary) was introduced in 2013 [23]. The concept utilized computer modeling in EnergyPRO software [24] to compare two different scenarios of the city development. The first model reflected current status of the city, highly dependent on traditional energy, and anticipated to feed growing demands with energy produced from biological wastes and wood processing leftovers. The second model assumed supreme engagement of the renewables.

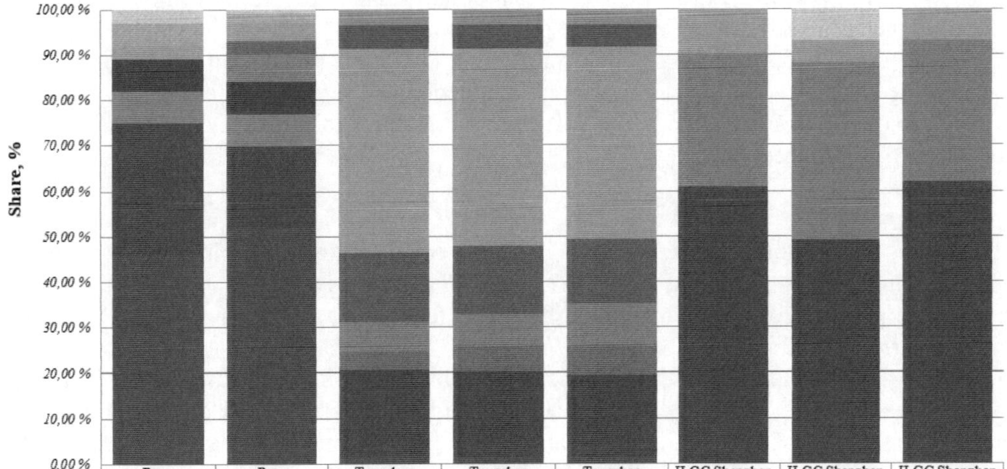


| | Pecs Scenario 1 | Pecs Scenario 2 | Tangshan Scenario 1 | Tangshan Scenario 2 | Tangshan Scenario 3 | ILCC Shenzhen Scenario 1 | ILCC Shenzhen Scenario 2 | ILCC Shenzhen Scenario 3 |
|---|---|---|---|---|---|---|---|---|
| Other sources | 3,00 % | 1,00 % | | | | | 7,00 % | |
| Associated gas | 5,00 % | | | | | | | |
| Geothermal Energy | | 2,00 % | | | | | | |
| Solar Energy | | 2,00 % | 0,60 % | 0,59 % | 0,73 % | 10,00 % | 5,00 % | 7,00 % |
| Domestic Waste | | | 2,84 % | 2,77 % | 2,67 % | | | |
| Industrial Wastes | | | 5,34 % | 5,20 % | 5,02 % | | | |
| Coal | 3,00 % | 2,00 % | 44,85 % | 43,68 % | 42,21 % | | | |
| Mixed: coal + natural gas | | | 15,24 % | 14,84 % | 14,34 % | | | |
| Biogas | | 3,00 % | | | | | | |
| Wind Energy | | 6,00 % | 6,35 % | 6,75 % | 9,24 % | 29,00 % | 39,00 % | 31,00 % |
| Hydroenergy | | | 4,12 % | 6,06 % | 6,35 % | | | |
| Nuclear Energy | 7,00 % | 7,00 % | | | | | | |
| Oil Products | 7,00 % | 7,00 % | | | | | | |
| Natural Gas | 29,00 % | 18,00 % | 18,93 % | 18,44 % | 17,81 % | 61,00 % | 49,00 % | 62,00 % |
| Biomass Energy | 46,00 % | 52,00 % | 1,73 % | 1,68 % | 1,62 % | | | |

**Figure 3.** Distribution of power generation in Pecs, Tangshan and Shenzhen cities in accordance with projects, %

The models computed optimal share of various energy sources in order to minimize $CO_2$ emissions and forecasted alternative energy as about 66 % in proportion to fossils in best case scenario by already 2020, reducing $CO_2$ emissions by 26 % which equals to 62 thousand tons of carbon dioxide.

The following can be borrowed from this project to develop a flowchart of transformations required for sustainable development of Ekaterinburg:

- It is necessary to develop energy strategy for the city, similar to the concept in Pecs, however making even a stronger accord on the renewables efficiency,
- Use of specific software to model and assess performance of the selected distribution between different energy sources.

### 4.2. *Tangshan*

The project around the Chinese city Tangshan has modeled three cases of alternative energy sources implementation in the city infrastructure until 2025–2029. The models have assumed different subsidy assistance and therefore acceptable emission levels of hazardous substances [25]. The first case suggested acceptable emissions in quantities corresponding to the current values, and complete absence of financial support. Second and third scenarios set 10 % and 15 % respective reductions in maximum acceptable contamination levels.

IOP Conf. Series: Earth and Environmental Science **72** (2017) 012024 doi:10.1088/1755-1315/72/1/012024

The research has indicated magnitude of investments required to effectively implement renewables in the city's power sector. It was also shown that subsidy assistance reduce CAPEX of the most expensive third case by only 7.15 %, which demonstrates that subsidizations are not the most necessary factor to successful development of a project.

Hence, the development of a similar project in Ekaterinburg can utilize Tangshan experience in modeling a range of possible economic and ecological indicators based on different shares of renewables. The result should outline precise requirements to an expected net project cost and unit cost of energy produced.

### *4.3. Shenzhen*

Shenzhen is one of the largest megapolises in China, concentrating approximately 40 % of the national GDP. As in many more similar big cities, air in Shenzhen is contaminated far above the tolerable levels. To improve ecological situation in the city, International Low Carbon City (ILCC) project was recently carried out [26].

Hybrid Optimization Model for Electric Renewable (HOMER), based on the software developed in the USA National Laboratory of Alternative Energetics, was selected as the main principle used in the project development. Similarly to Tangshan project, the model has analyzed different scenarios of ILCC execution and forecasted their economic and ecological effect with the following set of variable parameters: net project cost (NPC), cost of unit energy produced (COE) and fraction of energy produced by renewables, which by programming default can not fall below 30 %.

After simulating more than 32000 cases, three optimal scenarios were selected. They are presented on Figure 3. The first scenario, anticipating entire energy generation within the city, appeared as the most expensive with expected CAPEX of $ 653.2 mln. and COE of $ 0.187/kWh. The second case assumed part of the energy is produced outside of the city and delivered via existing infrastructure. The results showed to be having the best financial indicators – $ 511.7 mln. NPC and $ 0.128/kWh COE. The third scenario suggested to use city electrical transport as energy storage facilities and reached best ecological consequences – 78000 tons of carbon dioxide, as opposed to 97000 and 90000 tons for first and second cases. To develop this scenario the forecasted NPC and COE of $ 567.7 mln. and $0.150/kWh have fallen in between two first cases.

This example perfectly demonstrates that the highest investments do not necessarily allow to achieve desired results. And in order to maximize economic efficiency of Ekaterinburg city project, a thoroughly developed plan is required.

## 5. Work flow chart for RES implementation in Ekaterinburg city

It is necessary to designate the essential steps required for sustained and productive development of the city power system. Those are identified as follows:

1. Building up energy strategy on a city level, similar to the one developed in the Pecs city. This will require a team of people specialized in energy sector, ecology and economics. The result of their work will form specific goals and limitations of the strategy.
2. Strategic planning – a step when the expert team will set the acceptable levels for variable parameters and delineate project execution timeframe.
3. Economic analysis should reveal key financial figures – NPC and COE, similarly to Tangshan and Shenzhen projects. The analysis should work through a large number of modeled cases, with and without possible subsidy assistance, in order to reveal just a few rational solutions.
4. Technical analysis will further determine feasibility of alternative energy sources to maximize project efficiency. Results will be used as a basement for technical scope of work to build or modify energy generation and transportation systems.
5. Feasibility study of the planned transformations, using a dedicated software (EnergyPRO or similar), to assess performance of a suggested distribution between various energy sources. The study will allow to accurately determine characteristics of the city energy demands and supply, and prove efficiency and sustainability of the renewables share in the total energy production system.

To successfully develop the project on a short term, and accomplish all steps above on a long term planning, the authors recommend to carry out further detailed researches on every stage above. In addition it is suggested to evaluate long-term stability of the ecological, economic and social characteristics which are seen as dynamic and unstable in some conditions under the urban development process [27], which may influence the project, sensitive to those parameters.

## 6. Conclusions

In response to the experience of existing projects and analyses carried out in this study the following conclusions can be made:

1. Renewable energy sources is a valid alternative to partially displace traditional fossil energy to sustain Ekaterinburg power demands,
2. To a robust RES development in Ekaterinburg, key lessons were learnt from similar projects, having success in other parts of the world,
3. Working flow chart was developed to outline an overall plan and set required minimum for transformations of existing infrastructure to achieve sustainable progress of the metropolis,
4. A foundation for further scientific works required to increase share of the RES in Ekaterinburg's energy sector was laid.

## References

[1] York R 2007 Demographic trends and energy consumption in European Union Nations 1960–2025 *Social Science Research* **36 (3)** p 855–72

[2] Peters G P, Marland G and Quéré C L 2012 Rapid growth in CO2 emissions after the 2008–2009 global financial crisis *Nature Climate Change* **2 (1)** p 2–4

[3] 2014 *UN Department of Economic and Social Affairs Population Division* World Urbanization Prospects The 2014 Revision (New York: United Nations)

[4] Bhatt V, Friley P and Lee J 2010 Integrated energy and environmental systems analysis methodology for achieving low carbon cities *Journal of Renewable and Sustainable Energy* **2 (3)** p 1–19

[5] González M O A, Gonçalves J S and Vasconcelos R M 2017 Sustainable development: Case study in the implementation of renewable energy in Brazil *Journal of Cleaner Production* **142 (1)** p 461–75

[6] Sharma R and Goel S 2016 Stand-alone hybrid energy system for sustainable development in rural India *Environment, Development and Sustainability* **18 (6)** p 1601–14

[7] Lee C T 2017 Sustaining the low-carbon emission development in Asia and beyond: Sustainable energy, water, transportation and low-carbon emission technology *Journal of Cleaner Production* **146** p 1–13

[8] Demiroren A and Yilmaz U 2010 Analysis of change in electric energy cost with using renewable energy sources in Gökceada, Turkey *Renewable and Sustainable Energy Reviews* **14 (1)** p 323–33

[9] Bernal-Agustín J L and Dufo-López R 2009 Simulation and optimization of stand-alone hybrid renewable energy systems *Renewable and Sustainable Energy Reviews* **13 (8)** p 2111–18

[10] Zhou W and Lou C 2010 Current status of research on optimum sizing of stand- alone hybrid solar-wind power generation systems *Applied Energy* **87 (2)** p 380–89

[11] Shiroudi A and Rashidi R 2012 Case study: Simulation and optimization of photovoltaic-wind-battery hybrid energy system in Taleghan- Iran using homer software *Journal of Renewable and Sustainable Energy* **4** p 1–11

[12] Sen R and Bhattacharyya S C 2014 Off-grid electricity generation with renewable energy technologies in India: An application of HOMER *Renewable Energy* **62** p 388–98

[13] Pacesila M Burcea S G and Colesca S E 2016 Analysis of renewable energies in European Union *Renewable and Sustainable Energy Reviews* **56** p 156–70

[14] Nastasi B and Matteo U D 2016 Solar energy technologies in Sustainable Energy Action Plan of Italian big cities (Italy: Turin) *71st Conf. of the Italian Thermal Machines Engineering Association* p 14–16
[15] Karpova A and Tredler I 2010 *Ekaterinburg – a power efficient city* (Moscow: Siemens) p 12–18
[16] 2016 *Federal State Statistics Service of Sverdlovsk Region* (Electronic Materials) Available from: http://sverdl.gks.ru/wps/wcm/connect/rosstat_ts/sverdl/ru/statistics/sverdlStat/ [Accessed 18th Febuary 2017]
[17] Elansky N F, Lavrova O V, Skorokhod A I and Belikov I B 2016 Trace gases in the atmosphere over Russian cities *Atmospheric Environment* **143** p 108–19
[18] Mingaleva Z and Shpak N 2015 Possibilities of Solar Energy Application in Russian cities *Thermal science* **19 (2)** p 457–66
[19] Starikov E, Pahaluev V, Shcheklein S and Nikitin A 2016 Computational and experimental research of the technological steam generation using a solar energy concentrator *16th Int. Multidisc. Scient. GeoConf. (SGEM)* (Bulgaria: Albena) **1** p 237–44
[20] Gryaznov O N, Gulyaev A N Ruban N V Savintsev I A and Cherkasov S A 2015 Factors of engineering-geological conditions of the city of Yekaterinburg *News of the Ural State Mining University* **3** p 5–20
[21] Fuentes E, Waddicor D A and Salom J 2016 Improvements in the characterization of the efficiency degradation of water-to-water heat pumps under cyclic conditions *Applied Energy* **179** p 778–89
[22] Udalov S N and Zubova N V 2013 Simulation of Wind Speed in the Problems of Wind Power *Journal of Siberian Federal University. Engineering & Technologies* **2** p 150–65
[23] Kiss V M 2015 Modelling the energy system of Pecs – The first step towards a sustainable city *Energy* **80** p 373–87
[24] Connolly D Lund H Mathiesen B V and Leahy M 2010 A review of computer tools for analysing the integration of renewable energy into various energy systems *Applied Energy* **87 (4)** p 1059–82
[25] Zhen J L, Huang G H, Li W, Wu C B and Wang S 2016 Electric power system planning with renewable energy accommodation for supporting the sustainable development of Tangshan City, China *Journal of Cleaner Production* **139** p 1308–25
[26] Bin Y, Jingjing J and Lixin M 2015 Sustainable energy options for a low carbon demonstration city project in Shenzhen, China *Journal of renewable and sustainable energy* **7 (2)** p 1–17
[27] Shen L, Shuai C, Jiao L, Tan Y and Song X 2017 Dynamic sustainability performance during urbanization process between BRICS countries *Habitat International* **60** p 19–33

IOP Conf. Series: Earth and Environmental Science **72** (2017) 012025 doi:10.1088/1755-1315/72/1/012025

# Mathematical Approach to Identification of Load Structure at the Nodes of the Distribution Grids 6–10 kV and 0.4 kV

**T Nizamutdinova and N Mukhlynin**

Department of Automated Electrical Systems, Yeltsin UrFU, Ekaterinburg, Russia

e-mail: domptf16@gmail.com

**Abstract**. A significant increasing energy efficiency of the full cycle of production, transmission and distribution of electricity in grids should be based on the management of separate consumers of electricity. The existing energy supply systems based on the concept of «smart things» do not allow to identify the technical structure of the electricity consumption in the load nodes from the grid side. It makes solving the tasks of energy efficiency more difficult. To solve this problem, the use of Wavelet transform to create a mathematical tool for monitoring the load composition in the nodes of the distribution grids of 6–10 kV, 0.4 kV is proposed in this paper. The authors have created a unique wavelet based functions for some consumers, based on their current consumption graphs of these power consumers. Possibility of determination of the characteristics of individual consumers of electricity in total nodal charts of load is shown in the test case. In future, creation of a unified technical and informational model of load control will allow to solve the problem of increasing the economic efficiency of not only certain consumers, but also the entire power supply system as a whole.

## 1. Introduction

Currently the reduction costs on the production, transmission and distribution of energy resources is an important trend around the world. The increasing importance of energy efficiency is determined by several factors: total increasing needs of industry in electrical energy [1]; limitation, high cost and non-renewable nature of conventional energy resources [2]; ecological requirements toughening to power generation processes [3]. All these factors lead to the necessity of creation of complex management systems that control the process of consumers power supply. One of such systems is the "Internet of Things" system which has been developed for power consumers. The concept of the Internet of things ("smart things") of consumer interaction algorithm development is aimed to increasing the energy efficiency of individual devices. The concept presents internetwork relationship of physical devices, vehicles, buildings and other elements with inbuilt technologies, sensors and actuating devices [4]. Relationship between devices is carried out by means of software system using internet technologies. The main idea of the "Internet of Things" is to automate various processes through the exchange of information between devices by software without human intervention [5]. Thus, the optimization task of energy resources use is solved locally.

Further development of the "smart things" concept will allow to create a process control system of transmission, distribution and consumption of power energy. The main idea of developing such system should be based on a complex management of the individual electric plant of energy exchange participants. Existing systems based on the concept of "smart things", have not the technical capacity to identify the consumers composition in the load nodes on the network side. Also the systems have not a function of two-way information exchange between the subjects of electric power industry. These aspects are challenges to determine the load structure, which is necessary to solve the tasks of increasing energy efficiency.

This paper is organized as follows. The applicability of the proposed approach and application area of the developed algorithm are described in Section II. Section III is devoted to the mathematical description of the signal analysis tasks. Conducted experiment and its results are presented in section IV. The conclusions obtained from the study are presented in section V.

## 2. Application area

The information accessibility on the load composition in the network nodes will allow to solve the problems of increasing the work efficiency for all participants of energy exchange [6]. For example, the electricity grid company can use this information to change the service cost on power transmission from the utility to the consumer. Reducing electricity consumption volumes in the desired time intervals can be achieved through informing the consumer about the change cost. It will allow to increase efficiency of the electric utilities and to reduce losses by load levelling in the grid nodes. The process of full or partial control of the operating mode for each consumer on the network side is called network monitoring. Participant of energy can use network monitoring technology to their advantage, but on the condition that it leads to increasing the efficiency of generation, transmission and distribution of electricity. Energy exchange participant should provide electricity to consumers in contractual relationships. The structure of the proposed organization of the information collection and exchange is presented in Figure 1.

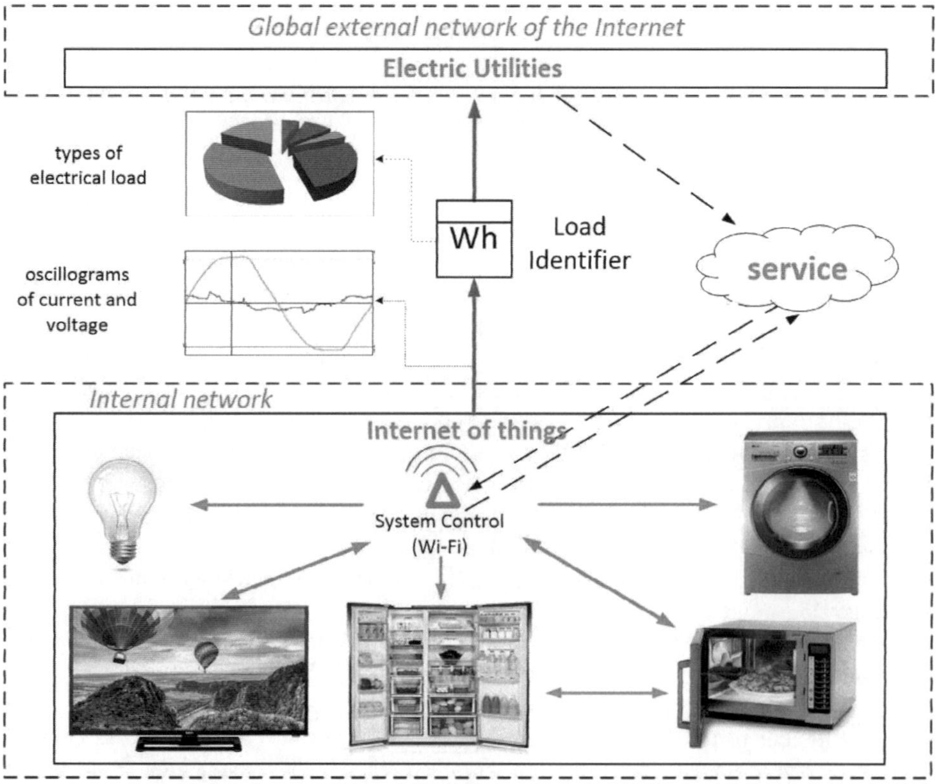


**Figure 1.** Organization of information collection and exchange

Figure 1 shows that the interaction of the various smart electricity consumers is carried out within a private area of the consumer. The internal system of information exchange is organized through the network of the smart things with required access to the external Internet network. The connection between the internal network of the consumer, which is organized by smart electricity consumers (Internet of things), and the global external network of the Internet is absent at this level. Thus, information on the composition and operating mode of a single consumer can't be accessible to external recipients. The network monitoring technology will be implemented in a device such as an electricity meter. It will allow to partially or completely monitor the work of consumers in each node of the network load without violating the information security of the consumer.

Reverse interaction with energy exchange participants can be carried out through services by transferring control information signals from the network to the consumer. Services provide recommendations for electricity consumers connected to the Internet of things. Signals can be the cost indicators of electricity by means of which the schedule of switching on and off some particularly power-consuming electric consumers is built. It is necessary to develop a mathematical algorithm in digital electricity meters that will allow to determine the composition of energy consumers for the realization of such opportunities.

Therefore, the main goal of the research described in the article is to propose a new approach for solving the problem of distinguishing different types of consumers from the total load signal.

## 3. Mathematical algorithms

To date, the conventional method of the analysis of electrical signals is the Fourier transform. However, this method has several disadvantages which limit its application in tasks of load composition analysis [6]. Fourier analysis is not capable to give an accurate analysis of the signals in the frequency as well as in the time domain simultaneously. Moreover, signals often include non-periodical components which are imposed on the fundamental frequency signal. This presents a problem for the Fourier transform, because its use requires periodic signals [6, 7]. Therefore, in latter days wavelet transform is widely used for the analysis of digital signals. Mechanisms of signal decomposition by Fourier and wavelet transformations are mathematically similar. Basis, where the original signal is decomposed, is used in both transformations. The strictly deterministic trigonometric basis is used in the Fourier transform, the arbitrary basis is applied in wavelet transform. The basis of wavelet transformation should satisfy certain conditions [7]. Wavelet analysis is a method based on a large-scale transformation, translation of basis (mother) wavelets and comparison of these wavelets with the original signal [8]. Wavelet-coefficients at different scales are obtained as a result of signal comparison with mother wavelet function. Figure 2 shows the principle of Wavelet transform.

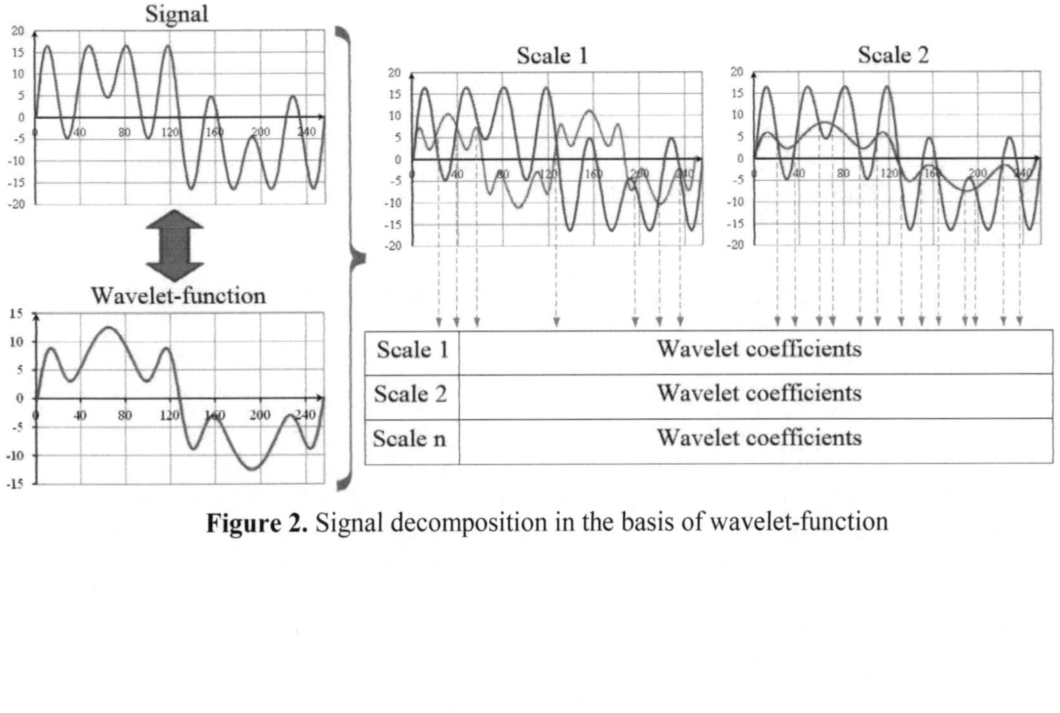


**Figure 2.** Signal decomposition in the basis of wavelet-function

In addition wavelet transformation is capable to localize the information in the time-frequency domain with a high quality. These characteristics and advantages of the described method will form the basis of a mathematical algorithm of analysis and identification of electrical load in the network nodes. Direct Wavelet transform is defined as the sum over the entire duration of the signal multiplied by scalable shifted versions of wavelet functions [9]:

$$W(a,b) = C = \int_{-\infty}^{+\infty} s(t) a^{1/2} \psi\left(\frac{t-b}{a}\right) dt \tag{1}$$

where $b$ is scale factor or parameter extension; a is the translation parameter (translation step); the function ψ(t) is transformation function which is called the basis function or mother wavelet; $a^{1/2}$ is the constant, it is necessary for normalization in order to the signal had the same energy at each scale.

It should be noted that the mother wavelet function must satisfy several conditions [9, 10]: it should be short and oscillatory, i.e. it must have zero average and decay quickly at both ends. Several examples of wavelets are shown in Figure 3.

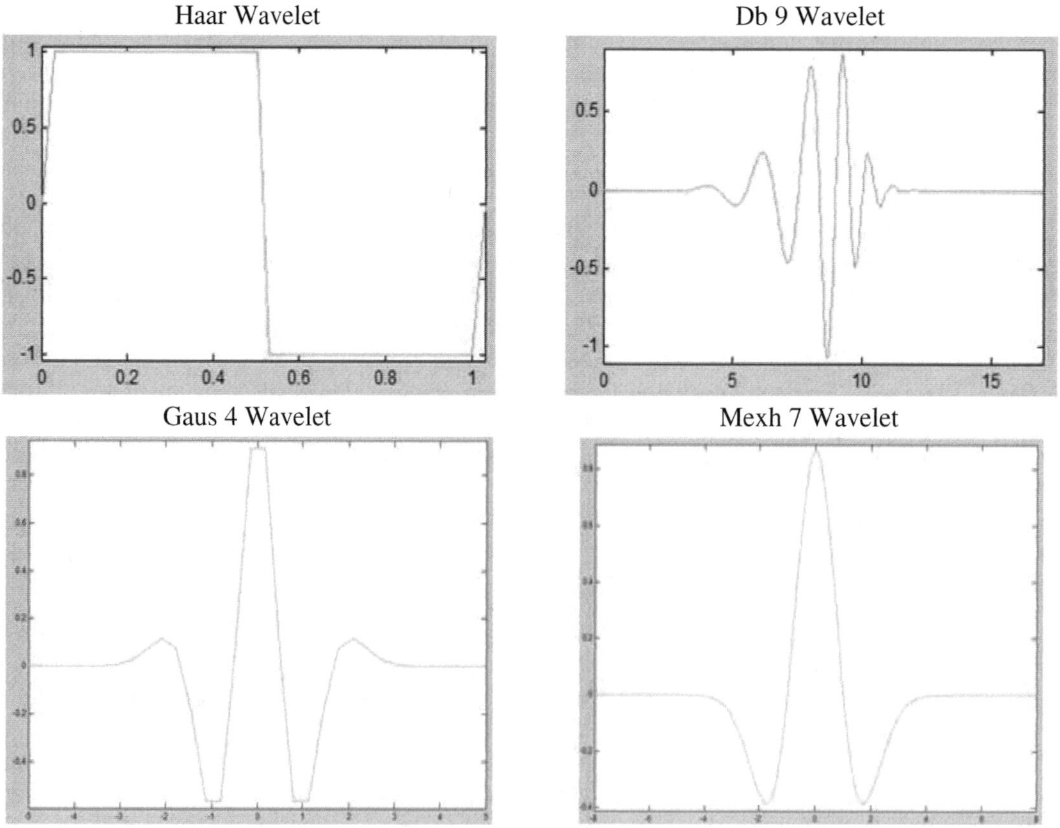


**Figure 3.** Different types of standard basic wavelet functions

Adequacy of signals analysis is significantly dependent on the chosen mother wavelet [10].This is particularly important to solve the problem of identifying specific types of load from the total load signal. Application of existing wavelets cannot provide a qualitative signal analysis in the context of the load identification, because the total load signals are to be analysed from side of each load.

Therefore, the application of wavelet functions based on an individual load signal is considered in this paper. This will allow to analyze the consumers' composition more accurately and correctly.

## 4. Experiment and results of analysis

### *4.1. Experiment description*

To prove the possibility of Wavelet transform use for solution of load identification task the experiment has been performed. It has included the follows:

1) Obtaining current and voltage oscillograms of various electric devices by type and power to clearly show the efficiency of Wavelet transform in identifying specific features of particular electricity receivers from the total load signal.

   The following power consumers oscillograms have been considered:

- Light-emitting-diode lamp ($P_{nom}$ = 6 W);
- Laptop ($P_{nom}$ = 28 W);
- Electric kettle ($P_{nom}$ = 1850 W).

2) Obtaining current and voltage oscillograms with different combinations of operating devices:

- LED Lamp and Laptop;
- Lamp and Electric kettle;
- Laptop and Electric kettle.

3) Creation of unique mother wavelet for each type of device under consideration;
4) Obtaining wavelet coefficients of the spectrum for each measured signal;
5) Analysis of coefficients obtained and formulation of criteria for load structure identification in the node of the distribution network by the type of power consumers.

Measurements have been performed by Chauvin Arnoux CA 8335 power quality analyzer. Oscillograms were recorded with one period of industrial frequency 50 Hz. Frequency of measurement was 256 point per period ($f$ = 12.8 kHz).

Figure 4 shows the protocol of the measurements and the experiment scheme.

**Figure 4**. Scheme of the experiment and measurements report

The analysis of current and voltage oscillograms has shown that current signals are more revealing in solution of load identification task. Figure 5 shows current signals which were measured.

From the Figure 5 it is clear that current signals of laptop and lamp are significantly distorted by higher harmonic components. This is due to the fact that there are line adapters in these electricity receivers. Also it should be mentioned that laptop has a line adapter of impulse type (dotted line in the Figure 5). The latter explains the "sawtooth" waveform of the signal. The graph of current signal of the kettle is close to the pure sinusoid. This is due to the fact that electric kettle is the pure active load.

From the common load signal it is seen that the more the power of the electric receiver the common graph form looks more like a graph form of powerful electric device. This is well traced for current signals where there is a kettle signal (lamp and electric kettle, laptop and electric kettle).

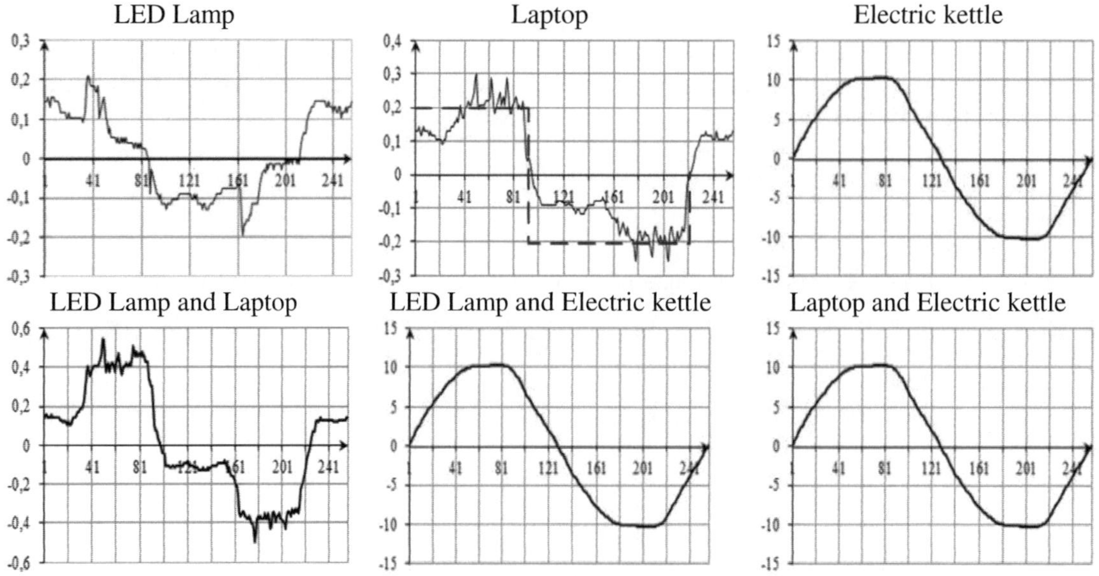


**Figure 5.** Current Signals
(Current characteristic according to the quantity of measurements)

However, identification of specific features of separate power consumers in the total nodal load schedules is necessary for increasing energy efficiency in electric networks. Therefore, the unique basis wavelet functions were obtained on the basis of the measured load current signals for electric devices under consideration (LED Lamp, Laptop, Electric kettle).

### *4.2. Creation of unique basic wavelet function*

Wavelet Toolbox from Matlab software package was used to create the unique wavelet function adapted to specific load current signal. Matlab software package allows to create wavelet functions for continuous Wavelet transform. Current oscillograms of electricity receivers under consideration (LED lamp, laptop and electric kettle) were used both in named and relative units as original signals to create unique basic wavelet-functions.

It was assumed that significant difference between current signals scales of LED Lamp, Laptop and Electric kettle will strongly influence the values of wavelet-coefficients. Therefore, initially measured oscillograms were considered in relative units. However, it was found by the experimental way that wavelet-functions created on the basis of current signals in named units are more effective for solution of load identification task. For this reason, the unique wavelets created on the basis of current signal in named units were used in these studies.

Unique mother wavelets for LED lamp, laptop and electric kettle are presented in the Figure 6.

Figure 6 shows that graphs of unique basic wavelet functions significantly differ in shape from graphs of current signals of electricity receivers. Graph of lamp's wavelet function has obvious "noise". At the same time, kettle's wavelet function is smoother and more symmetrical. Thus, it can be assumed that unique wavelet function for any active load will be symmetrical and smooth. This feature of Wavelet transform will be useful in identification of electricity receivers and their classification by the type.

Created basic wavelet functions adapted to particular load type were used to obtain wavelet coefficients for each measured load signal.

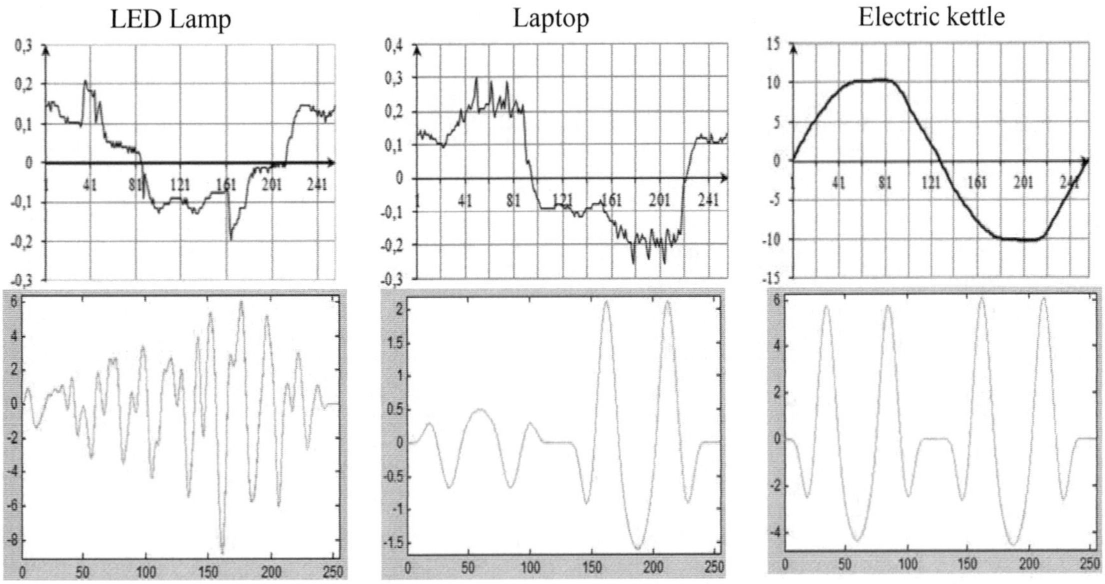


**Figure 6.** Unique mother wavelets

### *4.3. Obtaining wavelet-coefficients and results*

Wavelet Toolbox in Matlab software package allows to obtain wavelet-coefficients of analyzed signal at different scales. Wavelet-coefficients shows signal features: signal "noise" is traced at small scales and wavelet-coefficients reflects a behavior of measured signal with an increase in scale. It is difficult to visually observe signal of LED lamp and laptop in total measured oscillograms (Figure 5) where there is a signal of electric kettle. Therefore, it has been assumed that we should identify electric devices of small power by wallet-coefficients at small scales.

Values of wavelet-coefficients extremely depend on the type of the chosen wavelet on which basis the analyzed signal is decomposed. The choice of mother wavelet takes a significant part in identification of electric devices. So, wavelet-functions used in the present study were created for their own load type on which basis wavelet coefficients have been created [11].

The algorithm shown in the Figure 7 in the form of block-diagram was used to identify electric devices of total measured signals. It includes the following steps:

- Signals of LED lamp, laptop and electric kettle decomposition in the basis of wavelet-function of the same load and obtaining standard values of wavelet-coefficients;
- Wavelet-transform of total load signal (LED lamp and laptop signal, LED lamp and electric kettle signal, laptop and electric kettle signal) by wavelet-function of lamp, laptop and electric kettle.
- Calculation of deviations of total signal wavelet-coefficients from standard coefficients.
  Deviations are calculated by the formula [12]:

$$\sigma = \sqrt{\frac{\sum_{1}^{n}\left(k_{\mathrm{W}}^{STANDART} - k_{\mathrm{W}}\right)}{n}} \tag{2}$$

where $k_W^{STANDART}$ – value of standard wavelet-coefficient, $k_{\mathrm{W}}$ – value of total signal wavelet-coefficient in each measured signal, $n$ – frequency of measurements per period (in the present study it is equal to 256 point per period).

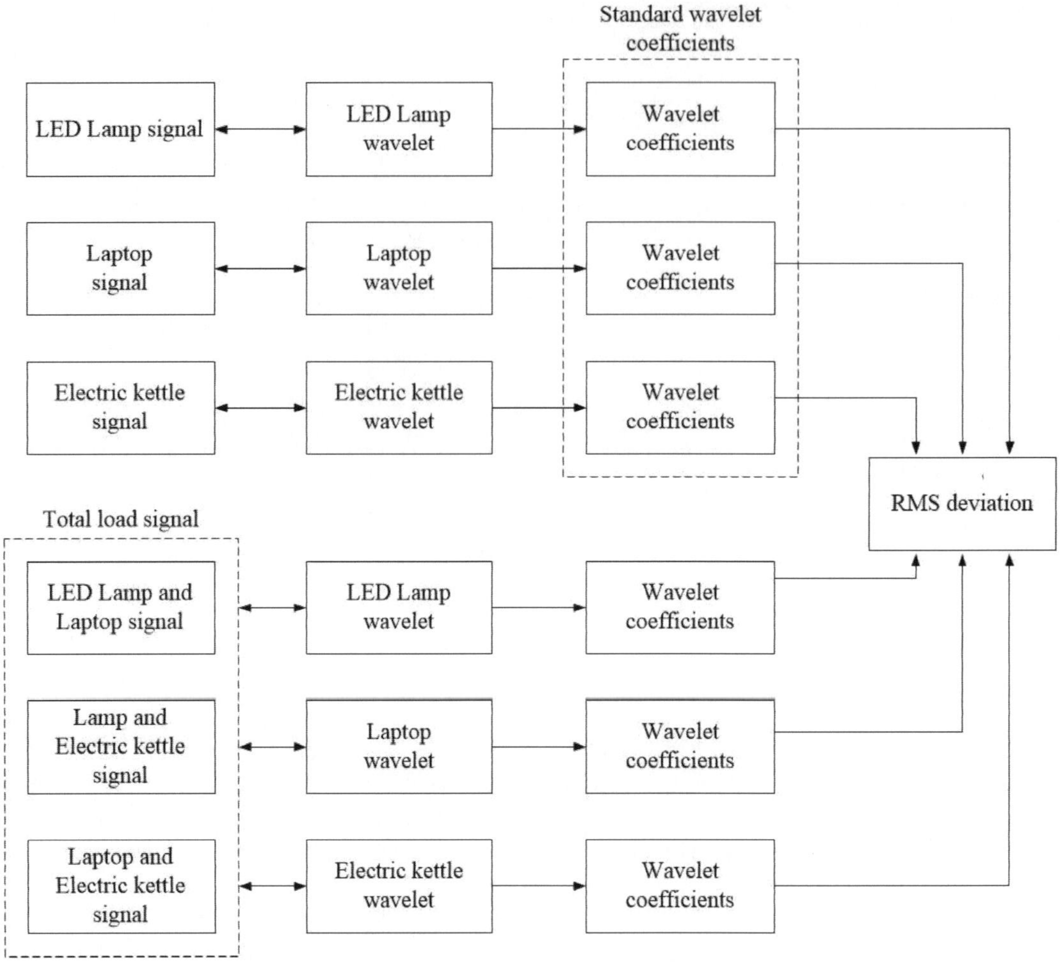


**Figure 7.** Block-diagram describing algorithm of electricity receivers identification

Calculation of total signal wavelet-coefficients is presented in Table 1.

**Table 1**. Deviations of wavelet-coefficients from standard

| LED Lamp wavelet-function | | Laptop wavelet-function | | Electric kettle wavelet-function | |
|---|---|---|---|---|---|
| | | LED Lamp and Laptop signal | | | |
| Scale 1 | 7090618291 | Scale 1 | 0,60641333 | **Scale 1** | **6,0341E+48** |
| | | Lamp and Electric kettle signal | | | |
| Scale 1 | 8420385789 | **Scale 1** | **5,38924183** | Scale 1 | 6,1113E+47 |
| | | Laptop and Electric kettle signal | | | |
| **Scale 1** | **9336881166** | Scale 1 | 5,38579427 | Scale 1 | 1,8538E+48 |

The results of coefficients deviations calculation presented in Table 1 shows that the total load signals decomposed in the basis of wavelet-function adapted for electric receiver which signal is presented in the total load signal gives smaller value of the deviation. For instance, in case of total load signals decomposed in the basis the lamp's wavelet-function lowest values of deviations were obtained for LED Lamp, Laptop and Electric kettle signals. The highest value of deviation (bold) was obtained for Laptop and Electric kettle signal. The same results were obtained in decomposition of total load signals by another basis wavelet-functions.

## 5. Conclusions

1. The algorithm of Wavelet-transform was proposed as mathematical technique for signal identification. Wavelet-transform is able to allocate local features of signals due to multiscale analysis. The latter is a significant advantage for load identification task.

2. The unique mother wavelet was created for each measured load signal. Using created unique mother wavelet function in signal decomposition it is possible to identify electric device for which the unique wavelet was created.

3. The algorithm for load identification was proposed. It includes 3 basic steps: determination of standard wavelet-coefficients values, obtaining coefficients for total load signals, calculation of coefficients deviation at small scales.

4. The further work will be devoted to improvement of the possibilities of load structure identification using Wavelet-transform by the increase of sampling frequency of measured signals.

**Acknowledgment**
The work was supported by Act 211 Government of the Russian Federation, contract No. 02.A03.21.0006 and the Ministry of Education and Science of the Russian Federation (in the framework of state assignment, No. 13.1928.2014/K (project No. 1928).

## References

[1] Huan K, Zhu Y and Shen H 2012 A new evaluation system for energy-saving effect based on extended energy consumption and extended energy efficiency *International Conference on Electricity Distribution* (Shanghai)
[2] Coroiu M and Chindris M 2014 Energy efficiency indicators and methodology for evaluation of energy performance and retained savings *49th International Universities Power Engineering Conference (UPEC)* (Romania)
[3] Jeandel A 2013 Energy Efficient Solutions for Existing Communities *World energy council* annex II p 1–38
[4] Mattern, Friedemann, Floerkemeier and Christian 2016 From the Internet of Computers to the Internet of Things *Distributed Systems Group, Institute for Pervasive Computing* (Zurich)
[5] Cisco.com 2016 *An Introduction to the Internet of Things* (*IoT*) (San Francisco: Lopez Research)
[6] Berger L T and Iniewski K 2012 *Smart Grid Applications, Communications, and Security* p 488
[7] Perera N, Rajapakse A D and Jayasinghe R P 2007 On-line Discrete Wavelet Transform in EMTP Environment and Applications in Protection Relaying *International Conference on Power Systems Transients (IPST'07)* (Lyon)
[8] Haibo He, Xiaoping S and Janusz A Sc 2009 Power quality disturbances analysis based on EDMRA method *Electrical Power and Energy Systems* **31** p 258–68
[9] Gu X, Li G, Zhou M and Lo K L 2011 Wavelet Transform Based Approach to Harmonic Analysis *11th International Conference on Electrical Power Quality and Utilisation* (Lisbon)
[10] Daubechies I 1992 *Ten Lectures on Wavelets* (Philadelphia, Pennsylvania: SIAM)
[11] Misiti M, Misiti Y, Oppenheim G and Poggi J M 1996–1997 *Wavelet Toolbox for use with MATLAB*
[12] Dodge and Yadolah 2003 *The Oxford Dictionary of Statistical Terms* (Oxford University Press)

# The use of urban wood waste as an energy resource

**G I Khudyakova*, D A Danilova and R R Khasanov**

Heat power plants Department, Ural Federal University, 19 Mira str., Ekaterinburg, 620002, Russia

*E-mail: uge87@mail.ru

**Abstract**. The capabilities use of wood waste in the Ekaterinburg city, generated during the felling of trees and sanitation in the care of green plantations in the streets, parks, squares, forest parks was investigated in this study. In the cities at the moment, all the wood, that is removed from city streets turns into waste completely. Wood waste is brought to the landfill of solid household waste, and moreover sorting and evaluation of the quantitative composition of wood waste is not carried out. Several technical solutions that are used in different countries have been proposed for the energy use of wood waste: heat and electrical energy generation, liquid and solid biofuel production. An estimation of the energy potential of the city wood waste was made, for total and for produced heat and electrical energy based on modern engineering developments. According to our estimates total energy potential of wood waste in the city measure up more 340 thousand GJ per year.

## 1. Introduction

In Russia, the problem of utilization of wood waste in the forestry and woodworking industry, as well as wood waste generated during pruning trees and bushes on city streets and park zones, is one of the most urgent. It is so because at present more than a half of the tree biomass is lost with the existing methods of processing. But in the cities at the moment, all the wood waste is brought to the landfill of solid household waste. It is indicates a low level of technological processes of woodworking and ignoring existing opportunities for using wood waste from cities.

Wood waste can be used as an energy resource to provide the urban consumer with heat and electrical energy. Such an approach will improve environmental performance by reducing the waste amount at urban landfills and production of useful energy from renewable sources of wood waste.

## 2. Research area – Ekaterinburg city

Ekaterinburg – one of the largest economic centers of the world, is on the list of City-600, compiled by the research organization Mc Kinsey Global Institute. Ekaterinburg is located in a zone moderately a continental climate, so characterized by sharp variability of weather conditions with pronounced seasons. According to the landscape zoning the city is located within the south-taiga forest zone, near its boundary with the subzone of mixed forests [1].

The vegetation cover of the city is represented by native south-taiga aspen, birch and pine forests, and grown on site cutting down the aspen-birch forests; and flooded meadows, tree and shrub thickets, vegetation along the on the coasts of transitional sedge-sphagnum marshes (on the city outskirts). In general, urban vegetation is characterized by high floristic wealth. Pine trees, birches, rare and relic plants grow on the territory of the Shartash forest park and in the mountains of Uktus.

The greening system in Ekaterinburg was formed over two centuries. It was influenced both by natural factors (location within the city of natural forests, mostly coniferous), and urban planning. The main stages of development: the beginning of the XIX century characterized by private, manor gardening; the end of XIX – the beginning of the XX century, the appearance of the first public landscaping objects (boulevards, a square on an urban dam, Weiner Garden and Kharitonovsky Garden); the second half of the twentieth century – the expansion of forestry functions.

The modern stage of the city planting is characterized by the compacting of the building, as a result the gardening area in the residential zone is reduced. The city has a rather tense ecological situation, conditioned by air pollution. High rates of construction lead to a reduction of some parked massifs area. The provision of green plantations in recent years is kept at the level of 17 square meters per person [2].

According to the city administration, in 2015 the area of green plantations of common use was 24 544 ha, the parks, boulevards and squares area – 2493 ha [3]. Planted trees and shrubs in the amount of 28 279 pieces in 2015. The total volume of waste entering municipal landfills in the same year amounted to 554 thousand tons.

Morphological composition of waste in cities, which collected to landfills is shown in Figure 1.

a

b

c

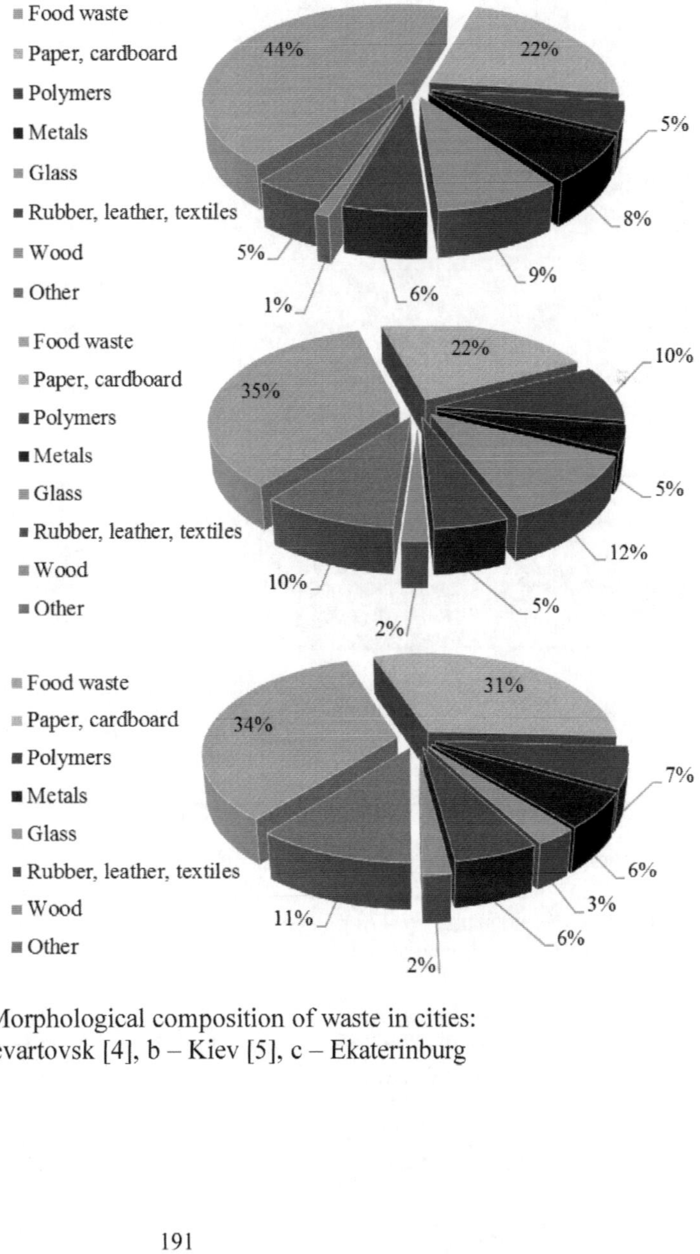


**Figure 1**. Morphological composition of waste in cities:
a – Nizhnevartovsk [4], b – Kiev [5], c – Ekaterinburg

For comparison, take the city:
a) Nizhnevartovsk (population 271 thousand people, 61° N, area 271.3 km²), located in the Sredneobsk lowland of the West Siberian Plain in the middle reaches of the Ob River on its northern coast;
b) Kyiv (population 2 927 thousand people, 50° N, area 847.7 km²), located in the center of Eastern Europe on the Dnieper River;
c) Ekaterinburg (population 1 456 thousand people, 56.5° N, area 468 km²), located on the eastern macroslope of the Middle Urals, along the coasts of the Iset River.

In the municipal economy at the moment, all the wood that is removed from city streets, parks and squares is brought to the landfill of solid household waste, that is, it turns into waste completely. According to the information of the organization MUE "Zelenstroy", which deals with the trees cutting, sorting and evaluation of the quantitative composition of wood waste is not carried out.

Figure 2 are shows photographs of the pruning process of woody and shrubby vegetation on the streets of Ekaterinburg.

**Figure 2.** Pruning trees and shrubs in the streets of Ekaterinburg

According to the municipal program "Improving the improvement of the territory of the municipality city of Ekaterinburg for 2017–2020" (Decree No. 2353 of 29.11.2016) [6] the purpose of the subprogram "Improvement and gardening of the territory of the municipality city of Ekaterinburg" is the organization of improvement and landscaping of the territory of the municipality, use, conservation, protection reproduction of urban forests and forest areas, are located within the boundaries of the municipality. The plans for 2017–2020 increase the number of planted trees and shrubs by 470 pieces per year, which will provide an average growth in wood material from 140 m³ per year.

## 3. The situation in foreign countries

The total wood stock in Russia reaches almost 82 billion m³. This is four times more than in the US, forty times more than in Sweden and sixteen times more than in Finland. The volume of wood waste in the domestic forest industry is proportionally potentially much higher. According to experts' estimates, it is technically possible to use up to 800 million tons of wood biomass annually in Russia for energy purposes only [7]. Almost all developed foreign countries have a common approach to integrated wood processing, including waste use [8].

The proportion of forest area in relation to the total area in northern European countries [9], with a high degree of urbanization of the population from 76% in Norway to 92% in Iceland similar in terms of climatic conditions and type of terrain to Ekaterinburg, is shown in Figure 3. This comparison is possible because the authors [9] consider define urban woodland as located in close to urban agglomerations.

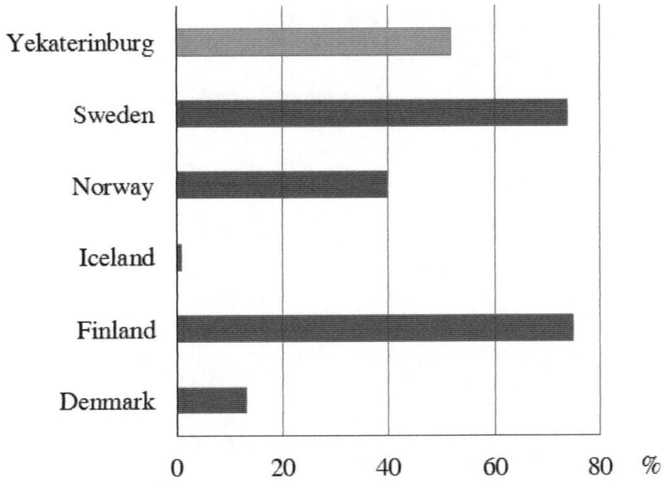


**Figure 3.** Area of forest and other wooded land

More complete use of wood in the US was made possible through the introduction of debarking machines and chippers, to produce clean chips of higher quality (without bark). In Canada, the bulk of waste is used in the production of pulp, with one quarter of sawdust due to the introduction of the continuous pulping process. In Sweden, wood chips (about 60%) prevail among the wastes, which compose the raw material base for the production of fiberboard and chipboard. In Finland, more than 85% of waste is consumed in the pulp and paper industry [8]. In Norway, in equal shares waste is directed to the production of plates and pulp.

Foreign companies in Germany, Austria, Finland and other countries offer equipment for the energy use of wood waste to generate heat and electrical energy.

The countries with the highly developed sawmilling and woodworking industry, for example, the USA, Canada, Japan and the countries of Northern and Central Europe, achieved the greatest results in the use of wood waste.

## 4. Energy use of wood waste

In large cities, it is a problem to the utilization of wood waste, generated during the felling of trees and sanitation in the care of green plantations in the streets, parks, squares, forest parks, etc. These waste are a low-quality wood of medium size, e.g. boughs, twigs peaks, bark fragments of coniferous and hardwoods.

In the greening of urban areas, green spaces are being reconstructed. It is a complex of agro-technical activities for replacing diseased and shrinking trees and shrubs, improving the species composition, and pruning woody and shrubby vegetation. Sawn trees and wood waste formed during pruning are exported by organizations that are engaged pruning branches in crowns and the removal of deadly, accidental trees that have lost their decorative value. The average composition of waste when cutting a tree [10] for minimum and maximum values is given in Figure 4.

**Figure 4**. The specific waste composition of a sawn tree

The values of the lowest energy of wood waste combustion [11] are shown in Figure 5.

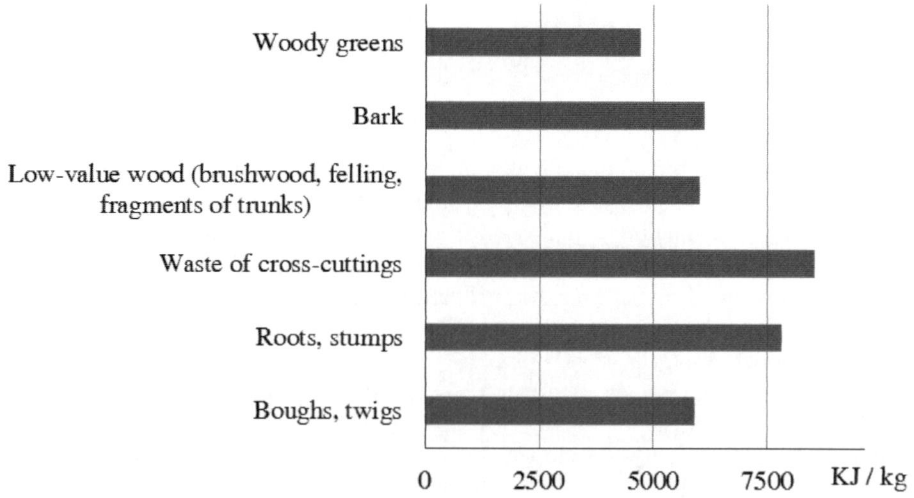


**Figure 5.** Lowest calorific value of wood processing waste

Figure 6 shows scheme of possible methods for the energy use of wood waste. In addition to waste disposal in landfill, technical solutions exist for converting them into heat and electrical energy, as well as into other types of products (pellets, briquettes, synthetic biofuels, etc.).

Wood waste can be used as fuel, for example sawdust and bark to increase the calorific value are briquetted. In this case, the calorific value of dry briquettes is much higher than that of raw sawdust, and wood pellets produce more heat energy in the process of burning.

Boilers on wood waste are designed to produce heat energy in the form of steam and (or) hot water by burning the crushed wood. Boilers of this type include a fuel bunker, a device for transporting waste to the combustion chamber, which can provide continuous and periodic fuel supply [12], a solid fuel boiler, ash removal equipment, chimneys equipped with a combustion gas purification system, an automated control and safety system, also other units and systems. The pellet as fuel are increasingly used in boilers at the present time [13]. The received thermal energy can be used for maintenance of the industrial enterprises, social objects or household consumers.

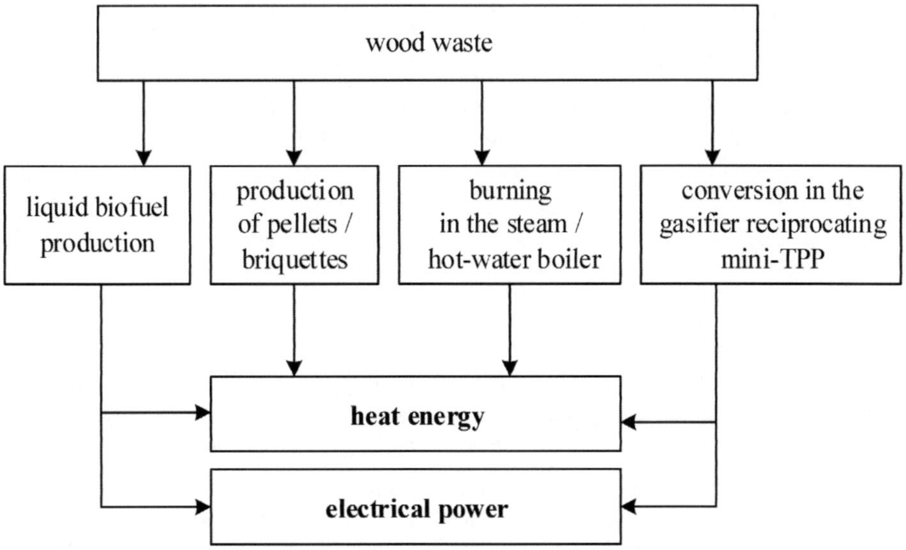


**Figure 6.** Possible ways of energy use of wood waste

The effective conversion of wood waste including bark is possible with using the gasification plants, which allows to have in addition to energy a significant amount of associated technological products [7]. In the world, technological solutions are also used to gasification wood waste with the generation of heat energy and electricity when the synthesis gas is fed into the internal combustion engine. The article [14] compares different gasification plants, e.g. two stage [15] and the one stage method of the gasification, that use biomass as a fuel. In their experiments the authors were able to get gas with a calorific value of dry gas (LHV) 4.64–5.33 MJ/Nm$^3$.

In the work [16] the theoretical results obtained using the engine's mathematical model are also in a satisfactory agreement with experimental data obtained on the engine as a part of investigations of coupled modelling and experimental research of the gasifier and the engine as a whole biomass energy system.

If are comparing the gasification and combustion of wood waste, it may be noted most gasification efficiency for small capacity plants for electrical power generation, since in the same time also heat energy is produced as well as in boilers.

The questions of costs for transporting wood waste to power installations in the general economic evaluation of their energy use in this case do not arise. The costs of wood waste transporting exist and when cutting down tree plantations are transported for disposal at the landfill. Thus, the total costs of shipping waste to the site energy recovery are equal to those that are available in the process of cutting down tree plantations and it transportation to the landfill.

The energy potential calculation of the wood waste in the Ekaterinburg city for 2015 by types of produced fuel and energy sources should begin with an estimate of the total energy potential (GJ) of wood waste [11], which is determined by the equations:

$$E_{ww} = m \cdot Q_i^r \cdot 10^{-3} \tag{1}$$

$m$ – potential resource (mass) used type of wood waste, t;

$Q_i^r$ – lower calorific value of wood waste on the working mass of fuel, MJ/kg.

Based on the technology for heat energy production described above, is determined the quantitative potential of the heat energy generated from waste taking into account the efficiency of the boiler according to the following formula:

$$Q_b = E_{ww}\eta_b \tag{2}$$

$\eta_b$ – efficiency of the plant for wood waste burning (boiler water).

The amount of electrical energy produced from wood waste during their gasification with further combustion of synthesis gas in the internal combustion engine is defined as:

$$W_g = E_{ww}\eta_{gf}\eta_{ICE}\eta_{eg} \tag{3}$$

$\eta_{gf}$ – efficiency of gasifier;

$\eta_{ICE}$ – efficiency of internal combustion engine;

$\eta_{eg}$ – efficiency of electric generator.

In addition to electric power, such a plant can provide consumers with heat energy together, that is, it work in the cogeneration cycle. When transferring the obtained values into GCal for heat and in MWh for electric energy, are obtained the final values given in Table 1.

**Table 1.** Annual energy potential of wood waste in the Ekaterinburg

| Type of energy source | Units | The resulting value |
|---|---|---|
| Wood waste | GJ | 341 825 |
| Heat energy | GCal | 71 894 |
| Electrical power | MW·h | 23 798 |

Equally important is the ecological aspect of the problem, wood fuel is practically sulfur-free and has a high reactivity. Therefore in exhaust gases by burning wood there is no sulfur dioxide and sulfur gas, and the carbon monoxide content with rationally designed combustion devices is minimal. In the same way, wood is considered to be $CO_2$ neutral fuel, because of the carbon dioxide is captured in the process of wood growth, and when sawed wood is used for energy production, the same amount of carbon dioxide is produced per kilogram of material as when it landfilled.

## 5. Conclusions

In the Ekaterinburg city there is a program for improvement and gardening of the territory of the municipality by planting more trees and shrubs. But, at the same time resources of wood waste, generated during the felling of trees and sanitation in the care of green plantations in the streets, parks, squares, forest parks are not being used, and a technologies for useful disposal are not being worked out now. In the municipal economy at the moment, all the wood that is removed from city streets, parks and squares is brought to the landfill of solid household waste, that is, it turns into waste completely.

The capabilities use of wood waste in the Ekaterinburg city was investigated in this study. According to our estimates total energy potential of wood waste city measure up more 340 thousand GJ per year. In addition to waste disposal in landfill, technical solutions exist for converting them into heat and electrical energy, as well as into other types of products (pellets, briquettes, synthetic biofuels, etc.).

To improve environmental in the site and production of useful energy, the wood waste can be used as an energy resource. The calculated potential of wood waste in Ekaterinburg to provide the urban consumer with heat energy is 71 894 GCal, and electrical energy is 23 798 MW·h per year. This use of wood waste renewable sources can reduce the waste amount at urban landfills.

**References**

[1] Zaitsev O B and Polyakov V E 2015 *Especially protected natural areas of the city of Ekaterinburg* (Ekaterinburg: Publishing House "Azhur") p 51
[2] Srodnykh T B 2009 *J. Forests of Russia and their economy* **3** p 48
[3] Results of the social and economic development of Municipality "city of Ekaterinburg" in 2015 Decree of the Administration of the Ekaterinburg City of 16.06.2016 No. 1203
[4] On approval of the city target program "Improvement of the ecological situation in the city of Nizhnevartovsk in years" Available from: http://pandia.ru/text/78/166/14201-2.php [Accessed 12th May 2017]
[5] Recovery of solid domestic wastes Available from: https://aebgroup.com.ua/services/removal-household-waste [Accessed 12th May 2017]
[6] Municipal program "Improving the improvement of the territory of the municipality "city of Ekaterinburg" for 2017–2020 Available from: http://www.ekburg.ru/ [Accessed 12th May 2017]
[7] 2006 *Bioenergetics* (St. Petersburg: Bell)
[8] Vasilyeva T V J 2002 The review of currently formed trends in wood residue utilization in overseas industries *J. Forestry bulletin scholar* **4(2002)**
[9] Gundersena V, Frivoldb L H, Lo I, Bruno Bilde Jorgensend B B, Falcke J and Oyena B H 2005 *J. Urban Forestry & Urban Greening* **3** p 189
[10] 1999 *Collection of specific indicators of generation of production and consumption wastes* (Moscow: State Committee of the Russian Federation for Environmental Protection) p 55
[11] Dubinina N N, Orlov A A, Korchuk Y A and Lyakh N I 2016 *The generation of energy using wood waste from the sawmill* Available from: http://science-bsea.bgita.ru/2016/les_2016/dubinina_energ.htm [Accessed 12th May 2017]
[12] Juszczak M 2016 *J. Environment Protection Engineering* **42** p 95
[13] Gehrig M, Jaeger D, Pelz S K, Weissinger A, Groll A, Thorwarth H and Haslinger W 2016 *J. Atmospheric Environment* **136** p 61
[14] Sulc J, Stojdl J, Richter M, Popelka J, Svoboda K, Smetana J, Vacek J, Skoblja S and Buryan P 2012 *J. Waste Management* **32** p 692
[15] Brand P, Larsen E and Henriksen U 2000 *J. Energy and Fuels* **14** p 816
[16] Centeno F, Mahkamov K, Silva Lora E E and Andrade R V 2012 *J. Renewable Energy* **37** p 97

IOP Conf. Series: Earth and Environmental Science **72** (2017) 012027 doi:10.1088/1755-1315/72/1/012027

# Alternative power supply systems for remote industrial customers

**N V Kharlamova, A I Khalyasmaa and S A Eroshenko**

Ural Federal University named after the first President of Russia B. N. Yeltsin

E-mail: ninakharlamova@gmail.com, lkhalyasmaa@mail.ru, stas_ersh@mail.ru

**Abstract.** The paper addresses the problem of alternative power supply of remote industrial clusters with renewable electric energy generation. As a result of different technologies comparison, consideration is given to wind energy application. The authors present a methodology of mean expected wind generation output calculation, based on Weibull distribution, which provides an effective express-tool for preliminary assessment of required installed generation capacity. The case study is based on real data including database of meteorological information, relief characteristics, power system topology etc. Wind generation feasibility estimation for a specific territory is followed by power flow calculations using Monte Carlo methodology. Finally, the paper provides a set of recommendations to ensure safe and reliable power supply for the final customers and, subsequently, to provide sustainable development of the regions, located far from megalopolises and industrial centres.

## 1. Introduction

In view of the small capacities developed at the powerhouses with alternative energy in use, renewable energy sources (RES) [1] are most often considered in power industry together with a concept of the distributed generation [2]. Such generation allows providing small consumers including ones working autonomous from the centralized.

The enterprise often needs implementation of additional production capacities for further development. In most cases the enterprise increases the external electric network consumption after receiving specifications of the electric grid company concerning connection of additional electroreceivers. However, in certain cases own source of electrical power is installed to reduce costs of production power supply. Such installations as wind generators, solar batteries and also the generators consuming biofuel as well as with geothermal power in use can act as energy resources.

Expediency of the power sources choice is explained by the fact that there is no fuel component in structure of expenses and also there is no payment for energy transmission on electric network which provides low cost of electric energy production even in case of high capital investments into the power equipment. To choose an optimum resource of renewable power for the objective solution it is expedient to investigate the potential of different types of alternative energy for the considered location area of the generating object.

The limited amount of Russian regions has the sufficient potential for realization of commercial use of sunlight energy. So, for instance, by information from Hevel Solar [3] only in the Far East of Russia, in the areas adjoining to China, the insolation level exceeds 4.5 kWh/sq.m/day. Only Kamchatka and the Kuril Islands territories have sufficient potentials for geothermal energy use due to its terrestrial heat flow corresponding to a geothermal gradient 20 °C/100 m in comparison with 6 °C/100 m in the Caucasus and in the Crimea [4] which are also considered as perspective territories for geothermal power exploitation. In different regions this indicator is much lower therefore now use of geothermal energy in Russia in areas besides Kamchatka and the Kuril Islands is inexpedient. According to the national atlas of Russia [5] the average wind speed is sufficient for operation of low-speed wind turbines on the most part of the Russian Federation territory. The paper deals with questions of application of various RES for creation of power supply systems for remote industrial consumers having electrical shortage which is one of the main limiting factors of sustainable development.

## 2. RES geography in Russia

To suggest a solution of the problem of partial autonomous enterprise power supply it is necessary to consider already available achievements in the field of renewable energy use in Russia.

### *2.1. Geothermal power*

Despite the presence of a certain potential in the North Caucasian, Volga region, East and West Siberian areas the most perspective from the point of view of geothermal resources the Kamchatka Region is considered. In Russia there are only five geothermal generating stations with a overall installed capacity of 80.1 MW and all of them are located in the Kamchatka Region or in the Sakhalin region [6]. At the same time nearly 60% of this value is made by the power of Mutnovsky geothermal power plant of 50 MW [7]. However, this area experience concerning installation of geothermal power plants [8] which have allowed to reduce significantly the region dependence on imported fuel oil and, as a result, to reduce energy prices is rather an exception to the rules.

### *2.2. Biofuel use*

Russia is included into the three of the biofuel exporting countries, however, only 20 % of produced fuel pellets are consumed in domestic market [9]. As biofuel production is closely connected with agriculture, its use is mostly widespread in territories which are historically more populated. Power plants of "Baytsury", "Luchiki" and the first Russian bioreactor in the Doshino village can be examples. Biogas power plants are generally used by agricultural complexes for ensuring own power needs. It should be noted that it is unprofitable for enterprises to sell the energy produced by biofuel in the retail or wholesale markets as unlike the European countries in Russia there are no special tariffs for "bioelectric power".

### *2.3. Solar power*

According to [10] the overall installed capacity of solar and wind power facilities in Russia has to make 1600 MW in ten years [11]. Nevertheless, for today there are only eight large solar power stations which are mainly located at the southern borders of Russia except for solar powerhouse of Batagai in Yakutia with only 1 MW capacity. Practically all solar objects have been implemented within the last several years and average quantity of installed capacity makes about 5 MW [12]. On some solar power stations new stages construction is planned but the maximum installed capacity will not exceed 10 MW after expansion [13] which is in many respects caused by features of a geographical location of Russia and, as a result, the specific climate which is not suitable for mass solar power assimilation.

### *2.4. Wind power*

The major wind power stations of Russia are in the Crimea. These are Donuzlavsky, Ostaninsky, Tarkhankutsky and East Crimean wind generating stations. Similar objects are operating in the Kaliningrad region (the Zelenograd WDPP), at Chukotka (the Anadyr WPS), in the Republic of Bashkortostan (Tyupkilda WPS), Kalmykia, at Bering Island of the Commander Islands and at the Kola Peninsula [14]. There is also a number of projects in other regions but they are in a development stage. Design capacities of the developed WPS can reach 75 MW [15] nonetheless the capacity of already operating power plants, except for the Crimea, does not exceed 5 MW [16].

Nevertheless, many steppe regions of Russia have sufficient potential from the point of view of speed and constancy of wind flows (for example, Kurgan and Tyumen regions) in total with high concentration of the industrial enterprises in these areas. These factors combination provides profitable wind power use for power supply of remote industrial consumers.

## 3. Assessment of expediency of wind power generation use

The expediency of the suggested solution is estimated on a practical example. The first step is the choice of the optimum place for WPS placement within the Kurgan region. This area was chosen because of the combination of a plain terrain, high concentration of the industrial enterprises and relative remoteness from the large cities, which had been described above.

### *3.1. The settlement choice*

The first step is the choice of the optimum place for placement of wind power station within the Kurgan region on the map based on meteorological data [17, 18]. This area was chosen because of the combination of a plain terrain, high concentration of the industrial enterprises and relative remoteness from the large cities, which had been described above.

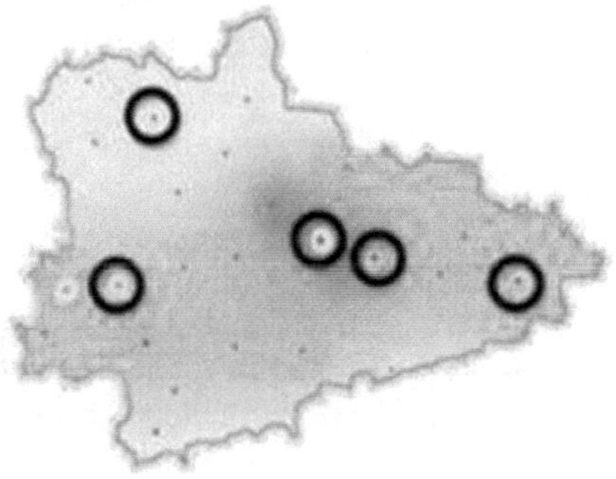

**Figure 1**. The map of average wind speed at the height of 10 m above ground, m/s:
– more than 5.5; – from 3 to 5.5; – less than 3.
The number of the industrial enterprises in the settlement:
– more than 5; – from 3 to 5; – 2 and less.
• – designation of the settlement, O– the chosen location

### *3.2. Assessment of expediency of wind power plants construction*

The first step is the choice of the optimum place for placement of wind power station within the Kurgan region on the map based on meteorological data [17, 18]. This area was chosen because of the effective combination of a plain terrain, high concentration of the industrial enterprises and relative remoteness from the large cities.

Calculation is made on the basis of Weibull distribution function [19, 20] as the way of an assessment of wind speed change in the medium-term period, i.e. in one year. Integral Weibull distribution function represents the probability density distribution of wind speed presented by the expression:

$$P(v)=1-e^{-\left(\frac{v}{c}\right)^k}, \quad (1)$$

where $P$ is the probability of occurrence of a wind flow with a speed not lower than $v$; $c$ is the scale factor, $k$ is the shape factor.

In each case the system of nonlinear equations with $P$ and $v$ found in result of averaging of observation results [17, 18] was solved for $c$ and $k$ by means of the Mathcad software package. Calculation for Kurgan is given as an example.

**Table I**. Probability of wind occurrence in the given speeds range, where $v_0$-$v$, m/s is the range of wind speeds

| $v_0$-$v_i$, m/s | *Jan.* | *Feb.* | *Mar.* | *Apr.* | *May* | *June* | *July* | *Aug.* | *Sept.* | *Oct.* | *Nov.* | *Dec.* |
|---|---|---|---|---|---|---|---|---|---|---|---|---|
| | **Possibility of wind speed accessory to this interval $v_0$-$v$,%** | | | | | | | | | | | |
| 0-3 | 19 | 21 | 3 | 0 | 3 | 0 | 6 | 3 | 7 | 6 | 0 | 13 |
| 3-6 | 52 | 54 | 37 | 20 | 19 | 47 | 58 | 55 | 33 | 36 | 23 | 42 |
| 6-10 | 26 | 21 | 37 | 53 | 75 | 40 | 33 | 39 | 53 | 39 | 50 | 36 |
| 10-14 | 3 | 4 | 23 | 20 | 3 | 13 | 3 | 6 | 7 | 19 | 24 | 3 |
| 14-16.5 | 0 | 0 | 0 | 7 | 0 | 0 | 0 | 0 | 0 | 0 | 3 | 6 |

The probability of average daily wind speed $v$ in the range ($v_0$,$v_i$) for calendar year is calculated as follows:

$$P(v=v_i)=\frac{\sum_{1}^{12} P(v\in(v_0,v_i))}{12}\cdot 100\% \qquad (2)$$

The probability of average daily wind speed $v$ in the range ($v_0$,$v_i$) for calendar month is calculated as follows:

$$P(v\in(v_0,v))=\frac{n(v\in(v_0,v_i))}{n} \qquad (3)$$

where $n(v \in (v_0, v_i))$ is the number of days in a month when average daily wind speed $v$ belongs to the range ($v_0$,$v_i$), $n$ is the number of days in a month.

Calculation of probability that the average daily wind speed $v$ is less than speed $v_i$ is made according to the formula (Results of calculations are presented in the table II):

$$P(v_i>v)k=\sum_{1}^{k} P(v_i=v) \qquad (4)$$

**Table II**. Probability of wind distribution

| **v, m/s** | Probability of wind speed | |
|---|---|---|
| | **P ($v_i$=v), %** | **P ($v_i$<v),%** |
| 0-3 | 6.750 | 6.750 |
| 3-6 | 39.667 | 46.417 |
| 6-10 | 41.833 | 88.200 |
| 10-14 | 10.667 | 98.917 |
| 14-16 | 0.833 | 99.750 |
| >16 | 0.500 | 100 |

On the basis of the obtained data the system of equations is set:

$$\begin{cases} 0.0675 = 1 - e^{-\left(\frac{3}{c}\right)^k} \\ 0.4640 = 1 - e^{-\left(\frac{6}{c}\right)^k} \end{cases}$$

The equations for two other settlements have been set in the same way. As a result of its solution the values presented in the table have been received III.

**Table III**. Determination of the scale factor $c$ and shape factor $k$

| Name of the settlement | Kurgan | Shchuch'ye | Lebyazh'ye | Kataysk | Zverinogolovsk |
|---|---|---|---|---|---|
| Scale factor, $c$ | 7.603 | 4.49 | 5.057 | 3.501 | 4.149 |
| Shape factor, $k$ | 1.993 | 2.903 | 2.768 | 2.326 | 2.347 |

On the basis of parameters from the table III and a formula (1) the Weibull function diagram is created. In the same coordinates the diagram displaying meteorological observations for the same period is constructed. Blue color is used to indicate temperature change according to weather services, light-blue is used for Weibull distribution on wind speeds.

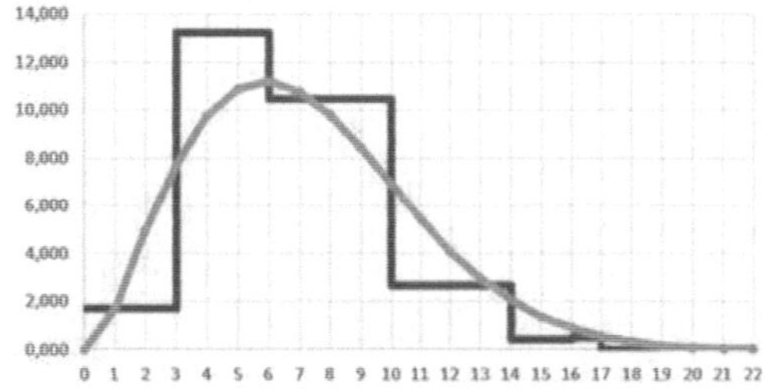


**Figure 2**. Weibull distribution and weather data, Kurgan

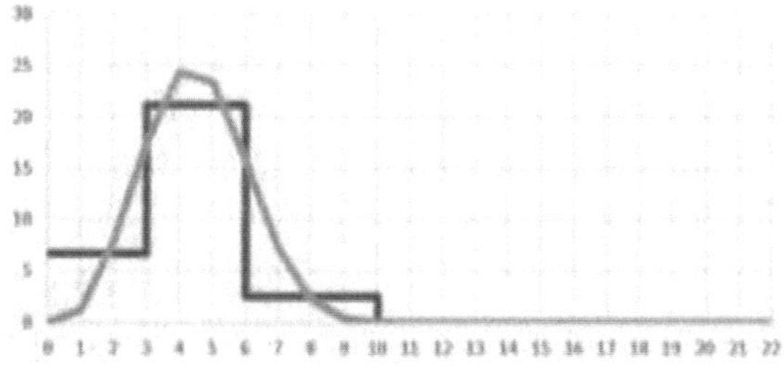

**Figure 3**. Weibull distribution and weather data, Shchuch'ye

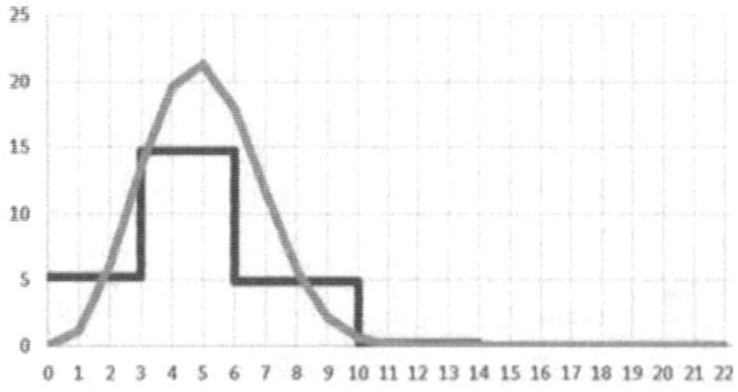


**Figure 4**. Weibull distribution and weather data, Lebyazh'ye

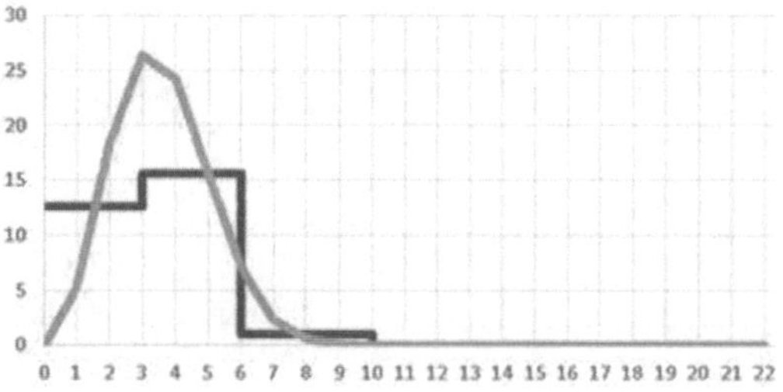


**Figure 5**. Weibull distribution and weather data, Kataysk

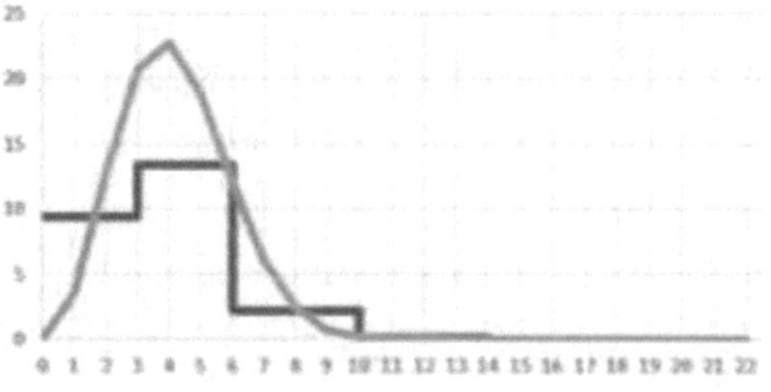

**Figure 6**. Weibull distribution and weather data, Zverinogolovsk

*3.3. Calculation of wind parameters for the considered area*

On the basis of the obtained data calculation of parameters which allowed estimating expediency of wind power stations use in these areas has been made. Calculations were made on the formulas listed below and the received values are given in the table IV. Calculation for Kurgan is given as an example.

1. Determination of wind speed at $z$ height knowing wind speed $v_0$ at $z_0$ height can be made according to a formula:

$$v = v_0 \cdot \frac{z}{z_0} \quad (5)$$

2. Expectation of average wind speed at the height of 10 m for Kurgan:

$$\bar{v} = c \cdot \Gamma(1+1/k) = 7.603 \cdot 0.891 = 6.774 \text{ m/s} \quad (6)$$

3. The most probable wind speed at the height of 10 m (mode) at $k > 1$ is calculated on the expression:

$$v_{p\max} = \frac{c \cdot (k-1)^{(1/k)}}{k^{(1/k)}} = \frac{7.603 \cdot (1.993-1)^{(1/1.998)}}{1.993^{(1/1.998)}} = 5.360 \text{ m/s} \quad (7)$$

4. Wind turbulence at the height of 50 m (a mean square deviation) is defined as follows:

$$\sigma_v = \sqrt{c^2 \cdot \Gamma(1+2/k) - (\bar{v})^2} = \sqrt{7.603^2 \cdot 1.001 - (8.485)^2} = 4.300 \text{ m/s} \quad (8)$$

5. Probability of occurrence of a wind flow with the speed not exceeding 3 m/s:

$$\exp\left[-\left(\frac{v}{c}\right)^k\right] = 1 - \exp\left[-\left(\frac{3}{7.603}\right)^{1.993}\right] = 14.414\% \quad (9)$$

6. Calculation of average capacity of wind power generation:

$$P(\bar{v}) = \Delta v \cdot \left(\sum_{n=0}^{50} p(n) \cdot P(n) - 0.5 \cdot p(50) \cdot p(50)\right) \quad (10)$$

**Table IV**. Results of calculation of wind power parameters

| **Name of the settlement** | Wind parameters | | | | | | | | | | |
|---|---|---|---|---|---|---|---|---|---|---|---|
| | ***c*** | ***k*** | ***$C_{exp.}$*** | ***$V_{av.}$*, m/s** | ***$V_{av.}$*, m/s** | ***P(v≤3)*, %** | ***$v_{p\ max}$*, m/s** | ***$σ_v$*** | ***$P_{av.}$*, %** | ***$V_{av.exp.}$*, m/s, *h=50 m*** | ***$v_{max}$*, m/s, *h=50 m*** |
| Kurgan | 7.603 | 1.993 | 7.295 | 6.500 | 6.774 | 14.410 | 5.360 | 4.340 | 23.355 | 8.485 | 20.043 |
| Shchuch'ye | 4.490 | 2.903 | 4.607 | 4.100 | 3.996 | 26.520 | 3.882 | 1.519 | 6.430 | 5.136 | 12.527 |
| Lebyazh'ye | 5.057 | 2.768 | 5.505 | 4.900 | 4.501 | 20.870 | 4.300 | 1.758 | 9.442 | 6.138 | 17.538 |

| Name of the settlement | Wind parameters | | | | | | | | | | |
|---|---|---|---|---|---|---|---|---|---|---|---|
| | $c$ | $k$ | $C_{exp.}$ | $V_{av.}$, m/s | $V_{av.}$, m/s | $P(v\leq 3)$, % | $v_{p\,max}$, m/s | $\sigma_v$ | $P_{av.}$,% | $V_{av.exp.}$, m/s, $h$=50 m | $v_{max}$, m/s, $h$=50 m |
| Kataysk | 3.501 | 2.326 | 4.606 | 3.865 | 3.967 | 50.020 | 2.750 | 1.217 | 3.244 | 5.219 | 12.731 |
| Zverinogolovsk | 4.149 | 2.347 | 5.505 | 4.600 | 4.589 | 37.126 | 3.275 | 1.442 | 5.658 | 6.089 | 16.155 |

### *3.4. Analysis of average capacities of wind power generation in the Kurgan region*

The obtained data have been plotted on the map for simplification of the analysis. According to this map it is possible to draw a conclusion on what part of installed capacity of the wind generator will be developed on average in standard conditions taking into account Weibull distribution and terrain features.

In practice the received results can be used for both the preliminary analysis of expediency of wind power stations use at the existing enterprises and the choice of optimum location of the new enterprise taking into account a possibility of its partial power supply by wind power.

For descriptive reasons the calculation is made to define what average capacity will be given by one wind generator with the installed capacity of $P_{inst}$ = 2500 kW according to data for Kurgan [16].

$$P_{av.} = \frac{P_{inst.} \cdot P_{av.\%} \cdot \cos\beta}{100} = \frac{2500 \cdot 23.355 \cdot 0.819}{100} = 478.194 \text{ kW}, \quad (11)$$

where $\beta = 270°-10°-C_{av.}=270°-10°-225° = 35°$. (12)

At that β is the same for all the area as the prevailing wind direction according to data [10] is southeast for all territory.

Thus it is possible to calculate average efficiency of power use:

$$K_{\exp.av.} = \frac{P_{av.}}{P_{inst.}} = \frac{P_{av.\%} \cdot \cos\beta}{100}, \quad (13)$$

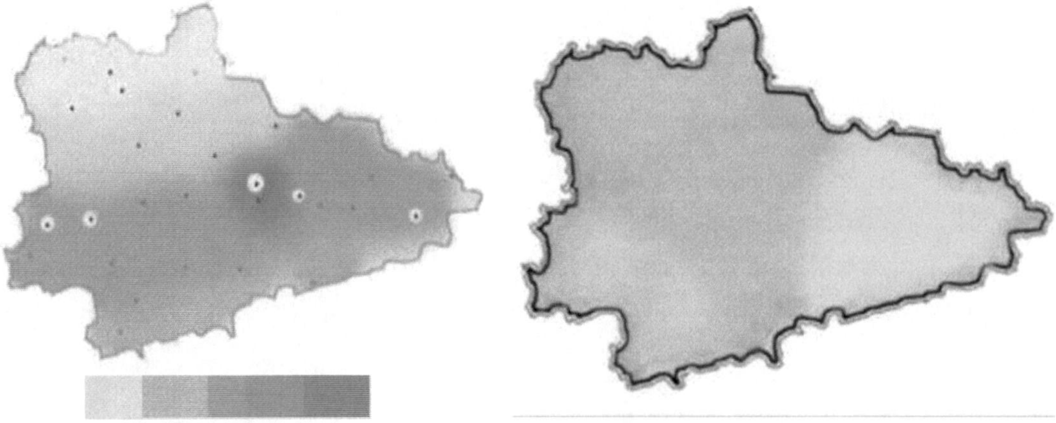


**Figure 7**. Average efficiency of power use

**Figure 8**. The map of electric power deficiency and surplus in Kurgan

East area is the most suffering from power shortages with power shortage making about 150 MW [21]. At the same time in the Kurgan energy area power surplus of 382 MW is observed by the data [22] for 2016. Nevertheless, in the Kurgan area there is a significant increase in power consumption due to implementation of the new energy-using equipment. Similar tendencies are observed also in East and Shumikhinsky energy areas.

Comparing maps in figure 2 and figure 3 it might be concluded that there is a possibility of wind power stations use as an alternative connection to the Unified energy system in all the energy areas, except for Shadrinsk as the average output there is insufficient for ensuring steady power supply of consumers.

## 4. Influence of wind generators operation on electric network

While using the generator as the power supply it is necessary to take into account inconstancy of the generated capacity. It is connected with its dependence on the wind speed [23]. For an assessment of this influence by means of a software package of @Risk Palisade Decision Tools Monte Carlo method [24] and the InorXL program unit [25] were used for calculations of the steady state modes.

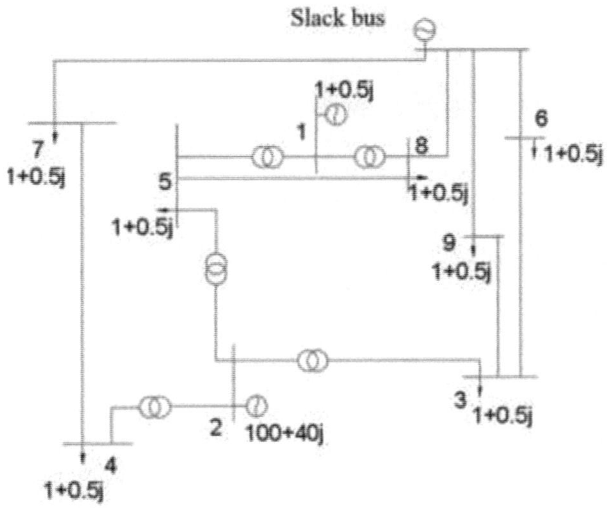


**Figure 9**. Single-line diagram of the network

One thousand iterations was made for which the capacity of the wind generator P has been presented by the Weibull function received on the basis of data from Table III. Herewith $c$ and $k$ also change in the ranges (0.959; 1.172) and (0.383; 0.469) respectively on formulas (14) and (15).

$$c = x^2 - 0.85 \cdot x + 0.179 \tag{14}$$

$$k = x^2 - 2.218 \cdot x + 1.120 \tag{15}$$

Weibull distribution for wind speeds is presented by a formula (16):

$$P = 1 - \exp\left[-\left(\frac{P_i}{c}\right)^k\right] \tag{16}$$

### *4.1. Analysis results*

Analysis results are shown in the table V and in figure 10 where maxima (max), minima (min) and the most expected (m. e.) modules of voltage in nodes of the considered electric network of 10 kV are displayed.

**Table V**. Voltage parameters

| **Busbar** | **$U_1$** | **$U_2$** | **$U_3$** | **$U_4$** | **$U_5$** | **$U_6$** | **$U_7$** | **$U_8$** | **$U_9$** |
|---|---|---|---|---|---|---|---|---|---|
| **Max** | 116.113 | 117.312 | 10.618 | 10.640 | 10.607 | 10.108 | 10.112 | 10.521 | 10.108 |
| **Min** | 115.540 | 116.745 | 10.567 | 10.588 | 10.555 | 10.069 | 10.073 | 10.470 | 10.108 |
| **M. e.** | 115.858 | 117.063 | 10.596 | 10.617 | 10.584 | 10.095 | 10.099 | 10.500 | 10.095 |

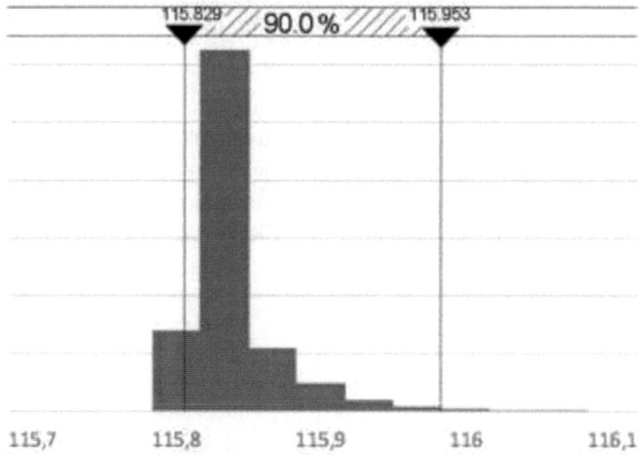


**Figure10**. Change of voltage on substations busbars

On the basis of each iteration data on the minimum and maximum voltage in each node and also the most probable values of voltage have been obtained. Also nodes voltage and *c*, *k* и *P* parameters relationship have been defined.

Analyzing the obtained data it is possible to conclude that on voltage classes of 110 and 10 kV the difference among the minimum and maximum values of voltage makes about 1 %. Therefore it is possible to conclude that change of the wind generator capacity influences the mode slightly and does not make the voltage in network nodes exceed admissible limits.

## 5. Conclusion

The research presents practical approbation of a method determining the potential of wind generators use for the choice of the most suitable generator taking into account climate, landscape and influence on external electric network. Criteria for such assessment are the required values of power and voltage and also its acceptable deviations. Besides an algorithm, recommendations for its use simplification and also the map for decisional process visualization are provided in the paper.

Though the preliminary choice might be made on the basis of only values of average wind speeds, such assessment will not provide exact results about expediency degree and a real possibility of use of wind power production in general for concrete electric network. In spite of the fact that installment of wind power generation will require more deep meteorological studies, the simple and evident algorithm accelerates and simplifies process of calculations.

Currently it is impossible to call wind power in Russian Federation widespread. Creation of transparent express methods for definition of expediency of wind power generation use in the long term will allow to improve a situation and to develop domestic market of this type of the generating capacities.

Wind power can become one of real alternatives of power supply from the centralized system and it is possible to make wind energy and renewed power in general more attractive to the end user only by simplifying the process of its use.

**References**

[1] Adefarati T and Bansal R C 2016 Integration of renewable distributed generators into the distribution system: a review *IET Renewable Power Generation* **10** (7) p 873–84 *references*

[2] Alimardani A, Therrien F, Atanackovic D, Jatskevich J and Vaahedi E 2015 Distribution System State Estimation Based on Nonsynchronized Smart Meters *IEEE Trans. Smart Grid* **6 (6)** p 2919–28

[3] *Official site of the company "Hevel"* Available from: http://www.hevelsolar.com

[4] Wind mode *National Atlas of Russia Volume 2* Available from: http://xn--80aaaa1bhnclcci1cl5c4ep.xn--p1ai/cd2/172/172.html [Accessed 08th May 2017]

[5] *Earth: Global Map of Winds, Weather and Sea Currents* Available from: https://earth.nullschool.net/

[6] Kolesnikov D V, Lyubin A A and Shulyupin A N 2015 Geothermal Power Generation in Kamchatka, Russia *Proceedings World Geothermal Congress Melbourne Australia* Available from: https://pangea.stanford.edu/ERE/db/WGC/papers/WGC/2015/26058.pdf [Accessed 08th May 2017]

[7] *Kamchatka Krai SIDP on 2016–2020* Approved by Order of the Governor of the Kamchatka Territory 7/14/2016 number 790-P The Ministry of Housing and Communal Services and Energy Kamchatka Territory

[8] Lappalainen H K, Kulmala M and Zilitinkevich S 2015 *Pan-Eurasean experiment peex science plan* Available from: http://www.atm.helsinki.fi/peex/images/PEEX_SP__27052015.pdf [Accessed 08th May 2017]

[9] 2016 *Russian Federation Biofuels Annual report* FAS Available from: https://gain.fas.usda.gov/Recent%20GAIN%20Publications/Biofuels%20Annual_Moscow_Russian%20Federation_6-23-2016.pdf [Accessed 08th May 2017]

[10] Meyers G 2016 *Renewable Energy Report For Russia* Available from: https://cleantechnica.com/2016/05/09/renewable-energy-report-russia/ [Accessed 08th May 2017]

[11] Solar power plants in Russia *Power industry Oil and Gas Website for suppliers* Available from: http://energybase.ru/en/power-plant/solar [Accessed 08th May 2017]

[12] Kajaste R 2008 *NEFCO Renewable Energy Projects in Russia, Experiences and Challenges* (Alterenergy, Moscow 23–24 October 2008) Available from: http://s3.amazonaws.com/zanran_storage/www.nefco.org/ContentPages/16787039.pdf [Accessed 08th May 2017]

[13] 2008 *Russia's Solar Energy given the all clear to develop 90 MW of solar plants* (Alterenergy, Moscow 23–24 October 2008) Available from: https://www.pv-magazine.com/2016/08/08/russias-solar-energy-given-the-all-clear-to-develop-90-mw-of-solar-plants_100025697/ [Accessed 08th May 2017]

[14] *The WindPower Wind Energy Market Intelligence* Available from: http://www.thewindpower.net/country_windfarms_en_59_russia.php [Accessed 08th May 2017]

[15] Askerov E 2016 Building of wind power industry in the Russian Federation *IX the National Conference in 2016*

[16] *The list of Russian WPS* Available from: http://energybase.ru/power-plant?PowerPlantSearch%5Bcountry_id%5D=&PowerPlantSearch%5Bdistrict_id%5D=&PowerPlantSearch%5Bname%5D=&PowerPlantSearch%5Btype_name%5D=wind&PowerPlantSearch%5Bcity_id%5D=&PowerPlantSearch%5Bstate%5D [Accessed 08th May 2017]

[17] Diary of the weather *Forecast Gismeteo* Available from: https://www.gismeteo.ru/diary/ [Accessed 08th May 2017]
[18] Wind speed and Stats Available from: http://www.atlas-yakutia.ru/weather/wind/stat_weather_28661_wind.php [Accessed 08th May 2017]
[19] Eroshenko S, Pasderin A and Samoilenko V 2014 *The report of research and development work – Development and implementation of autonomous power supply systems to improve the quality and reliability of remote power supply up to 100 kW*
[20] Khan M S A, Khalique A and Abouammoh A M 1989 On estimating parameters in a discrete Weibull distribution *IEEE Transactions on Reliability* **38** (3) p 348–50
[21] Wilson M P, Given M J, Timoshkin I V, MacGregor S J, Wang T, Sinclair M A, Thomas K J and Lehr J M 2012 Weibull statistical analysis of impulse-driven surface breakdown data *Pulsed Power Conference (PPC) 2011 IEEE*
[22] 2014 The disposal of the Kurgan Region Government Available from: http://www.ind.kurganobl.ru/assets/files0/rasporyajenie_28.04.14.pdf [Accessed 08th May 2017]
[23] Gupta D and Sadiq M 2008 Software Risk Assessment and Estimation Model *International Conference on 2016 Computer Science and Information Technology*
[24] Raychaudhuri S 2008 Introduction to Monte Carlo simulation *Simulation Conference WSC 2008*
[25] Chowdhury A A, Bertling L, Glover B P and Haringa G E 2007 A Monte Carlo Simulation Model for Multi-Area Generation Reliability Evaluation *International Conference on Probabilistic Methods Applied to Power Systems PMAPS 2006*

# The use of solar energy for residential buildings in the capital city

**V Velkin, S Shcheklein and V Danilov**

Ural Federal University named after the first President of Russia Boris Yeltsin, Ekaterinburg, Russia

E-mail: v.i.velkin@urfu.ru

**Abstract**. Taking into account the conditions of sharply continental climate of Russia the implementation of the project of heating for an apartment house can be structured on the basis of solar vacuum collectors. A diagram of the operation of the vacuum solar collectors can be considered for heating and hot water. The calculations of effective angle of solar collectors also vary for stationary use in winter and summer. Consumption of centralized heat in the spring and autumn is reduced by 30 % due to the use of solar collectors. In summer the main problem of application of solar collectors is to protect the tubes from overheating. In winter, the use of solar heating is able to provide not more than 25 % of the needs regarding the utility and the results of experimental exploitation. It is shown that the main problem of using solar energy in Russia relies not in technology, but in the legislative field. The use of a vacuum manifold in Russia will be widely implemented in areas with a cold climate and in the modern houses after solving the issues of legislative support from the state and municipal authorities.

## 1. Introduction

Involvement of renewable energy sources in energy production is a modern trend in all industrialized and in many developing countries [1]. The extent of the practical use of solar, wind, biological energy and low-grade heat reached in some countries is extremely high level, commensurate with the level of the traditional installations on organic and nuclear fuel [2]. The modern installations of renewable energy use many of the achievements of science and technology of the 21st century and stimulate a new knowledge, supporting the creation of equipment of the 6th technological structure. Additive and information technologies that enable the creation of new mass and low-cost products, remote control and monitoring, efficient system of storage and conversion of energy are increasingly used in machines and renewable energy systems [1, 2]. Commercial products of the energy range, especially power plants that use renewable energy becomes one of the most popular entries of goods, covering consumers in all countries of the world and with maximum added value. The Russian Federation is currently not among the world leaders in the development of renewable energy (unlike nuclear industries, which managed to keep the leading position). One of the problems hindering the use of renewable energy in Russia is the harsh climate of the greater part of the country [3].

In this work we present the results of application of solar energy in the harsh climate of the Ural regions for heating and hot water for an apartment house that allows us significantly (20–25 %) to reduce the consumption of fossil fuels and environmental impact on densely populated areas of the city. It is shown that in the period of summer repairs and installation of district heating systems solar heat generation is able to provide 100 % amount of power supply for the purposes of hot water. Experimentally proved the necessity of using special heat-generating equipment with a high degree of repossessions. Practical use of solar energy in Russia is currently the greatest application in the Krasnodar region [4] consisting a high level (DDHP degree-days of heating period) – 3500. Sverdlovsk region has much harsher climate (6000 DDHP) therefore, there exists a high risk of using solar collectors.

To investigate the efficacy of using solar energy in the Ural region the scheme has been developed and executed by the experimental installation of solar collectors with vacuum tubes [5] for heating and hot water. The obtained experimental data clarifies the effectiveness of the system of production of thermal energy for the conditions of sharply continental climate, characteristic of the Ural-Siberian climate zone. In the course of the study- the advent of solar radiation, change in thermal characteristics of elements of the unit; calculated energy performance and efficiency of the installation, the general scheme have been shown in Figure 1.

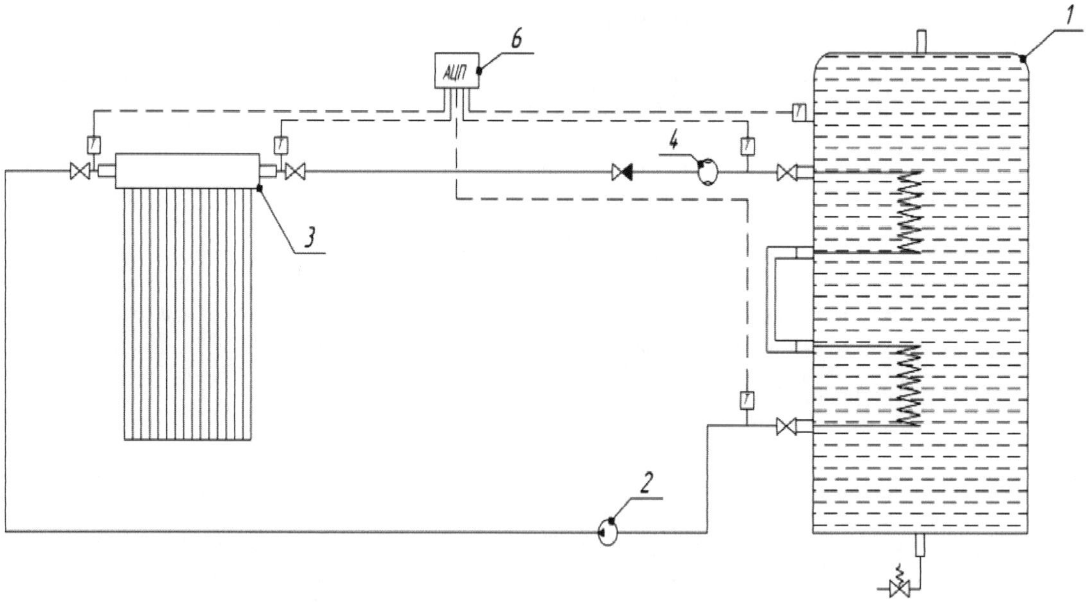


**Figure 1.** The schematic diagram of the setup to investigate the efficacy of vacuum solar collector under natural conditions of Ural region (Urfu, Ekaterinburg)

1 – double circuit storage tank capacity of 250 liters; 2 – circulation pump Wilo 48 W with three-way adjustment-power; 3 – vacuum solar collector Ariston Kairos VT-15; 4 – flowmeter; 6 – analog-to-digital Converter

## 2. Research setting

Studies of the energy characteristics of the long term (several months) installation, collection and accumulation of information - both the solar radiation and temperature characteristics have been automated.

For the study the radiation characteristics were applied to two independent automatic weather complex (made in USA), recording direct and scattered radiation, temperature and humidity environment, the strength and direction of wind in the automatic mode saving the data in the computer's memory. Solar collectors had a southern orientation and fixed tilt angle to the horizon of 45°.

## 3. Installation

In April 2012, on the basis of calculations and experimental data the scheme and installation of vacuum solar collectors on a 10-storied apartment building in Ekaterinburg [6]. Diagram of the vacuum system of the IC is presented in Figure 2.

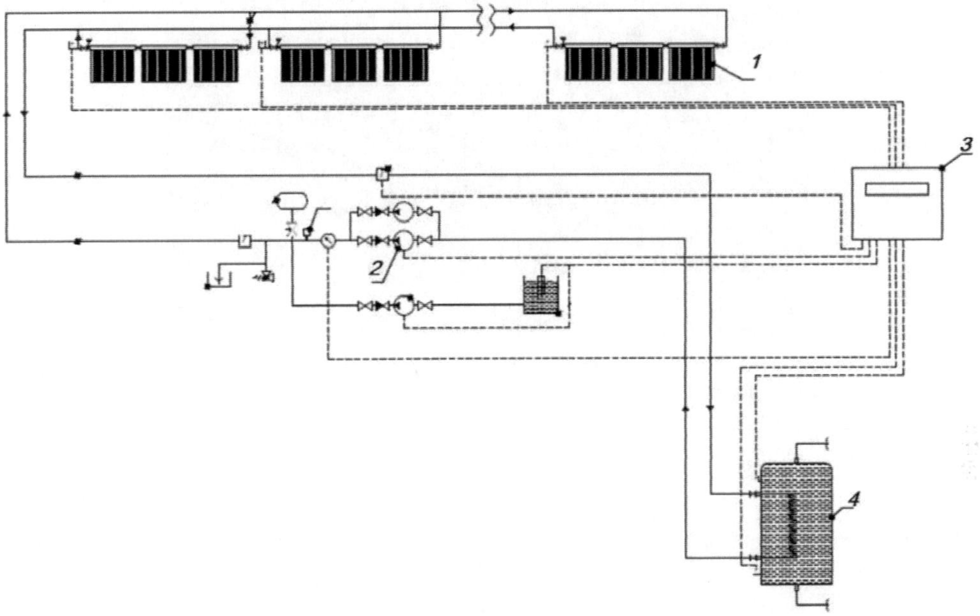


**Figure 2.** System diagram of a solar heat supply of apartment houses in the metropolis (Ekaterinburg, Rhodonite str., 8)

1 –Solar collectors with vacuum tubes; 2 – a pump for supplying coolant; 3 – controller; 4 – accumulator

Flat SK, as it is known, lose more heat into the environment in comparison with vacuum solar collectors [7]. These losses grow up with the increase in temperature difference. Vacuum solar collectors have a lower area ratio of absorption to total area (60–80 %) in comparison to flat plate collectors. Based on the values of effective area per square meter, vacuum collectors are more efficient. This allows their use in tight spaces on the roof of multistoried residential building.

The selected installation type (solar collector) is determined taking into account climatic conditions. In regions of temperate and cold climate (if any solar radiation) flat collectors will carry significant heat loss due to convection. This fact makes their use impractical. On the other hand, the layer of vacuum tube collector allows to save thermal energy.

From a technological point of view, the process of manufacturing a vacuum manifold (especially with the heat pipes) is more complex and therefore more expensive compared to flat systems.

System with flat plate collectors are widely used in countries with large values of daily sums of solar radiation (Israel, China, India) [8]. In Russia the flat solar collectors are used in the Krasnodar region. On the other hand, they provide hot water (mostly individual households) and are the most cost-effective investment, since they have a relatively low cost.

To solve simultaneously the tasks of providing hot water and heating, it is necessary to have a high temperature coolant and the opportunity to work in winter conditions, and in cloudy weather.

Largely these requirements can be met by the system vacuum collectors with heat pipes. Heat transfer from the heat pipes is more efficient and allows the installation to operate in a wide range of weather conditions. In Figure 3 shows a part of vacuum installation SC on an apartment building in Ekaterinburg.

**Figure 3**. Photo frame for installation of solar vacuum collectors at an apartment house in Ekaterinburg

The dependence of thermal energy (*E*) obtained per day against the temperature difference between the coolant in the reservoir and the ambient temperature (Tm-Ta) for different values of the advents of solar energy presented in Figure 4 ("SRCC").

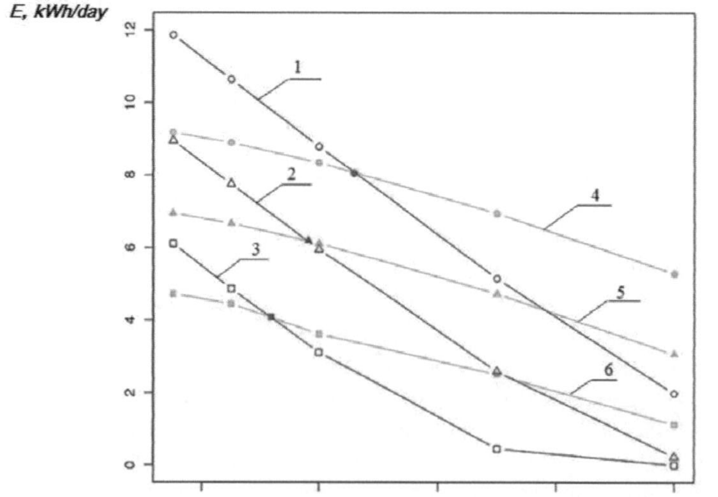


**Figure 4.** The dependence of thermal energy (*E*) obtained per day against the temperature difference, between the coolant in the reservoir and the ambient temperature
1 – flat SK, Sunny, Asun = 6,4 kWh/($m^2$·day) ; 2 – flat SK, cloudy, Asun = 4,7 kWh/($m^2$·day) ; 3 – flat SK, cloudy, Asun = 3 kWh/($m^2$·day) ; 4 – vacuum SK, Sunny, Asun = 6,4 kWh/($m^2$·day) ; 5 – vacuum-UK cloudy Asun = 4,7 kWh/($m^2$·day) ; 6 – vacuum-UK cloudy with a chance of Asun = 3 kWh/($m^2$·day)

The presented dependence figure illustrates the main differences between two types of systems. So, the flat solar collectors (curves 1–3) have large energy service with small temperature difference. However, with the increase of temperature in a solar collector, heat losses grow up, consequently useful power output is reduced. On the other hand, the curves of vacuum solar collectors (4–6) have a more flat look. This gives an indication of lesser heat losses with increasing temperature.

Thus, taking into account the advantages and disadvantages of both systems, for the project of using solar energy for apartment building the system with vacuum solar collectors consisting heat pipes are selected as the main elements

## 4. Calculations

Figure 5 presents the power comparison of solar flux for 4 different values of angles of inclination to the horizon system vacuum of solar collectors. The maximum amount of energy during the summer period is observed for angles of inclination less than the latitude (in our geographic location lat 56°), whereas for the winter period larger angles.

For solar energy systems whose main task is heating, the most appropriate angles for "winter" are (inclination from 60 to 70 degrees). For example, for the angle 65° in January observed solar insolation is 400 W/m², while at the same time for angle 10 degrees $I_R$ = 180 W/m². Further increase in angle is impractical because when the angle to 90 ° $I_n$ = 415 W/m² for January, but in the summer there is a significant shortage of energy, because solar radiation falls on a vertical surface only partially.

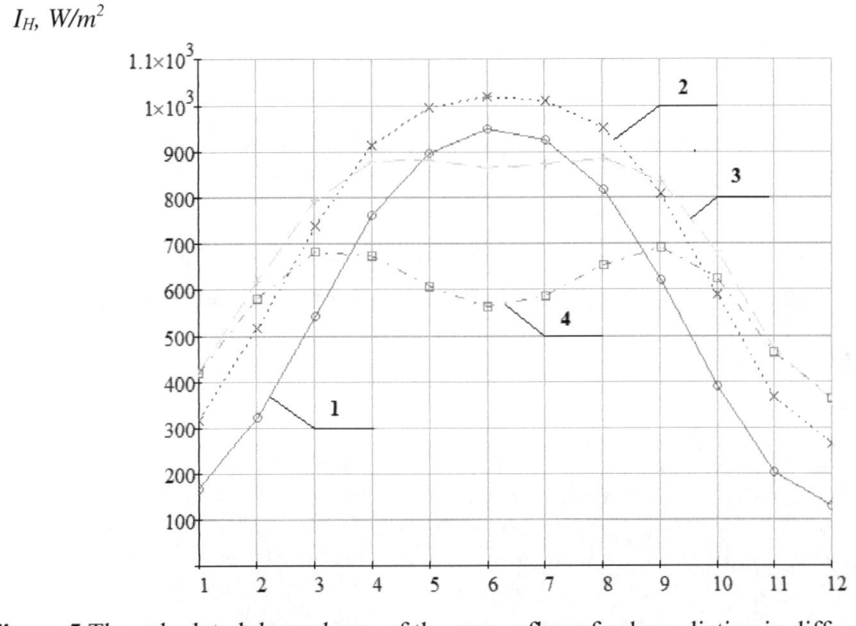


**Figure 5.**The calculated dependence of the power flux of solar radiation in different months for different tilt angles of solar collectors: 1 – 10; 2 – 45°; 3 – 65°; 4 – 90°

Thus, for the system of solar collectors used for heating and DHW for a residential apartment house in the conditions of the Urals Federal district, the considered effective angle was 65°.

## 5. Flow control

In addition to the automatic control system of performance controlled by temperature, the implementation system of data collection (Figure 6) which allows us to display the control screen data for the temperatures of the individual sections of the vacuum SK as well as other parameters, such as pressure and temperature of straight pipe. The availability of such information allows us to monitor the performance of the system in whole and its separate elements. Further information is supplied to the computer, where it is processed and stored in a database that allows us to collect statistical information about the whole system in different periods of the year.

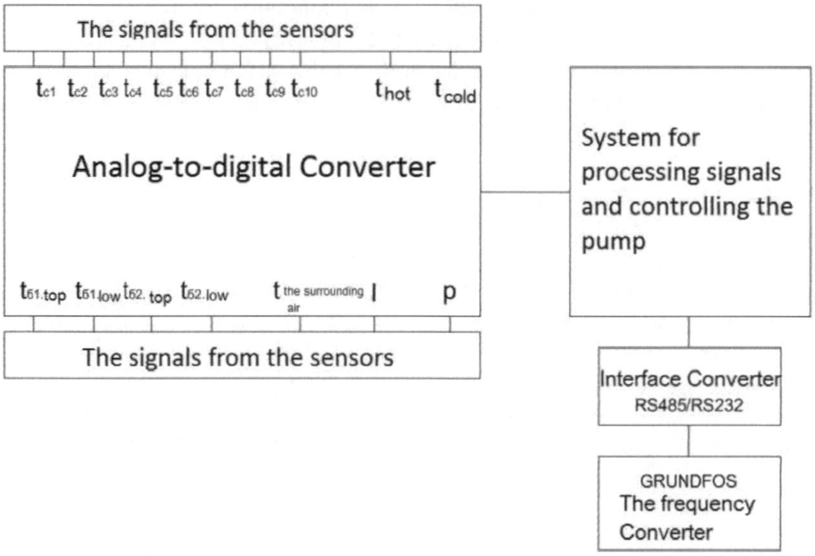


**Figure 6.** A block diagram of a data collection system,
where $t_{c1}$, $t_{c2}$,...,$t_{cn}$ the temperature of each section headers;
$t_{б1 \cdot top}$, $t_{б1 \cdot bottom}$, $t_{б2 \cdot top}$, $t_{б2}$.
The bottom temperature in tanks-accumulators;
$t_{hot}$., $t_{cold}$. – temperature in pipelines, p – system pressure,
I – power flux of solar radiation

Vacuum installations SC on a block of flats in the metropolis are presented in Figure 7.

**Figure 7.** Installation of vacuum solar collectors on multi-10-storied building in Ekaterinburg (Students of the UrFU on the practical training for installation)

## 6. Conclusions

1. Experimentally confirmed the possibility of using solar energy for heating of residential buildings in the metropolis for the conditions of sharply continental climate in the Ural region.

The Operation of vacuum tube solar collectors in an apartment building for 1.5 years has demonstrated the efficiency of hot water and heat supply in autumn and spring.

In the summer, coolant and anti-solar protection are over heated because of no necessity of heating there were required vacuum tubes.

In winter, the system is unable to cope with peak loads (outdoor air temperature from -15 °C and below) and requires the supply of heat from centralized heat supplier.

2. The main problem of the use of solar collectors in the metropolitan areas is the lack of a legislative base for renewable energy and the associated relationship between the consumer (tenant) and a supplier (heat Supply Company). The heat supplier raised a question about the complete shutdown of the centralized system of heat supply and removing it from service or the elimination of a system for solar heating and domestic hot water, as there could be financial losses of up to 30 % in autumn and spring periods.

3. The use of renewable energy sources reduces the city budget advantageously by strategical saving of fuel resources. However, the gaps in Russian Legislation and lack of will of the leadership is not yet possible to overcome the barriers for implementation of solar collectors in the metropolitan areas.

4. The use of vacuum collectors in Russia will start to be introduced widely in areas with a cold climate and in smart houses after resolving the issues of legislative support from the state and municipal authorities. It will open wide opportunities for the construction of smart city with the use of solar energy for heating purposes.

**References**

[1] Bushuev V V 2014 *Energy and the fate of Russia* (Moscow: publishing house "Energy") 292 p

[2] Directive 2009/28/EC of the European Parliament and of the Council of 23 April 2009 on the promotion of the use of energy from renewable sources and amending and subsequently repealing Available from: https://www.energy-community.org/portal/page/portal/ENC_HO-ME/DOCS/2938033/0633975AD78C7B9CE053C92FA8C06338.PDF [Accessed 08th May 2017]

[3] Popel O S, Frid S E, Kolomiets Y T, Kiseleva S V and Terekhov E N 2010 *Atlas of solar energy resources on the territory of Russia* (Moscow) 83 p

[4] Butuzov V A, Bryantseva E V and Gnatyuk I S 2011 A solar power plant in Krasnodar region *Industrial Power engineering* **7** p 45–7

[5] Danilov V S, Korzhavin S A, Shscheklein S E and Velkin V I 2012 Experimental research of efficiency of the combined system of solar heat *Alternative energy and ecology* **3** p 77–81

[6] Velkin V I 2015 *Methodology of calculation of integrated renewable energy systems for use in autonomous objects* (Ekaterinburg: UrFU) 226 p

[7] Elistratov V V 2016 *Renewable energy* (St. Petersburg: Polytechnical University Publishing House) 424 p

[8] Butuzov V A and Butuzov V V 2015 *The use of solar energy for heat production: reference-methodical edition* (Moscow: Interenergo-Izdat) p 304

# Implementation of energy efficient smart technologies at the urban territories of the Arctic zone of Russia

**M V Tsurkan[1], M A Liubarskaia[2, 3]*, A M Vorotnikov[4, 5] and S V Maiorov[6]**

[1] Tver State University, Tver, Russian Federation
[2] Saint-Petersburg State University of Economics, Saint-Petersburg, Russian Federation
[3] Baltic Academy of Tourism and Entrepreneurship, Saint-Petersburg, Russian Federation
[4] Russian Presidential Academy of National Economy and Public Administration under the President of the Russian Federation, Moscow, Russian Federation
[5] System Consulting LLC
[6] Ministry of Economic Development of the Russian Federation, Moscow, Russian Federation

* E-mail: lioubarskaya@mail.ru

**Abstract**. Special climate conditions and out-of-date energy system at the urban territories of the Arctic zone of Russia reinforce the need for the development of a new concept for the implementation of energy efficient smart technologies at these territories. In this regard, the article is focused on identifying the prospects for the development of the energy sector of the urban territories of the Arctic zone of Russia with the involvement of smart technologies. Methods of the study include system analysis and modeling, which allow authors to consider the prospects of implementing energy efficient smart technologies in a comprehensive manner. Article suggests the interpretation of the **concept** of "energy efficient smart technologies" determines the priorities and introduces the models for the implementation of these technologies at the urban territories of the Arctic zone of Russia. The proposed concept and models were developed taking into account the modern information and communication infrastructure of the given territories. Particular attention was paid to the complex of characteristics influence the implementation of energy efficient smart technologies at the urban territories of the Arctic zone of Russia.
**Keywords**: energy efficiency, smart technologies, Arctic zone of Russia, biofuel, renewable energy source

## 1. Introduction

Russia can become the first country where energy efficient smart technologies will be implemented by the order of the government, and not at the initiative of municipalities or IT companies. According to Otis O., Ponamareva L., Isahi A., legislative and technological prerequisites for this have been created [1].

At the same time, the process of ubiquitous implementation of energy efficient smart technologies, using common tools, may not be effective, taking into account the climatic and environmental specifics of the urban territories of the Arctic zone of Russia. This zone includes 8 subjects of Federation: Murmansk and Arkhangelsk regions; Republic of Komi and Yakutia; Krasnoyarsk region; Nenets and Yamalo-Nenets Autonomous Districts, Chukotka Autonomous District.

All cities of the Arctic zone has harsh climate. The average summer temperature is about 0 °C, and average annual temperature is negative. Most of these cities are located on the coast of the Arctic seas or in the immediate vicinity of it, as well as in the lower part of the rivers flowing into the Arctic Ocean.

Authors consider some cities as the most prospective for the implementation of energy efficient smart technologies. Among them Murmansk (325,000 inhabitants), Norilsk (205,000 inhabitants), Vorkuta (85,000 inhabitants), cities of Yakutia and the Krasnoyarsk region, mono-cities of the Arkhangelsk region (Severodvinsk, Novodvinsk, Onega), and the city of Monchegorsk, Murmansk region.

Currently, these cities are the objects of research of many scientists from the prospective for the sustainable development, and some projects are already under implementation.

One of the large scale projects is "Arctic Partnership for International Research and Education: Promoting Urban Sustainability in the Arctic" (2015–2021), funded by the National Research Foundation of the United States.

Research in this project is already being provided by a number of US universities (The George Washington University, University of Northern Iowa, Cedar Falls, University of Alaska, MIT, University of Virginia) with the participation of many partners around the world, including the University of Helsinki, University of Saskatchewan, Norwegian Institute of International Relations, University of Tromse, St. Petersburg State University, Kola Scientific Center of the Russian Academy of Sciences, Institute of Geography of Moscow State University, SOPS, State Hydrological Institute of the Russian Federation.

In the next two years Russian scientific center "Skolkovo" together with the Department of Urban Planning and Research of the MIT International Science and Technology Initiative will implement a project focused on the Urban Sustainable Development Index of the Arctic.

This Index will provide an opportunity to assess different aspects of the development of the cities of the Far North in a uniform way, such as the effects of climate warming, the cyclical nature of the resource economy development, the level of social tension, and others; compare the Arctic cities among themselves on a single scale and make recommendations on the application of city policies.

Many of the existing studies consider the possibility of implementation of energy efficient smart technologies for the development of Arctic cities, but the concept of this process remains virtually undeveloped.

Authors of this paper developed new definition of the category "energy efficient smart technologies" as a set of technical decisions and methods aimed at:
- optimization of energy consumption;
- automation of an intelligent energy network;
- formation of a flexible energy distribution system;
- intellectualization of the control and regulation of demand;
- integration of renewable energy sources.

Also, term "energy efficient smart technologies" includes the software and hardware for managing an intelligent power grid or energy efficient buildings and structures.

## 2. Priorities for the implementation of energy efficient smart technologies at the urban territories of the Arctic zone of Russia

The research company Frost & Sullivan for 15 years has been analyzing the practice of creating the smart cities around the world. In 2016 Frost & Sullivan formed the list of the main city improvements reached through smart technologies implementation. These list includes reducing harmful emissions, creating a scalable and flexible urban environment, increasing security level and level of resource saving and energy efficiency, providing social support and universal access to the information, improving logistics and transportation, formation of rational supply chain management, etc. [1].

The analysis of the provisions of the Federal "Strategy for the development of the Arctic zone of the Russian Federation and ensuring national security for the period until 2020" suggests that the urban infrastructure of the Arctic regions is in need of smart technologies in the sphere of:
- ensuring environmental safety (this aspect also included in the National Public Standard, a public agreement of the population of cities, businesses and authorities on preserving the ecological structure of territories [2]);
- ensuring the safety of the population, including the development of a system for monitoring the geophysical situation in Arctic cities;
- monitoring the use of renewable and alternative energy sources (taking into consideration local recourses);
- power supply management;
- management of energy distribution, including nuclear power plants energy (for example, the Kola NPP with VVER-440 units in the Murmansk region [3]);
- advancing telecommunications;
- formation of information intelligence centers in the cities.

The strategy for the development of the Arctic zone implies a transition from centralized power supply systems to a distributed networked intellectual energy systems, which is impossible without implementation of smart technologies at the energy sector.

Nowadays the most promising for the urban territories of the Arctic zone of Russia looks the implementation of energy efficient smart technologies at multi-apartment buildings.

## 3. Characteristics of energy infrastructure of the urban territories of the Arctic zone of Russia

The Arctic region of Russia as a whole could be described with the following characteristics:

- high energy intensity and low efficiency of extraction of natural resources, high cost of manufacturing, and lack of effective compensation mechanisms, low labor productivity;
- underdevelopment of the energy system, as well as irrational structure of generating capacities, high production costs of generation and transportation of electricity.

At the same time, according to the estimates of the Ministry of Energy of the Russian Federation, the Murmansk region is one of the leaders in the energy efficiency rating of the Subjects of Federation. Basically, such results have been achieved through extra budgetary sources, using the mechanism of the energy service contracts. Now the cities of the Murmansk region develop bioenergy sector, focused on using local biofuels (wood waste and peat) [4].

The rating of energy efficiency in the regions of the Arctic zone is presented in Figure 1.

The city of Monchegorsk will perform as a pilot territory for the implementation of bioenergy sector development projects in the Murmansk region.

At the same time, another related projects are under implementation in Kodinsk, Krasnoyarsk Region. The city heating infrastructure predominantly using electricity, one of the most costly and scarce energy sources in the Arctic zone. At the moment, an electric boiler-house with a total heat capacity of more than 110 GCal per hour is operating in Kodinsk at a peak load of not more than 55 GCal per hour. This project allowed a partial replacement of the existing electric boiler house with a bio-boiler that operates on wood waste, secondary unclaimed product of sawmilling and timber processing.

New project with total investments of 354 500 thousand rubles is aimed at the construction of a biofuel water-heating boiler with capacity of 20.0 MW. Expected useful heat supply (from collectors) is 105 055 GCal per year. The cost of raw materials (chips) is 63.5 rubles per cubic meters, including VAT. This is an example of successful financing of an energy efficient project at the expense of credit resources.

According to researches of Russian scientists, development of renewable energy sources at the urban territories of the Arctic zone of Russia has big prospective. First of all, we are talking about solar and wind energy. Considering the remoteness of cities in the Arctic zone, from each other, autonomous energy generation infrastructure, including alternative energy sources, becomes especially important. These can be autonomous combined power plants, urban nuclear power-technological complexes, etc. For these objects implementation of smart technologies for the automation of data collection from alternative energy systems becomes urgent.

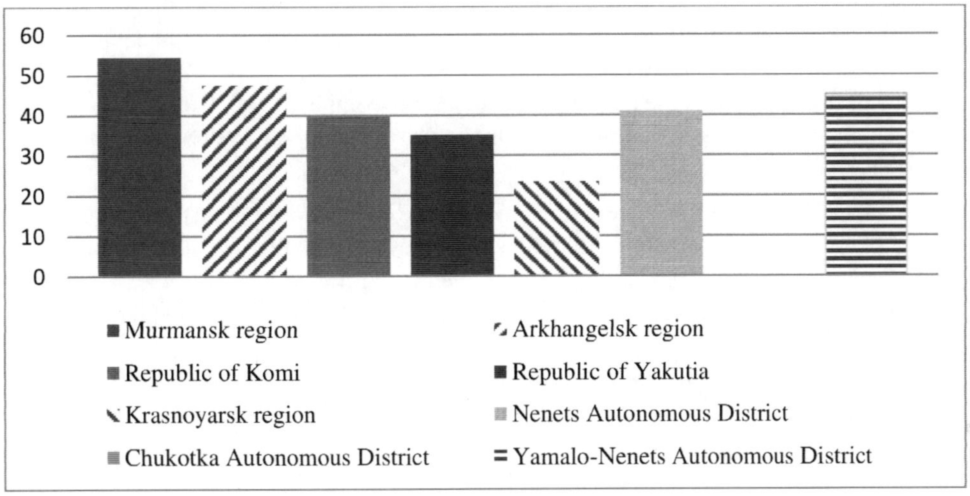


**Figure 1.** Rating of energy efficiency in the regions of the Arctic zone as of January 1, 2017
*Source:* compiled by authors based on the official data at www.energy.s-kon.ru

Belugin A.V., Lagunov A.Y. and Pozdeev V.A. emphasize the need to introduce a smart monitoring system of alternative energy sources in the Arctic [5].

Currently, Russia has developed a system of intelligent control with advanced adaptation and diagnostic functions. This system implements elements of artificial intelligence in the form of blocks for predicting the states of jointly operating subsystems, and has been recommended by experts for the urban territories of the Arctic zone of Russia [6].

The experience of implementation of energy efficient smart technologies in multi-apartment buildings has already been obtained in Yakutia. Three-storey "smart homes" have increased insulation; energy-efficient windows are installed everywhere. Six of these homes use heat from three autonomous block-modular gas boiler houses, the rest of them get heat centrally, but they have automatic electronic units that clearly react to temperature changes and support it in the set parameters inside the house. Hot water is heated through solar collectors, and the emergency power system operates on solar panels, all houses also have a ventilation system. Another important detail is that the tenants themselves can regulate the microclimate in the apartment.

Residents of "smart homes" pay only for cold water, saving on hot water about 50 %, and 25 % on heating. The payment goes according to the actual consumption of resources, not to average norms as in the city of Yakutsk or in Yakutia in general [7].

## 4. Models of implementation of energy efficient smart technologies at the urban territories of the Arctic zone of Russia

The analysis of market of energy efficient smart technologies based on theoretical and methodological sources allowed the authors to develop two models of implementation of those technologies at the urban territories of the Arctic zone of Russia:

1. A basic model, when energy efficient smart technologies are implemented locally to solve a specific problem. In this case, there is a traditional way of planning and financing at the level of energy sector enterprises or multi-apartment buildings. Smart projects are carried out independently. The purpose

of the model is to prove the value of smart solutions, to show the benefits of investing in pilot energy efficient smart projects.
2. The growth model, when the state set up standards for the development of the energy efficient smart systems are approaches to data processing and management. The purpose of the model is to improve the results of the implementation of smart initiatives in the energy sector and to coordinate them at the city level.

Today, we can say that the basic model is implemented in Yakutsk, at the level of apartment buildings.

As part of the analysis, the authors developed complex of characteristics influence the implementation of energy efficient smart technologies at the urban territories of the Arctic zone of Russia (Table 1).

**Table 1.** Characteristics influence the implementation of energy efficient smart technologies at the urban territories of the Arctic zone of Russia

| Characteristics | Suggested groups | Notes and examples |
|---|---|---|
| Localization level | Apartment building | Measuring instruments and devices (first of all, smart meters and smart sensors) |
| | City section | The experience was gained in Yakutsk, using alternative energy sources (solar energy) |
| | City | Provided at a single autonomous energy supply facility |
| Model of implementation | Basic model | Recommended by the authors for implementation at the level of apartment building or a city section |
| | Growth model | Recommended by the authors for implementation at the city level |
| Core of Energy concept | Renewable energy sources | |
| | Small Nuclear Power Engineering | |
| | Optimization of the use of traditional fuels | |

*Source:* compiled by authors

Traditional devices, unlike smart meters and traditional counters with connected smart sensors, measure only the total amount of energy consumed and do not provide information of schedule of consumption. Using intelligent meters we can easily get such information. It allows responsible agencies to introduce groups of tariffs for consumption depending on the time of the day and season, and to organize monitoring of consumption. This leads to the reduction of unnecessary consumption of resources by end-users. It is known that a decrease in consumption for 1 kWh by the consumer saves up to 4–5 kWh of energy from the manufacturer. In the residential sector of Russia, the consumption of hot water and heating energy is up to 80–90 % of the total energy consumption. Therefore, smart meters and systems for monitoring consumption and saving water and heat are most relevant in smart homes and public buildings [8–10].

Practically in all Arctic cities, which are the objects of research, the tariff for energy is much higher than average in Russia. Climatic conditions, the need for fuel delivery lead to a natural rise in price, and most of the energy tariffs are subsidized. The introduction of smart meters and smart sensors will partly help to reduce the corresponding tariffs.

Based on analysis of the Russian market the best technical characteristics for implementation at the urban territories of the Arctic zone of Russia have the following devices: Counters "Matrix", AIIS KUE "Smart IMS", "Compact V", Automated Smart Metering System "Alpha smart AS1440", Counter "Light S" and others. Their difference lies in the fact that they fix and normalize significant differences in the supply of energy, operate at low temperatures, do not require a high-speed connection.

An example of an engineering solution that can be implemented in the cities of the Arctic zone is smart energy technology, the implementation of which is justified by Kolosov R.V., Titov V.V., Titov V.G. [11].

As part of the state order, scientists analyzed smart technologies that allow to generate energy in parallel on the basis of wind turbines and solar batteries and from small-scale power systems.

This approach allows to obtain constant energy, regardless of the climatic conditions of a particular period. The intellectual energy system in this case predicts the energy consumption in the network, estimates the economic benefit from the use of a particular source in a certain period of time, and automatically decides on the choice of that one.

This smart technology is based on the use of the algorithm of the increment of conductivity. The main advantage of the method of additional conductivity over the random perturbation algorithm is that it can calculate the direction to change "working point" for absorbing solar energy. This change could be short-term or long-term. Therefore, with a sharp change in illumination, the operation of such an algorithm will not lead to an erroneous choice of the direction of the changes.

The scheme of operation of the conduction incremental algorithm used in the analyzed smart technology is shown in Figure 2.

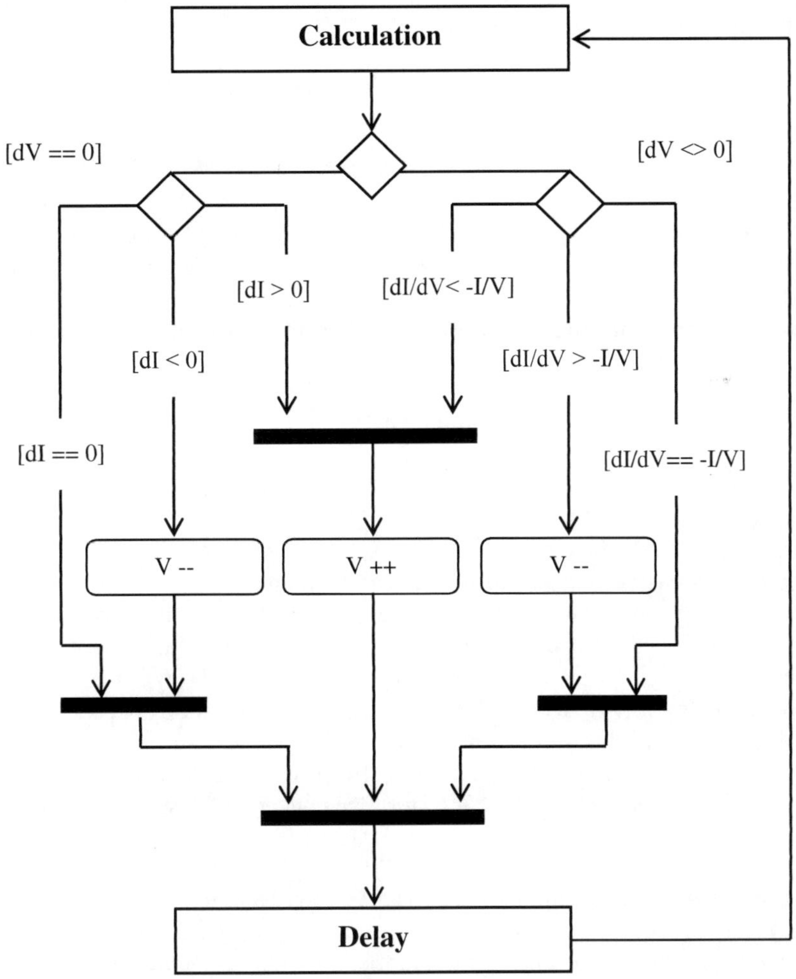


**Figure 2.** The scheme of operation of the conduction incremental algorithm
*Source:* compiled by authors based on the [11]

It should be noted that in the Arctic zone the daily intake of direct solar energy according to NASA SEE data varies from 3 to 5 kWh per sq. m per day or 0.7 to 1.8 kWh per sq. m per day (from 60 to 150 kg of fuel equivalent per sq. m per year). For comparison, we note that the average daily supply of solar energy in southern Germany, where solar installations are widely used, is only about 3.4 kWh per sq. m per day.

## 5. Efficiency of the use of the proposed energy efficient smart technologies for sustainable development of the urban territories of the Arctic zone of Russia

Under the concept of sustainable development of the cities, considered technologies assume either the effective use of traditional energy resources (resource-saving, energy efficiency) or the use of renewable energy sources, instead of traditional ones. This aspect allows us to speak about very positive environmental characteristics of energy efficient smart technologies, and about the observance of main principles of sustainable development.

In addition, the introduction of these technologies can contribute to the economic growth of cities, as their use will free up part of the budget funds spent on electricity and heat supply of public facilities.

To calculate the budgetary efficiency of the considered energy efficient smart technologies ($E_B$), the following formula was proposed and used:

$$E_B = \frac{\sum_{i=1}^{t}\left(BR_i + BS_i - BE_i\right)}{BI \cdot (1+\frac{r}{100})^t}, \quad (1)$$

where $BR_i$ – expected budget revenues for the urban territories of the Arctic zone of Russia from the implementation of energy efficient smart technologies for year "*i*" of the considered period, thousand RUR;

$BS_i$ – budget savings for the urban territories of the Arctic zone of Russia from the implementation of energy efficient smart technologies for year "*i*" of the considered period, thousand RUR;

$BE_i$ – expenses of the city budget for the implementation of energy efficient smart technologies for year "*i*" of the considered period, thousand RUR;

BI – initial budget investments in launching technology in a particular city, thousand RUR;

r – the refinancing rate established by the Central Bank of the Russian Federation for the time frame of expenses, %;

t – considered period for budget efficiency, from $t_{min}$ = 1 year to $t_{max}$ = planned period of use of energy efficient smart technologies in the territory of a particular city.

The use of energy efficient smart technologies at the Arctic Territories of Russia should be considered effective if, as a result of calculating the budgetary effectiveness, the result is $\geq 0$.

The technical efficiency in the framework of the given study is understood as the property of an energy efficient smart technology to convert costs into effects, taking into account the compliance of a technical solution with its goal.

Summary results of the evaluation of the effectiveness of the energy efficient smart technologies considered or effects assessed in the context of the concept of sustainable urban development are presented in Table 2.

## 6. Concept of implementation of energy efficient smart technologies at the urban territories of the Arctic zone of Russia

At the current stage of development of the information and communication infrastructure of Russian Arctic cities, it is possible to propose a concept for the implementation of energy efficient smart technologies (Figure 3).

IOP Conf. Series: Earth and Environmental Science **72** (2017) 012029 doi:10.1088/1755-1315/72/1/012029

**Table 2.** Assessment of the effectiveness of the implementation of energy efficient technologies at the Arctic Territories of Russia

| Name of energy efficient smart technologies | Budgetary efficiency for the city (the aspect of economic growth) | Improving the quality of life of urban population | Environmental characteristics | Technical efficiency |
|---|---|---|---|---|
| Biofuel water heating boiler house | + (1,23), calculated on the example of the city of Kodinsk | + uninterrupted supply of heat energy and hot water, regardless of the supply of traditional fuels | + "zero waste" | + expected useful heat output is 105,055 Gcal/year |
| Management of intra-urban energy distribution, including those allocated by nuclear power plants | + (≥ 1) | + | + preserving clean air and water | + reduction of energy use: ≥ 10%, but ≤ 17% |
| Smart-meters and Smart-sensors | + ( от 2-5), calculated for electricity meters "Matrix", AIIS KUE "Smart IMS", "Compact V", automated Smart Metering System "Alpha smart AS1440", electricity meters "Light S" | – | – | + reduction of energy use: ≥5 %, but ≤ 47% |
| Low-power power systems that combine wind turbines and solar panels | + (≥ 1) | + | + "zero waste", preserving clean air and water | + reduction of energy use ≥ 5 GW, но ≤ 22,5 GW |

*Source:* compiled by authors

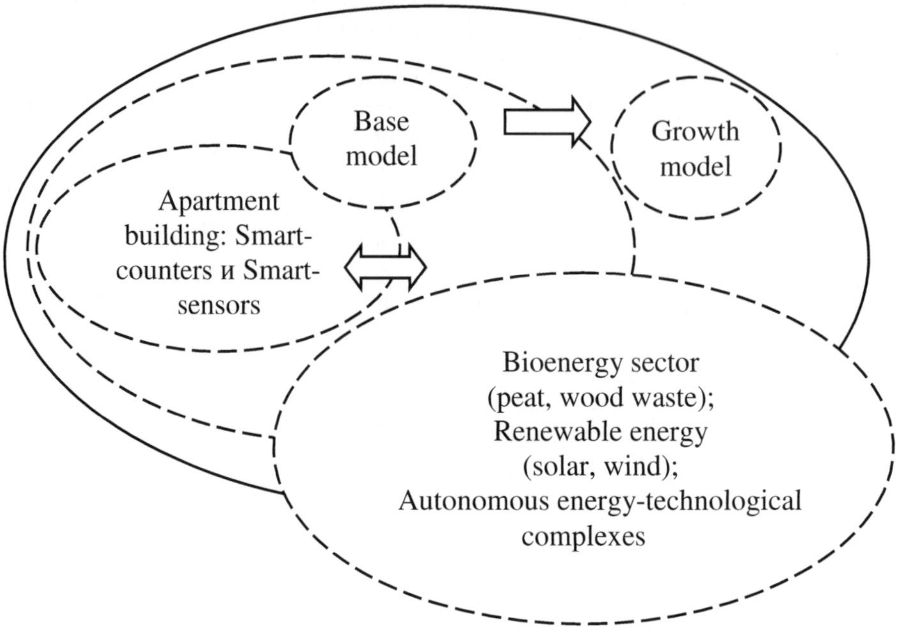


**Figure 3.** Concept of implementation of energy efficient smart technologies at the urban territories of the Arctic zone of Russia
*Source:* compiled by authors

## 7. Conclusion

The study allows authors to conclude that the implementation of energy efficient smart technologies at the urban territories of the Arctic zone of Russia is most effective through creation of autonomous energy-technological complexes, involving biofuels and renewable energy.

At the level of apartment buildings using traditional sources of energy could be optimized by implementation of smart meters and smart sensors.

For Murmansk, Norilsk, Vorkuta, Severodvinsk, Novodvinsk, Onega, Monchegorsk, the cities of Yakutia and Krasnoyarsk Region it is rational to gradually increase the use of energy-saving technologies in decisions to ensure the safety of communal infrastructure. Some experience of using energy efficient smart technologies, bioenergetics, wind and solar energy already gets the city of Murmansk, Yakutia and Krasnoyarsk Region. It can be adapted for other cities in the Arctic zone.

Nowadays at the Russian market we can find a lot of "products" to help rational generation, accumulation and consumption of energy suitable for the urban territories of the Arctic zone of Russia, but this process requires the political will of regional and local authority bodies.

## References

[1] Otis O, Ponamareva L and Isahi A 2016 City of the mind *J. Rubezh* **4** p 70–4
[2] Konygin E 2017 Ecology of the Arctic as a driver of innovation development *J. Reg. Energy and Energy Sav.* **1** p 33
[3] Kutovoy G 2017 The Arctic needs the Federal National Program for the Development of Small Nuclear Power *J. Reg. Energy and Energy Sav.* **1** p 36–7
[4] Kovtun M 2017 Murmansk region is one of the leaders in the energy efficiency rating *J. Reg. Energy and Energy Sav.* **1** p 48–53
[5] Belugin A, Lagunov A and Pozdeev V 2014 Automation of data collection from the alternative energy system *J. Scien. Works SWorld* 9 **1** p 90–7

[6] Tsereteli M 2017 Power and housing and communal services in the Arctic *J. Reg. Energy and Energy Sav.* **1** p 26–8
[7] "Smart home" systems allow you to save on heat to residents in Yakutia Available from: http://tass.ru/arktika-segodnya/4127534
[8] Wilson C, Hargreaves T and Hauxwell-Baldwin R 2017 Benefits and risks of smart home technologies *J. Energy Policy* **103** p 72–83
[9] Bhati A, Hansen M and Chan C M Energy conservation through smart homes in a smart city: A lesson for Singapore households *J. Energy Policy* **104** p 230–9
[10] Pocero L, Amaxilatis D, Mylonas G and Chatzigiannakis I 2017 Open source IoT meter devices for smart and energy-efficient school buildings *J. HardwareX* **1** p 54–67
[11] Kolosov R V, Puchenkin A V, Titov V V and Titov V G 2013 Renewable energy sources in small generation systems *Proc. Int. Conf. on Ren. Energy* **3** (Nizhny Novgorod: Nizhny Novgorod State Technical University named by Alekseev) p 207–11